Introduction to Polymers

Introduction to Polymers

ROBERT J. YOUNG

Department of Polymer Science and Technology
UMIST, Manchester

LONDON   NEW YORK

CHAPMAN AND HALL

First published 1981 by Chapman and Hall Ltd
11 New Fetter Lane, London EC4P 4EE
Published in the USA by Chapman and Hall
29 West 35th Street, New York NY 10001
Reprinted 1983 with additional material
Reprinted 1986

© 1981, 1983 Robert J. Young

ISBN 0 412 22170 5 (cased)
ISBN 0 412 22180 2 (paperback)

Printed in Great Britain
at the University Press, Cambridge

British Library Cataloguing in Publication Data

Young, Robert J
  Introduction to polymers
  1. Polymers and polymerization
  I. Title
  547.7   QD381   80-42133

  ISBN 0-412-22170-5
  ISBN 0-412-22180-2 Pbk

Contents

vi   *Contents*

Preface

Polymers are a group of materials made up of long covalently-bonded molecules, which include plastics and rubbers. The use of polymeric materials is increasing rapidly year by year and in many applications they are replacing conventional materials such as metals, wood and natural fibres such as cotton and wool. The book is designed principally for undergraduate and postgraduate students of Chemistry, Physics, Materials Science and Engineering who are studying polymers. An increasing number of graduates in these disciplines go on to work in polymer-based industries, often with little grounding in Polymer Science and so the book should also be of use to scientists in industry and research who need to learn about the subject.

A basic knowledge of mathematics, chemistry and physics is assumed although it has been written to be, as far as is possible, self-contained with most equations fully derived and any assumptions stated. Previous books in this field have tended to be concerned primarily with either polymer chemistry, polymer structure or mechanical properties. An attempt has been made with this book to fuse together these different aspects into one volume so that the reader has these different areas included in one book and so can appreciate the relationships that exist between the different aspects of the subject. Problems have also been given at the end of each chapter so that the reader may be able to test his or her understanding of the subject and practise the manipulation of data.

The textbook approaches the subject of polymers from a Materials Science viewpoint, being principally concerned with the relationship between structure and properties. In order to keep it down to a manageable size there have been important and deliberate omissions. Two obvious areas are those of polymer processing (e.g. moulding and fabrication) and electrical properties. These are vast areas in their own right and it is hoped that this book will give the reader sufficient grounding to go on and study these topics elsewhere.

Several aspects of the subject of polymer science have been updated compared with the normal presentation in books at this level. For example, the mechanical properties of polymers are treated from a mechanistic viewpoint rather than in terms of viscoelasticity, reflecting modern developments in the subject. However, viscoelasticity being an important aspect of polymer properties is also covered but with rather less emphasis than it has been given in the past. The presentation of some theories and experimental results has been changed from the original approach for the sake of clarity and consistency of style.

I am grateful to Professor Bill Bonfield for originally suggesting the book and for his encouragement throughout the project. I am also grateful to my other colleagues at Queen Mary College for allowing me to use some of their material and problems and to many people in the field of Polymers who have contributed micrographs. A large part of the book was written during a period of study leave at the University of the Saarland in West Germany. I would like to thank the Alexander von Humboldt Stiftung for financial support during this period. The bulk of the manuscript was typed by Mrs Rosalie Hillman and I would like to thank her for her help. Finally, my gratitude must go to my wife and family for giving me their support during the preparation of the book.

ROBERT J. YOUNG

Queen Mary College, London
October 1980

1 Introduction

1.1 Polymers as materials

Polymer Science is a relatively young discipline which developed rapidly following the recognition 30 or 40 years ago that polymers are made up of long molecules. Although it is now widely accepted that these materials contain macromolecules it was only with the pioneering work of people like Staudinger, Carothers and others that the chemical aspects of Polymer Science were established on a firm scientific footing.

Polymers have been around in a natural form since life began and naturally occurring biological polymers play an extremely important role in plant and animal life. Natural rubber has been known and used in practical applications for over 100 years and its unusual properties were studied by Scientists in the last century. This book is concerned mainly with synthetic polymers, the use of which has increased enormously over the past few decades. This has been due principally to their relative cheapness, ease of fabrication, low density, chemical inertness, and high electrical resistivity, but also to other factors. For example, polytetrafluoroethylene has a very low coefficient of friction and also very good 'non-stick' properties when used as a lining for cooking utensils. Poly(methyl methacrylate) has good optical clarity and superior mechanical properties to glass. It is therefore used in aeroplane windows where its lightness is an added bonus.

Polymers are now accepted as materials in their own right along with more conventional materials such as metals and ceramics. The study of the structure and properties of materials is known as *Materials Science*. This has developed relatively recently from metallurgy, which is only concerned with metals, to encompass now all materials used in practical or engineering situations. This book adopts a Materials Science approach to the study of polymers and is concerned principally with the relationship between structure and properties in these materials.

1.2 Basic definitions and nomenclature

In the discussion of the structure and properties of polymers there are several important words and concepts used which need to be defined. Most of these are discussed in detail when they are introduced, but at this stage there are several terms that must be defined.

A *polymer* is a long molecule which contains a chain of atoms held together by covalent bonds. It is produced through a process known as *polymerization* whereby *monomer* molecules react together chemically to form either linear chains or a three-dimensional network of polymer chains. The main characteristic of the chain is that the chemical bonding is strong and directional along the chains, but they are only bonded sideways by weak secondary van der Waals bonding or occasionally with hydrogen-bonding.

If only one type of monomer is employed to form the polymer the resulting molecule is called a *homopolymer*. Often superior properties are obtained by using different types of monomer species. In this case the polymer is termed a *copolymer*.

The naming of polymers or envisaging the chemical structure of a polymer from a knowledge of its name is an area which often gives students a great deal of difficulty. This problem is compounded by the variety of trade-names which are also used to describe particular polymers. In the case of addition polymers or in other cases where a single monomer species is used the prefix 'poly-' is added to the name of the monomer. So that ethylene polymerizes to give polyethylene, styrene to give polystyrene and methyl methacrylate forms poly(methyl methacrylate). In the last case brackets are used because the monomer name consists of two words. Some condensation polymers are made by reacting two types of monomer together which form an alternating sequence in the polymer chain. The polymer formed from the reaction between ethylene glycol and terephthalic acid is then termed poly(ethylene terephthalate). Attempts have been made to unify the nomenclature of polymers rather than using these trivial names. However, the trivial names are still widely used especially for simple polymers. The number of polymers in this book referred to by their chemical name has been kept as small as possible and it is hoped that the student will be able to recognize the more important polymers very quickly without being confused by a large number of different names.

1.3 Molecular architecture and classification of polymers

The main feature which sets polymers apart from other materials is that polymer samples are made up of long molecules. A typical sample of polyethylene may have molecules which contain an average 50 000 atoms and would be 25 000Å long. The presence of these macromolecules has a dominant effect upon the properties of the polymeric material. There can be a considerable variation in the architecture of the individual molecules in different polymer samples. Although by definition polymer molecules are long they can be linear, branched or even in the form of a three-dimensional network as shown in Fig. 1.1. The particular type of architecture the molecules possess controls the properties of the material.

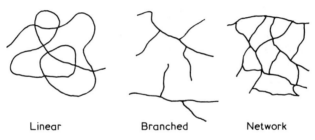

Linear   Branched   Network

Fig. 1.1 *Schematic representation of different types of polymer molecules.*

Any attempts to classify polymers into different categories tend to be somewhat arbitrary. One useful way is to put them into groups displaying similar properties which also has the advantage reflecting the underlying molecular structure. This classification is outlined in Fig. 1.2 where they are separated into three groups; *thermoplastics, rubbers* and *thermosets*. In addition thermoplastics are separated into those which are either crystalline or non-crystalline (amorphous).

Thermoplastics, which are often referred to just as 'plastics' are linear or branched polymers which can be melted upon the application of heat. They can be moulded and remoulded using conventional techniques and now make up the largest bulk of polymers used. Thermoplastics can be sub-divided into those which crystallize on cooling and those which do not and are normally used as polymer glasses. The ability of the polymers to crystallize depends upon many factors such as the degree of branching and the regularity of the molecules. However, crystalline thermoplastics are invariably only partly (semi-) crystalline and do not crystallize completely when cooled from the melt.

Rubbers are materials which display elastomeric properties, i.e. they can be stretched easily to high extensions and will spring back rapidly when the stress is released. This extremely important and useful property is a reflection of the molecular structure of the polymer which consists of a lightly cross-linked macromolecular network. The molecules slide past each other on deformation, but the cross-links prevent permanent flow and the molecules spring back to their original position on removal of the

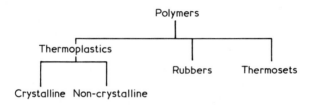

Fig. 1.2 *Classification of polymers.*

stress. The presence of the cross-links means that it is impossible to melt a rubber once it has undergone cross-linking.

Thermosets are heavily cross-linked polymers which are normally rigid and intractable. They consist of a dense three-dimensional molecular network and, like rubbers, degrade rather than melt on the application of heat. Common thermosetting polymers include phenol–formaldehyde or urea–formaldehyde resins and high-performance adhesives such as epoxy resins.

The chemical structure and uses of a variety of common thermoplastic polymers are listed in Table 1.1.

1.4 Molar mass and degree of polymerization

Since the characteristic properties of polymers depend intimately upon the size of the macromolecules present in the sample it is essential to have a method of describing their dimensions. This is normally done by quoting the *molar mass (M)* of the polymer which is simply the 'mass of one mole of the polymer' and so has units of g mol^{-1} or kg mol^{-1}. A *mole* of a substance being defined as 'the amount of that substance containing as many particles (i.e. polymer molecules) as there are atoms in 0.012 kg of carbon 12'. The term 'molecular weight' is often still used instead of molar mass, but this is not preferred because it can be somewhat misleading. It is really a dimensionless quantity, the relative molecular mass, rather than the weight of an individual molecule, which is of course a very small quantity ($\sim 10^{-20}$g).

An associated parameter which is often used to describe the size of a polymer molecule is the *degree of polymerization (x)* which is simply 'the number of chemical repeat units in the polymer chain'. It can be readily determined since

$$x = M/M_0 \tag{1.1}$$

where M_0 is the molar mass of the polymer repeat unit which can usually be taken as the molar mass of the monomer.

1.4.1 *Distribution of molar mass*

A normal polymer sample contains molecules with a variety of lengths and it is only possible to quote an average value of the molar mass. A typical distribution of molar mass is shown in Fig. 1.3. Often the exact form of the distribution is not known and it is conventional to describe it in terms of either a number average molar mass, $\bar{M}_n$ or a weight average molar mass, $\bar{M}_w$. Both of these parameters can be measured experimentally. In the definition of $\bar{M}_n$ and $\bar{M}_w$ it is envisaged that the distribution of molar mass

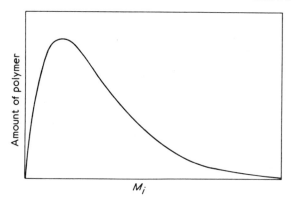

Fig. 1.3 *Typical molar mass distribution curve.*

can be split up into discrete fractions of molar mass M_i made up only of molecules which contain a discrete number of i repeat units.

(i) *The number average molar mass,* $\bar{M}_n$ is defined as 'the sum of the products of the molar mass of each fraction multiplied by its mole fraction'

i.e. $\quad \bar{M}_n = \sum X_i M_i$ (1.2)

where X_i is the mole fraction of molecules of length i, the ratio of the number of molecules of length i, N_i, to the total number of molecules, N. It follows therefore that

$$\bar{M}_n = \sum N_i M_i / N$$

or $\quad \bar{M}_n = \sum N_i M_i / \sum N_i$ (1.3)

It is often more convenient to deal in terms of mass fractions w_i rather than the numbers of molecules, the mass fraction w_i being defined as the mass of molecules of length i divided by the total mass of all molecules.

i.e. $\quad w_i = N_i M_i / \sum N_i M_i$ (1.4)

which can be written as

$$\sum (w_i / M_i) = \sum N_i / \sum N_i M_i$$ (1.5)

Combining Equations (1.3) and (1.5) gives $\bar{M}_n$ in terms of mass fractions as

$$\bar{M}_n = 1 / \sum (w_i / M_i)$$ (1.6)

It is possible to define a number average degree of polymerization as

$$\bar{x}_n = \bar{M}_n / M_0$$ (1.7)

TABLE 1.1 *Some common thermoplastic polymers*

Monomer	Polymer	Comments
Ethylene $CH_2=CH_2$	polyethylene (PE)	Moulded objects, tubing, film, electrical insulation, e.g. 'Alkathene'.
Propylene $CH_2=CH(CH_3)$	polypropylene (PP)	Similar uses to PE; lighter, stiffer, e.g. 'Propathene'.
Tetrafluoroethylene $CF_2=CF_2$	polytetrafluoroethylene (PFTE)	Mouldings or film. High-temperature polymer. Excellent electrical insulator. Low coefficient of sliding friction. Expensive, e.g. 'Teflon', 'Fluon'.
Styrene $CH_2=CH(C_6H_5)$	polystyrene (PS)	Cheap moulded objects. Polymerized with butadiene to make high impact polystyrene (HIPS). Expanded with pentane to make plastic foam.
Methyl methacrylate $CH_2=C(CH_3)COOCH_3$	poly(methyl methacrylate) (PMMA)	Transparent sheets and tubing. More expensive than PS. Aeroplane windows, e.g. 'Perspex', 'Lucite'. Low-molecular-weight grade 'Diakon' for transparent mouldings.

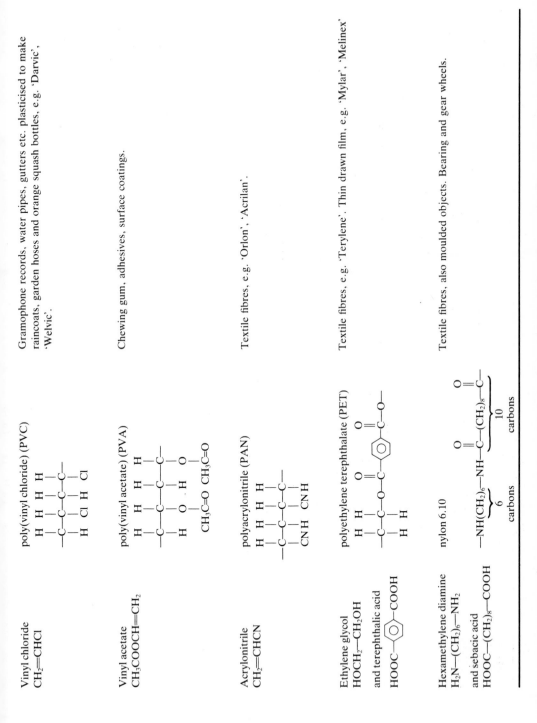

Vinyl chloride
$CH_2=CHCl$

poly(vinyl chloride) (PVC)

Gramophone records, water pipes, gutters etc. plasticised to make raincoats, garden hoses and orange squash bottles, e.g. 'Darvic', 'Welvic'.

Vinyl acetate
$CH_3COOCH=CH_2$

poly(vinyl acetate) (PVA)

Chewing gum, adhesives, surface coatings.

Acrylonitrile
$CH_2=CHCN$

polyacrylonitrile (PAN)

Textile fibres, e.g. 'Orlon', 'Acrilan'.

Ethylene glycol
$HOCH_2—CH_2OH$
and terephthalic acid
$HOOC—⬡—COOH$

polyethylene terephthalate (PET)

Textile fibres, e.g. 'Terylene'. Thin drawn film, e.g. 'Mylar', 'Melinex'.

Hexamethylene diamine
$H_2N—(CH_2)_6—NH_2$
and sebacic acid
$HOOC—(CH_2)_8—COOH$

nylon 6.10

Textile fibres, also moulded objects. Bearing and gear wheels.

(ii) *The weight average molar mass,* $\bar{M}_w$ is defined as 'the sum of the products of the molar mass of each fraction multiplied by its weight fraction, w_i'

i.e. $$\bar{M}_w = \sum w_i M_i \tag{1.8}$$

and combining this equation with Equation (1.4) gives

$$\bar{M}_w = \sum N_i M_i^2 \Big/ \sum N_i M_i \tag{1.9}$$

which is an expression for $\bar{M}_w$ in terms of the numbers of molecules. As before, a weight average degree of polymerization can be given as

$$\bar{x}_w = \bar{M}_w / M_0 \tag{1.10}$$

The breadth of the molar mass distribution or dispersivity is often described in terms of the *heterogeneity index,* $\bar{M}_w/\bar{M}_n$. Some types of polymerization lead to a value of $\bar{M}_w/\bar{M}_n = 2$ whereas a *homodisperse* polymer would have a value of $\bar{M}_w/\bar{M}_n = 1$. The value of the index gives an idea of the spread of the distribution for a *polydisperse* sample in which there is a range of molar mass.

2 Synthesis

2.1 Step-growth polymerization

It is conventional to divide the synthesis of polymers into two main categories. One is step-growth polymerization which is often also called *condensation polymerization* since it is almost exclusively concerned with condensation reactions taking place between multifunctional monomer molecules. The other category is *addition polymerization* where the monomer molecules add on to a growing chain one at a time and no small molecules are eliminated during the reaction. Step-growth polymerization is characterized by the gradual formation of polymer chains through successive reactions coupling monomers to each other to form dimers which can also react with other dimers or unreacted monomer molecules.

2.1.1 *Condensation reactions*

Since step-growth polymerization is mainly concerned with condensation reactions it will be useful first of all to look at what is meant by a condensation reaction. In simple terms it is a reaction between an organic base (such as an alcohol or amine) with an organic acid (such as a carboxylic acid or acid chloride) in which a small molecule like water is eliminated (condensed out).

A typical condensation reaction is the reaction between acetic acid and ethyl alcohol

$$CH_3COOH + C_2H_5OH \rightarrow CH_3COOC_2H_5 + H_2O$$

acetic   ethyl   ethyl acetate
acid   alcohol

Two important points to note are that a small molecule, water, is produced by the reaction. Also the product ethyl acetate is known as an *ester*. There are many examples of condensation reactions in organic chemistry and other products as well as esters are produced. For example, amides can be produced by reactions between organic acids and amines.

2.1.2 *Functionality and polycondensation reaction*

In order to produce polymer molecules by step-growth polymerization it is essential that the monomer molecules must be able to react at two or more sites. In the condensation reaction shown in Section 2.1.1 each of the two reactants can only react at one point. Because of this once the initial reaction has taken place the ester produced is incapable of taking part in any further reactions. Now, if reactants are used which are capable of reacting at two sites polymerization can start to take place. A reaction between a diacid and a diol (di-alcohol) yields an ester which still has unreacted end groups.

$$HOOC-\text{⬡}-COOH + HO-(CH_2)_2OH \rightarrow$$

terephthalic　　　　　ethylene
acid　　　　　　　　glycol

$$HOOC-\text{⬡}-COO-(CH_2)_2-OH + H_2O$$

reactive ester

This reaction can be written in a general form,

$$HOOC-R_1-COOH + HO-R_2-OH \rightarrow$$

diacid monomer　　　　diol monomer

$$HOOC-R_1-COO-R_2-OH + H_2O$$

Dimer

where R_1 and R_2 are general groups. The ester produced has an acid (COOH) group at one end and a hydroxyl (OH) group at the other and so can react with other monomer or dimer molecules to form a trimer or tetramer and go on through further condensation reactions to form long polymer molecules. The number of sites at which a monomer molecule is able to react is called its *functionality* and this is a concept which is of prime importance in step-growth polymerization. In order to form linear molecules it is necessary that the monomer molecules are bi-functional. If the monomers have a functionality of greater than two it is sometimes possible to form network polymers rather than linear molecules.

The general reaction for the production of a polyester from a diacid and a diol is

$$nHOOC-R_1-COOH + nHO-R_2-OH \rightarrow$$
$$HO[-OCR_1COOR_2-O]_nH + (2n - 1)H_2O$$

In the initial stages of the reaction diol monomer molecules react with diacid molecules. However, as the supply of monomer molecules is used up it is more likely that growing polyester molecules will also react with others of their own type.

e.g.　$$HO[-OCR_1COOR_2O-]_iH + HO[-OCR_1COOR_2O-]_jH \rightarrow$$
$$HO[-OCR_1COOR_2O-]_{(i+j)}H + H_2O$$

In the general reaction it is assumed that n molecules of diacid reacted with n molecules of diol. In fact it is usually necessary that equimolar mixtures (i.e. equal numbers of functional groups) are used otherwise the yield of high-molar-mass polymer is low. This condition is very difficult to achieve in the laboratory and is especially so in an industrial situation. If there is an excess of diacid, for example, intermediate short polymer molecules are produced which have acid end groups:

$$HO[—OCR_1COOR_2O—]_iOCR_1COOH$$

These intermediates are unable to react with further growing chains of the same type and so the reaction is effectively blocked. One of the ways of overcoming the problem is to use an ω-hydroxy carboxylic acid as monomer rather than the diacid and diol. The ω-hydroxy carboxylic acid molecule,

$$HOOC—R—OH \qquad (R— \text{general group})$$

has an acid group at one end and a hydroxyl group at the other. These molecules are able to undergo condensation reactions with others of their own type but they also have the great advantage of, if they are pure, guaranteeing equal numbers of functional groups in the reaction. Two monomer molecules will react to form a dimer

$$HOOC—R—OH + HOOC—R—OH \rightarrow$$

$$HOOC—R—OOCR—OH + H_2O$$

which is also an ω-hydroxy carboxylic acid. The dimers can react with either monomer or dimer molecules and so chain growth can proceed as before.

Another factor which can tend to reduce the yield of long polymer molecules, even when equal numbers of functional groups are employed, is the equilibrium which occurs between the reactants and products during condensation reactions. If the concentration of the condensate (such as water) is allowed to build up the reaction may stop and can even be forced in the reverse direction. This problem is usually readily overcome by removing the condensate and so driving the reaction in the forward direction. Also the kinetics of polyesterification can be rather slow at ambient temperature and so the reactants are often heated to a higher temperature and a catalyst is sometimes added. Both of these two modifications can speed up the reaction.

The principles outlined above for polyesterification reactions are equally applicable to step-growth polymerization reactions involving other types of monomer. For example, diamines and dicarboxylic acids combine together to give polyamides (nylons). Typical reactions which produce linear polymers by step-growth polymerization are given in Table 2.1 and in most cases they are basically condensation reactions. The most notable

TABLE 2.1 *Common reactions used to form linear polymers by step-growth polymerization.*

Reaction	Polymer
nHO—R—COOH → H[—O—R—CO—]$_n$OH + $(n-1)$H$_2$O	Polyester
nHO—R$_1$—OH + nHOOC—R$_2$—COOH → H[—O—R$_1$—OOC—R$_2$—CO—]$_n$OH + $(2n-1)$H$_2$O	Polyester
nNH$_2$—R—COOH → H[—NH—R—CO—]$_n$OH + $(n-1)$H$_2$O	Polyamide
nNH$_2$—R$_1$—NH$_2$ + nHOOC—R$_2$—COOH → H[—NH—R$_1$—NHOC—R$_2$—CO—]$_n$OH + $(2n-1)$H$_2$O	Polyamide
nHO—R$_1$—OH + nOCN—R$_2$—NCO → [—O—R$_1$—O—OCNH—R$_2$—NHCO—]$_n$	Polyurethane

exception is that certain step-growth polymerization reactions in which polyurethanes are formed do not produce any condensation products.

2.1.3 *Ring formation*

It has so far been assumed that during step-growth polymerization with bi-functional monomers only linear chains are formed. However, this may not always be so and in some cases rings can be formed rather than linear chains.

e.g. nHO—RCOOH

H[—ORCO—]$_n$OH + $(n-1)$H$_2$O
linear polyester

R
nO—C=O + nH$_2$O
lactone

It is possible that ω-hydroxy carboxylic acids can form lactones rather than linear polyesters and in principle any of the hydroxy carboxylic acids produced during polymerization reactions between diacids and diols may undergo a similar cyclization process.

The extent to which a hydroxy carboxylic acid tends to form a ring depends strongly upon the size of ring that would be produced. It is found that 5-, 6- and 7-membered rings are the most stable and will usually form in preference to a linear chain. γ-hydroxybutyric acid will always form γ-butyrolactone in preference to a polymer chain

HO(—CH$_2$)$_3$—COOH →
CH$_2$—CH$_2$ / CH$_2$ O / C ‖ O   + H$_2$O

3- and 4-membered rings are highly strained and do not form. The strain increases in 8- to 11-membered rings and decreases for very large rings. However, for long chains the probability of the two ends of the growing chain meeting each other during a reaction is so low that large rings are rarely formed.

It must be pointed out that early polymer chemists were under the impression that synthetic polymers were ring molecules, but modern analytical techniques have shown that this is only true under very special circumstances.

2.1.4 *Carothers equation*

Carothers, who was a pioneer of step-growth polymerization proposed that there was a simple relationship between the number average degree of polymerization, $\bar{x}_n$, and a parameter, p, known as the *extent of reaction*. The definition of p is best considered for the polymerization of an ω-hydroxy carboxylic acid. When the reaction takes place the number of molecules decreases as the average length of each molecule increases. If the number of molecules present initially is N_o and the number remaining at time t, is N, then the total number of COOH (or OH) groups reacted is ($N_o - N$). The extent of reaction p at time t is defined as the probability that any functional group present initially has reacted and so p is given by

$$p = \frac{\text{number of groups that have reacted}}{\text{number of groups present initially}}$$

and so

$$p = \frac{(N_o - N)}{N_o} \quad \text{or} \quad \frac{N_o}{N} = \frac{1}{(1 - p)} \tag{2.1}$$

As polymerization takes place the total number of molecules is reduced and their average length becomes longer. The number average degree of polymerization $\bar{x}_n$, is therefore given by

$$\bar{x}_n = \frac{\text{number of molecules present initially}}{\text{number of molecules remaining after time}, t.}$$

i.e. $\bar{x}_n = \dfrac{N_o}{N}$ $\tag{2.2}$

Combining Equations (2.1) and (2.2) gives the *Carothers equation.*

$$\bar{x}_n = \frac{1}{(1 - p)} \tag{2.3}$$

The equation is also valid for the reaction between a diacid and diol when there are also equal numbers of functional groups. In practice, it is found that values of p must be close to unity if polymers with useful mechanical properties are to be produced. Normally degrees of polymerization in excess of 50 are needed. Table 2.2 shows the relation between $\bar{x}_n$ and p.

TABLE 2.2 *Relationship between extent of reaction and number average degree of polymerization.*

p	0.50	0.90	0.99	0.999	0.9999
$\bar{x}_n$	2	10	100	1000	10000

Even when the percentage conversion of monomer into polymer is 50 percent the polymer molecules only have, on average, two repeat units.

If equal numbers of functional groups are not used a modified version of the Carothers equation must be employed which is

$$\bar{x}_n = \frac{(1 + r)}{(1 + r - 2rp)} \qquad (2.4)$$

where r is the ratio of the number of molecules of the two types of reactant and is always defined so that it is less than or equal to one. If r is equal to unity the equation becomes identical with the original Carothers Equation (2.3). The effect of not having equal numbers of functional groups is most easily shown with a numerical example. If there is 5 percent more diacid molecules than diol molecules in a linear polycondensation reaction then r is equal to $1/1.05 = 0.9524$. For an extent of reaction, p, of 0.999 Equation (2.4) gives a value of $\bar{x}_n$ of

$$\bar{x}_n = \frac{(1 + 0.9524)}{1 + 0.9524 - 2 \times 0.999 \times 0.9524} = 39.4$$

compared with a value of 1000 for the case where equal numbers of functional groups are employed (Table 2.2). As the reaction approaches completion

$$p \to 1 \quad \text{and} \quad \bar{x}_n \to \frac{(1 + r)}{(1 - r)}$$

and so for r equal to 0.9524 the maximum possible value of $\bar{x}_n$ is 41. This is a dramatic reduction in the maximum possible degree of polymerization which would theoretically be infinite (Equation 2.3) for equimolar quantities and it gives a good example of the importance of maintaining the stoichiometric balance in step-growth polymerization.

2.1.5 *Kinetics*

In theory the kinetics of step-growth polymerization could be very complicated as molecules of a variety of lengths are present once the reaction proceeds but in practice things are greatly simplified by the assumption, which has been shown experimentally to be valid, that the reactivity of a functional group is unaffected by the length of polymer chain to which it is attached. The kinetics of step-growth polymerization are normally exemplified through the kinetics of polycondensation reactions which depend upon whether or not a catalyst is used. The rate of reaction is normally defined as the rate of consumption of acid groups which is of course also equal to the rate of consumption of hydroxyl groups since every COOH group reacts with one OH group.

$$\text{i.e.} \quad \text{rate of reaction} = \frac{-d[COOH]}{dt} = \frac{-d[OH]}{dt}$$

where the square brackets refer to the concentration of COOH groups and OH groups in mol/m^3.

In the case in which *no catalyst is added* it is found experimentally that acid groups in the reaction mixture act as self-catalysts. The basic elementary reaction is then

$$\text{⋀COOH} + \text{HO⋀} + \text{catalyst} \rightarrow \text{⋀COO⋀} + H_2O + \text{catalyst}$$

and so the rate of reaction is given by

$$\frac{-d[COOH]}{dt} = k[COOH][OH][\text{catalyst}]$$

But since the [COOH] groups catalyse the reaction then

$$\frac{-d[COOH]}{dt} = k[COOH]^2[OH] \tag{2.5}$$

where k is the rate constant for the reaction. If equal numbers of functional groups are employed in the reaction as would be the case for an ω-hydroxy carboxylic acid then

$$[COOH] = [OH] = c$$

and Equation (2.5) becomes

$$\frac{-dc}{dt} = kc^3$$

This equation can be readily integrated since if at $t = 0$, $c = c_o$

$$\int_{c_o}^{c} \frac{-dc}{c^3} = \int_{o}^{t} k dt$$

and so

$$\frac{1}{c^2} - \frac{1}{c_o^2} = 2kt \tag{2.6}$$

Equation (2.1) can be rewritten in terms of the concentration of reactants, c, rather than the numbers of molecules, N, by considering unit volume of the mixture

$$\text{i.e.} \quad \frac{N_o}{N} = \frac{1}{(1-p)} = \frac{c_o}{c}$$

and so the concentration of reactants can be related to p and Equation (2.6) becomes

$$2ktc_o^2 = \frac{1}{(1-p)^2} - 1 \tag{2.7}$$

For the polycondensation reaction without added catalyst a plot of $1/(1-p)^2$ versus reaction time should yield a straight line of slope $2kc_o^2$ and intercept 1. Fig. 2.1(a) gives some experimental data which are in good agreement with Equation (2.7).

Often the uncatalysed reactions are rather slow and if a *catalyst is added* there can be a considerable acceleration of the rate. Equation (2.5) can also be applied in this case and it becomes

$$\frac{-d[COOH]}{dt} = k'[COOH][OH] \tag{2.8}$$

where k' is a new rate constant which is in fact a composite quantity

$$k' = kf([catalyst])$$

where k is the fundamental rate constant and $f([catalyst])$ some function of the catalyst concentration. k' can only be a constant if the substance added is a true catalyst and does not alter in concentration as the reaction proceeds. If equal numbers of function groups are employed Equation (2.8) becomes

$$\frac{-dc}{dt} = k'c^2$$

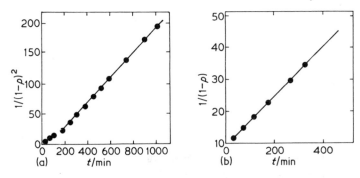

Fig. 2.1 (a) *Plot of $1/(1-p)^2$ as a function of time for the polymerization of ethylene glycol with adipic acid at 459 K. (b) Plot of $1/(1-p)$ as a function of time for the polymerization of decamethylene glycol with adipic acid as a function of time using p-toluene sulphonic acid as a catalyst at 369 K. (Both sets of data taken from Flory).*

and this can be integrated as before if k' remains constant with time to give

$$\frac{1}{c} - \frac{1}{c_o} = k't$$

or $\quad c_o k't = \dfrac{1}{(1-p)} - 1$ $\hfill$ (2.9)

This equation has been verified experimentally as shown in Fig. 2.1(b) by plotting $1/(1-p)$ versus t for an acid catalysed polycondensation reaction.

2.1.6 *Size distribution in linear polymers*

The creation of long polymer chains is a random process and this leads to a distribution in the lengths of molecules that are formed. However, the distribution of chain lengths can be determined through statistical calculations. If we consider a general ω-hydroxy carboxylic acid

HO—R—COOH

undergoing self-polymerization the problem is to calculate the probability of finding a polymer chain composed of i basic repeat units H[—ORCO—]$_i$OH at a time, t, during the reaction. This can be readily calculated if the number of ester linkages is considered

H—ORCO—ORCO—ORCO— .... —ORCO—OH
$\quad\quad\quad$ 1 $\quad\quad$ 2 $\quad\quad$ 3 $\quad$ $(i-1)$

In a polymer containing i repeat units there are $(i-1)$ ester linkages. In Section 2.1.4 it was shown that the extent of reaction, p, is also the probability that any functional group has reacted and so it follows that after time t the probability that a particular molecule has been formed is the product of the probabilities of forming the individual bonds in that molecule. In the molecule drawn above containing i repeat units

prob. of forming 1st ester linkage	$= p$
prob. of forming 2nd ester linkage next to 1st	$= p^2$
prob. of forming 3rd ester linkage next to 2nd	$= p^3$
prob. of forming $(i - 1)$ ester linkage next to $(i - 2)$	$= p^{(i-1)}$

and so the probability that all the ester linkages in the molecule H[—ORCO—]$_i$OH have been formed is $p^{(i-1)}$. However for the molecule to exist, as well as having the ester linkages, it must also have an unreacted group at the end. The probability that a group has not reacted is $(1 - p)$ and so the probability that the polyester molecule containing i units exists is

$$P(i) = p^{(i-1)}(1 - p) \tag{2.10}$$

The problem is now to calculate the number of molecules of this type. If the total number of molecules present at time t is N and the number of molecules of length i is N_i then it follows that

$$N_i = P(i)N = Np^{(i-1)}(1 - p) \tag{2.11}$$

Often the number of molecules present at any time cannot be measured but normally the number of molecules present initially, N_o is known. In Section 2.1.4 it was shown that

$$N = N_o(1 - p)$$

and so Equation (2.11) becomes

$$N_i = N_o p^{(i-1)}(1 - p)^2 \tag{2.12}$$

which is an expression for the number of molecules of length i in terms of the initial number of molecules N_o and the extent of reaction, p.

It is often useful to know the mass fraction w_i of molecules of a particular length as well as the numbers and this can be readily calculated since w_i is given by

$$w_i = \frac{\text{mass of molecules of length } i}{\text{total mass of all the molecules}}$$

and so if M_o is the molar mass of a repeat unit or monomer molecule then

$$w_i = \frac{N_i \times (i \times M_o)}{N_o \times M_o} = \frac{iN_i}{N_o} \tag{2.13}$$

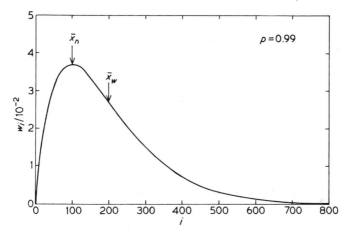

Fig. 2.2 *The mass fraction distribution of molecular lengths during step-growth polymerization according to Equation (2.14) for p = 0.99.*

if end-groups and the difference in mass between repeat units and monomer molecules are ignored. Combining Equations (2.12) and (2.13) gives

$$w_i = ip^{(i-1)}(1 - p)^2 \qquad (2.14)$$

Equation (2.14) is plotted in Fig. 2.2 for $p = 0.99$. It is found that there will be a maximum value of w_i at a particular value i for a given p. But again Fig. 2.2 shows that high degrees of conversion are needed before significant proportions of long molecules are obtained.

The example given here is for an ω-hydroxy carboxylic acid, but an identical result is obtained for other linear polycondensation reactions as long as equal numbers of functional groups are used.

2.1.7 *Molar mass distribution*

Once N_i and w_i are known the number average and weight average molar masses, $\bar{M}_w$ and $\bar{M}_n$ can be readily determined for a linear polycondensation reaction.

The *number average molar mass* $\bar{M}_n$ was given in Section 1.4.1 as

$$\bar{M}_n = \sum N_i M_i \Big/ \sum N_i$$

Now if a polymer sample contains N molecules then $\sum N_i = N$ and from Equation (2.11)

$$N_i = Np^{(i-1)}(1 - p)$$

and so if M_o is the molar mass of the polymer repeat unit it follows that

$$\bar{M}_n = M_o(1 - p) \sum ip^{(i-1)}$$

Using the standard mathematical relation

$$\sum_1^\infty ip^{(i-1)} = \frac{1}{(1 - p)^2} \quad \text{for} \quad p < 1$$

we have

$$\bar{M}_n = \frac{M_o}{(1 - p)} \tag{2.15}$$

Also since $\bar{x}_n = \bar{M}_n/M_o$ then it follows that

$$\bar{x}_n = \frac{1}{(1 - p)} \tag{2.16}$$

which is the Carothers Equation (2.3) that this time has been calculated using purely statistical considerations.

The *weight average molar mass*, $\bar{M}_w$ was given in Section 1.4.1 as

$$\bar{M}_w = \sum w_i M_i$$

Using Equation (2.14) it follows that

$$\bar{M}_w = M_o(1 - p)^2 \sum i^2 p^{(i-1)}$$

and another standard mathematical relation

$$\sum_1^\infty i^2 p^{(i-1)} = \frac{1 + p}{(1 - p)^3}$$

leads to

$$\bar{M}_w = M_o\frac{(1 + p)}{(1 - p)} \tag{2.17}$$

and

$$\bar{x}_w = \frac{(1 + p)}{(1 - p)} \tag{2.18}$$

which is the weight average degree of polymerization.

A parameter which is often quoted for polymer samples is the heterogeneity index, $\bar{M}_w/\bar{M}_n$ which can be determined from Equations (2.15) and (2.17) as

$$\frac{\bar{M}_w}{\bar{M}_n} = (1 + p) \tag{2.19}$$

As a step-growth polymerization approaches completion p tends to unity and so

$$\frac{\bar{M}_w}{\bar{M}_n} \to 2 \quad \text{as} \quad p \to 1$$

A polymer with a heterogeneity index of 2 is often said to have an *ideal distribution* of molar mass. However, care must be taken because it is possible to formulate polymer samples for which $\bar{M}_w/\bar{M}_n$ is 2 but the distribution of molar mass may not be ideal.

The relation between $\bar{x}_w$ and $\bar{x}_n$ and the distribution of chain lengths is shown in Fig. 2.2. Normally $\bar{x}_n$ is close to the peak value of w_i and $\bar{x}_w$ is somewhat higher. It must again be stressed that the calculations in this section are only for polycondensation reactions in which there are equal numbers of functional groups and no side reactions occur.

2.1.8 *Linear step-growth polymerization systems*

In order to produce polymer molecules by direct polyesterification reactions highly pure reactants are required and accurate control of the amount of each species in the reaction mixture is needed. Even so these reactions rarely go to conversions in excess of 98 percent and so there is not much high molar mass polymer produced. Because of the problems of using direct condensation reactions several specialized reactions have been developed for the formation of different step-growth polymers.

Ester interchange reactions are often employed in the production of polyesters. Poly(ethylene terephthalate) which is sometimes known commercially in the form of fibres as 'terylene' could be produced by direct esterification

$$n\text{HOOC}\!-\!\!\bigcirc\!\!-\!\text{COOH} + n\text{HO}(\text{CH}_2)_2\text{OH} \to$$

terephthalic acid   ethylene glycol

$$\text{HO}[-\text{OC}\!-\!\!\bigcirc\!\!-\!\text{COO}\!-\!(\text{CH}_2)_2\!-\!\text{O}]_n + (2n-1)\text{H}_2\text{O}$$

poly(ethylene terephthalate)

However, the yield in such a reaction is very low and there are several problems in using such a reaction in the industrial situation. Firstly, aromatic acids such as terephthalic acid are difficult to purify because of low solubility and a high melting point. Secondly, the reverse reaction can take place and the equilibrium does not strongly favour the formation of polyester even when water is removed. Finally, equal numbers of functional groups are essential in such a reaction if high values of $\bar{x}_n$ are to be achieved (Equation 2.4) and strict control of the amount of reactants is therefore essential. These problems may all be overcome by using ester-interchange reactions.

Impure terephthalic acid can be taken first of all and reacted with methanol to form a dimethyl ester.

$$HOOC-\langle O \rangle-COOH + 2CH_3OH \rightarrow$$

$$CH_3OOC-\langle O \rangle-COOCH_3 + 2H_2O$$

This ester can be easily purified by distillation. The ester will then undergo an ester-interchange reaction with an excess of ethylene glycol at a relatively low temperature

$$CH_3OOC-\langle O \rangle-COOCH_3 + 2HO-(CH_2)-OH \xrightarrow{\;400\ K\;}$$

$$HO(CH_2)OOC-\langle O \rangle-COO(CH_2)OH + 2CH_3OH$$

to form a low molar mass polyester with OH end groups. Moreover, the methanol can be distilled off and recirculated to be used to purify more terephthalic acid. If the low molar mass polyester is taken and heated to 530K it will undergo further ester-interchange reactions with other polyesters and produce reactions with other polyesters and produce longer polymer molecules.

$$2HO(CH_2)_2OOC-\langle O \rangle-COO(CH_2)_2OH \xrightarrow{\;530\ K\;}$$

$$HO(CH_2)_2OOC-\langle O \rangle-COO(CH_2)_2-OOC-\langle O \rangle-COO(CH_2)_2OH$$

$$+ HO(CH_2)_2OH$$

Further polymerization will take place as long as the ethylene glycol produced is removed (and recirculated). By using ester-interchange reactions the need for strict stoichiometric control is eliminated.

It should be noted that ester-interchange reactions can take place at any stage during polyesterification, but they do not affect the molar mass distribution. This is because two molecules of random length react together to produce two more molecules and the total number of molecules in the mixture remains unchanged.

Salt dehydration is a convenient reaction which can be used to produce the polyamide nylon 6.6 which is used widely as a textile fibre. The monomers used are hexamethylene diamine and adipic acid, but a normal condensation reaction is not employed. Instead the stoichiometric balance is ensured by utilizing the tendency for the two monomer molecules to form a 1:1 salt which can be crystallized from solution helping to purify the reactants. The salt melts at 456 K and can be converted into the polyamide by heating at 550 K under pressure

$$n\mathrm{H_2N(CH_2)_6NH_2} + n\mathrm{HOOC(CH_2)_4COOH}$$

hexamethylene   adipic acid
diamine

$$n\,^{+}\mathrm{H_3N(CH_2)_6NH_3}^{+}\; {}^{-}\mathrm{OOC(CH_2)_4COO}^{-}$$

550 K
under pressure

$$\mathrm{H[-NH(CH_2)_6NHOC(CH_2)_4CO-]_{\textit{n}}OH} + (2n-1)\mathrm{H_2O}$$

nylon 6.6

Interfacial polymerization can also be used to produce polyamides. A spectacular example of this is the 'nylon rope trick' where a continuous film of nylon 6.10 is drawn from the interface between immiscible solutions of sebacoyl chloride and hexamethylene diamine as illustrated schematically in Fig. 2.3. The reaction takes place only at the interface.

$$n\mathrm{H_2N(CH_2)_6NH_2} + n\mathrm{ClOC(CH_2)_8COCl} \rightarrow$$

hexamethyl   sebacoyl chloride
diamine

$$\mathrm{H[-HN(CH_2)_6NHOC(CH_2)_8CO]_{\textit{n}}Cl} + (2n-1)\mathrm{HCl}$$

nylon 6.10

and so it is controlled by diffusion and there is no need for strict control of stoicheiometry. In order to produce long chains and speed up the kinetics of the reaction the system can be stirred vigorously to ensure a constantly changing interface. Nylon 6.10 is widely used in the form of monofilaments such as for bristles or in sports equipment. It is less susceptible to water uptake than nylon 6.6 because it has a higher proportion of $-\mathrm{CH_2}-$ units on the polymer chain.

Ring-opening polymerization is another reaction that can be used to form polyamides and it can be contrasted with the ring-formation reactions described in Section 2.1.3. An important example is the polymerization of caprolactam to give nylon 6 which takes place at 520 K and is catalysed by water.

caprolactam   nylon 6

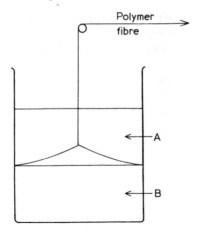

Fig. 2.3 *Schematic illustration of interfacial polymerization. Solution A is an aqueous solution of hexamethylene diamine and B a solution of sebacoyl chloride in carbon tetrachloride. The polymer can be drawn off in the form of a fibre.*

The ring in caprolactam contains 7 atoms and can be fairly readily opened. As in the case of ring formation reactions, 5- and 6-membered rings are more stable and are resistant to ring-opening reactions. The polyamide nylon 6 is used extensively in tyre cord and is finding increasing use in mouldings such as those for gears or bearings, where metal parts are often replaced.

2.1.9 *Non-linear step-growth polymerization*

If bifunctional monomers are used in step-growth polymerization linear polymer molecules are formed at sufficiently high degrees of conversion, but if molecules with a functionality of greater than two are introduced then the polymerization reaction can be very different. If a triol is reacted with a diacid a branched rather than a linear chain will form

$$
\begin{array}{c}
\mathrm{OH} \\
|
\end{array}
$$

e.g. $2\mathrm{HO{-}R_1{-}OH} + 3\mathrm{HOOC{-}R_2{-}COOH} \rightarrow$

$$\mathrm{HO{-}R_1{-}OOC{-}R_2{-}COO{-}R_1{-}OOC{-}R_2{-}COOH} + 4\mathrm{H_2O}$$

$$
\begin{array}{cc}
| & | \\
\mathrm{OH} & \underset{\underset{\mathrm{CR_2{-}COOH}}{\overset{\mathrm{O}}{}}}{\mathrm{O}}
\end{array}
$$

The branched chains can go on by further condensation reactions to form complex three-dimensional network structures which have physical and mechanical properties which are quite different from those of linear polymers. The point at which a three-dimensional network is formed is manifested by an abrupt change in the properties of the reaction mixture. At first the viscosity of the mixture increases steadily as the molecules become longer as is the case for linear step-growth reactions. When the network forms there is a dramatic increase in the viscosity and the mixture changes from a viscous liquid into a gel which shows no tendency to flow and in which bubbles cease to rise. This can occur at a relatively early stage in the reaction ($p < 0.8$) and it is known as the *gel-point* for the reaction.

2.1.10 *Prediction of gel-point*

Two main theories have been developed to predict the onset of gelation in step-growth reactions involving polyfunctional monomers. A simple method uses a *modified version of the Carothers equation*. In this case a parameter f_{av} is produced which is defined as the average functionality of all of the monomer units. If a system contains N_o molecules initially then total number of functional groups present is $N_o f_{av}$. If there are N molecules present at time, t, then the number of functional groups that have reacted to form these molecules is $2(N_o - N)$ since two different groups must react together to form one linkage. The extent of reaction p is the probability that any functional group has reacted and so

$$p = \frac{\text{number of groups reacted at time } t}{\text{number of groups present initially}}$$

i.e. $\quad p = \dfrac{2(N_o - N)}{N_o f_{av}}$

But since $\bar{x}_n = N_o/N$ the equation becomes

$$\bar{x}_n = \frac{2}{(2 - pf_{av})} \qquad (2.21)$$

If it is postulated that gelation occurs when $\bar{x}_n$ goes to infinity then it follows that the critical extent of reaction for gelation p_G is given by

$$p_G = 2/f_{av}. \qquad (2.22)$$

Increasing the average functionality has a dramatic effect upon the degree of polymerization as is demonstrated in Table 2.3.

TABLE 2.3 *Values of $\bar{x}_n$ for different values of p and f_{av} using Equation (2.21)*

p	0.5	0.7	0.9	0.95	0.99
$f_{av} = 2.0$	2	3.33	10	20	100
$f_{av} = 2.1$	2.10	3.77	18.18	400	∞
$f_{av} = 2.2$	2.22	4.35	100	∞	∞

It can be seen that at degrees of conversion where only moderate values of $\bar{x}_n$ are achieved with bifunctional monomers, slightly increasing the average functionality $f_{av.}$ greatly increases the degree of polymerization. Equation (2.21) is only valid for the case where the total numbers of each type of functional group are equal. It is possible to calculate an expression for p_G when this is not so but it is beyond the scope of this book.

It is also possible to derive a different expression for p_G using purely *statistical* considerations. This was done originally by Flory who took a system in which there are three types of monomer molecules.

bifunctional monomer (e.g. diol)   A—A

bifunctional monomer (e.g. diacid)   B—B

trifunctional monomer (e.g. triol)

These three types of molecule will react together to form a general linkage

and the terminal trifunctional units may or may not be joined to other B—B units. It is necessary to define a parameter α called the *branching coefficient* which is the probability that one trifunctional unit is connected to a chain which has a second trifunctional unit at the other end. In other words it is the probability that the general linkage drawn above exists and α can be derived using statistical arguments. In order to do this it is necessary to introduce another term γ which is defined as

$$\gamma = \frac{\text{number of A groups on branch units}}{\text{total number of A groups present}}$$

Using γ it is possible to calculate the probabilities that each of the bonds have been formed in the general linkage. If the extent of reaction is p then

prob. that any group has reacted $= p$

prob. that an A is linked to a B $= p$

prob. that a B—B is linked to a $= p\gamma$

prob. that a B—B is linked to a A—A $= p(1 - \gamma)$

and so for the bonds in the general linkage

$$\begin{array}{c}\rangle\!-\!A[B\!-\!BA\!-\!A]_iB\!-\!BA\!-\!\langle\\ \uparrow\qquad\uparrow\qquad\uparrow\qquad\uparrow\\ p\quad p(1-\gamma)\,p\qquad p\gamma\end{array}$$

the probability that the linkage has formed is $p[p(1-\gamma)p]^i p\gamma$. The branching coefficient α is the probability that linkages with all values of i are formed and so

$$\alpha = p^2\gamma \sum_0^\infty [p^2(1-\gamma)]^i$$

Using the standard mathematical relation

$$\sum_0^\infty x^i = 1/(1-x) \quad \text{for} \quad x < 1$$

then

$$\alpha = \frac{p^2\gamma}{1 - p^2(1-\gamma)} \tag{2.23}$$

In order to calculate the critical conditions for gelation the functionality of the branching units, f, must be introduced. f is equal to 3 for the triol units in the present example and is not the same as f_{av}. If one end of the linkage is connected to more polymer segments then the condition for network formation can be considered to be when more than one growing chain extends from the other end. The number of chains capable of emanating from the end is $(f-1)$ and so the probable number of chains emanating from the end is $\alpha(f-1)$. The condition for gelation is simply that this probability is not less than one i.e. $\alpha(f-1) > 1$ for network formation. The critical branching coefficient α_G is given by

$$\alpha_G = \frac{1}{(f-1)} = \frac{p_G^2\gamma}{1 - p_G^2(1-\gamma)}$$

and so

$$p_G = [1 + \gamma(f-2)]^{-1/2} \tag{2.24}$$

which is the critical extent of reaction for gelation using the statistical theory.

It is interesting to compare the predictions of the two theories with each other and with what happens in practice. Fig. 2.4 shows the variation of p, $\bar{x}_n$ and the viscosity η with time for a reaction between diethylene glycol (a diol), succinic acid (a diacid) and 1,2,3-propane tricarboxylic acid (a triacid). The mixture was found to gel after about 230 min and this was manifested by η becoming infinite. p_G is about 0.9 and $\bar{x}_n$ was only 25 at this point (and not infinite). For this system it was found when the total number of the two types of functional groups is the same, for a value of γ of 0.29 the modified Carothers Equation (2.22) predicts a p_G of 0.95 and the statistical theory (Equation 2.24) predicts a value of 0.88 whereas experimentally it was found that p_G was equal to 0.91. The discrepancy between the three values is usually found. The approach which uses the modified Carothers equation assumes that $\bar{x}_n$ is infinite at the gel-point, but Fig. 2.4 clearly shows that gelation takes place earlier in the reaction. It must be remembered that $\bar{x}_n$ is an average and molecules both longer and shorter than the average exist. The long molecules form a network at a stage in the reaction earlier than the theory would predict. The statistical theory does not suffer from these problems, but even so it predicts that gelation occurs at a value of p lower than actually found. This is because all of the branches do not help to form a network. Some branches are lost by the two ends reacting with each other to form a loop. The polymerization must proceed further to overcome these non-productive reactions.

2.1.11 *Network polymers*

Network or thermosetting polymers are widely used commercially because cross-linked structures sometimes have improved toughness and strength. They are normally produced in two stages. The first is a prepolymer which may either be liquid or solid and contains molecules with a low molar mass. The polymer will flow to fill a mould in this form and the material is then usually cured to give a heavily cross-linked product. Once they have been cured thermosetting polymers cannot be re-melted and tend to decompose on heating.

   Phenol–formaldehyde resins were developed by Baekeland in the early years of this century as 'Bakelite'. Depending upon the molar ratio of phenol to formaldehyde, two quite separate phenolic resins can be produced.

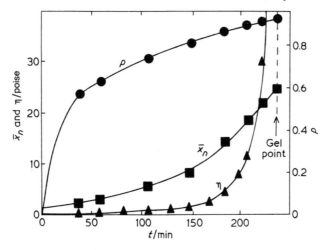

Fig. 2.4 *Variation of p, $\bar{x}_n$ and η with time for the polymerization of ethylene glycol with succinic acid and 1,2,3-propane tricarboxylic acid. (After Flory).*

If the ratio of phenol to formaldehyde is greater than one and an acid catalyst is used a Novolak can be formed. The phenol reacts with the formaldehyde through a methylolation reaction

OH   OH

+ HCHO →   CH₂OH

The methylol derivatives condense with phenol

OH CH₂OH   OH   OH   OH

+   →   CH₂   + H₂O

and go on through further methylolation and condensation reactions to form the novolak. These products contain no reactive —CH₂OH groups and so are not able to cross-link. They are normally dried and ground up, mixed with fillers (e.g. mica, glass fibre, sawdust) and colourants and cured in a hot mould using hexamine as a cross-linking agent and CaO as a catalyst. During the heating process the hexamine decomposes to form formaldehyde and ammonia by the catalytic action of the CaO and the ammonia acts as a catalyst for the final cross-linking by the formaldehyde to form a three-dimensional network. The fillers are added to improve the electrical or mechanical properties of the resin.

In the presence of an alkaline catalyst and with a lower phenol–formaldehyde ratio resoles are formed. These products will condense with each other on further heating to give a cross-linked structure and no separate curing agent is necessary.

Amino resins are formed from the reaction between formaldehyde and a bifunctional amine (urea) or a trifunctional amine (melamine)

$$
\begin{array}{cc}
\underset{\text{urea}}{O=C\diagup\!\!\!\!\!\!\overset{NH_2}{\underset{NH_2}{\diagdown}}} & \text{melamine}
\end{array}
$$

Polymerization takes place through a series of methylol and condensation reactions to form amino resins which are clear and colourless. This makes them easier to colour than the phenolic resins which are naturally dark coloured.

Epoxy resins have been developed principally as adhesives for the aero-space industry, but are now also used as electrical insulators and surface coatings. A low molar mass prepolymer is formed by the reaction between epichlorohydrin and bisphenol A in an aqueous alkaline environment.

$$(n+2)Cl\!-\!CH_2\!-\!CH\!-\!CH_2 + (n+1)HO\!-\!\!\bigcirc\!\!-\!\!\underset{CH_3}{\overset{CH_3}{C}}\!\!-\!\!\bigcirc\!\!-\!OH$$

$$\xrightarrow[\text{NaOH}]{\text{aqueous}} CH_2\!-\!CH\!-\!CH_2\!-\![O\!-\!\bigcirc\!\!-\!\!\underset{CH_3}{\overset{CH_3}{C}}\!\!-\!\!\bigcirc\!\!-\!OCH_2CHCH_2\!-\!]_n\!-$$

An excess of epichlorohydrin is used so that the epoxy group $(CH_2\!-\!CH\!-\!)$ is found at both ends of the chain. The prepolymer is usually a viscous liquid or low melting-point solid depending upon the value of n and can be cured with a variety of substances such as amines, acids and anhydrides. The principal curing agents used in practice are multi-functional amines which react with the terminal epoxy groups, or polyanhydrides which cross-link through the —OH group on the chain. The reaction with a diamine is illustrated schematically below

No water is produced in the reaction and the shrinkage on curing is small. Fillers are often added to improve the electrical and mechanical properties, and epoxy resins are becoming increasingly used as matrix materials for high performance carbon-fibre reinforced composites.

2.2 Free radical addition polymerization

Addition polymerization is the second main type of polymerization reaction. It differs from step-growth polymerization in several important ways. It takes place in three distinct steps *initiation, propagation* and *termination* and the principal mechanism of polymer formation is by addition of monomer molecules to a growing chain. Monomers for addition polymerization normally contain double bonds and are of the general formula $CH_2{=}CR_1R_2$. Some typical monomers are listed in Table 2.4. The double bond is susceptible to attack by either free radical or ionic initiators to form a species known as an *active centre*. This active centre propagates a growing chain by the addition of monomer molecules and the active centre is eventually neutralized by a termination reaction. Since the reaction only occurs at the reactive end of the growing chain long molecules are present at an early stage in the reaction along with unreacted monomer molecules which are around until very near the end of the reaction. It is possible to generalize about many features of addition polymerization, but there are sufficient differences between free radical and ionic initiated reactions for the two types to be treated separately.

TABLE 2.4 *Some monomers used in addition polymerization reactions and the repeat unit of the corresponding polymers.*

Monomer name	Monomer structure	Polymer repeat unit
Ethylene	$CH_2{=}CH_2$	$[-CH_2-CH_2-]_n$
Vinyl chloride	$CH_2{=}CH$ │ Cl	$[-CH_2-CH-]_n$ │ Cl
Styrene	$CH_2{=}CH$ │ (phenyl)	$[-CH_2-CH-]_n$ │ (phenyl)
Vinyl acetate	$CH_2{=}CH$ │ $OOCCH_3$	$[-CH_2-CH-]_n$ │ $OOCCH_3$
Acrylonitrile	$CH_2{=}CH$ │ CN	$[-CH_2-CH-]_n$ │ CN
Methyl methacrylate	$CH_2{=}C-CH_3$ │ $COOCH_3$	CH_3 │ $[-CH_2-C-]_n$ │ $COOCH_3$
Vinylidene chloride	$CH_2{=}CCl_2$	Cl │ $[-CH_2-C-]_n$ │ Cl

2.2.1 *Free radical initiators*

Free radicals are species which contain an unpaired electron and will be denoted as R·. They are extremely reactive and will react with monomers containing a double bond to form an active centre which is capable of reacting with further monomer molecules to give a macromolecular chain. In order for an initiator molecule to be effective it must be capable of breaking down into radicals which are stable long enough to react with a monomer molecule and form an active centre. Initiators are normally required to form radicals in a controlled way usually by the application of heat or electromagnetic radiation or during a chemical reaction as described below.

Free radicals may be readily produced by the *thermal decomposition* of certain peroxides and azo compounds. Benzoyl peroxide breaks down into benzoyloxy radicals which in turn break down into phenyl radicals and carbon dioxide.

$$\underset{\text{Benzoyl peroxide}}{\bigcirc\!\!\!\!-\overset{\overset{\text{O}}{\|}}{C}-O-O-\overset{\overset{\text{O}}{\|}}{C}-\bigcirc} \rightarrow 2\,\underset{\substack{\text{benzoyloxy}\\\text{radicals}}}{\bigcirc\!\!\!\!-\overset{\overset{\text{O}}{\|}}{C}-O\cdot} \rightarrow 2\,\underset{\substack{\text{phenyl}\\\text{radicals}}}{\bigcirc\!\!\cdot} + 2CO_2$$

This reaction takes place at 60°C in a benzene solution, but not all the radicals produced may go on to initiate chain growth. Secondary reactions can occur between the radicals produced because of the confining effect of solvent molecules (the 'cage' effect). Primary recombination can occur when the two benzoyloxy radicals produced are unable to diffuse away from each other fast enough

$$\bigcirc\!\!\!\!-\overset{\overset{\text{O}}{\|}}{C}-O\cdot\;+\;\cdot O-\overset{\overset{\text{O}}{\|}}{C}-\bigcirc \rightarrow \bigcirc\!\!\!\!-\overset{\overset{\text{O}}{\|}}{C}-O-O-\overset{\overset{\text{O}}{\|}}{C}-\bigcirc$$

Also two phenyl radicals or a phenyl radical and a benzoyloxy radical can combine together. There can also be wastage of initiator molecules through induced decomposition by phenyl radicals

$$\bigcirc \;+\; \bigcirc\!\!\!\!-\overset{\overset{\text{O}}{\|}}{C}-O-O-\overset{\overset{\text{O}}{\|}}{C}-\bigcirc$$

$$\rightarrow \bigcirc\!\!\!\!-\overset{\overset{\text{O}}{\|}}{C}-O-\bigcirc \;+\; \bigcirc\!\!\!\!-\overset{\overset{\text{O}}{\|}}{C}-O\cdot$$

These secondary reactions tend to reduce the efficiency of the initiator molecules since the decomposition would otherwise be quantitative.

Azo compounds will decompose both with the application of heat and by *photolysis*. Azobisisobutyronitrile breaks down into cyanopropyl radicals and nitrogen at ambient temperature under the action of ultra-violet radiation.

$$\underset{\text{Azobisisobutyronitrile}}{(CH_3)_2\!\!-\!\!\underset{\overset{|}{CN}}{C}\!\!-\!\!N\!\!=\!\!N\!\!-\!\!\underset{\overset{|}{CN}}{C}\!\!-\!\!(CH_3)_2} \overset{h\nu}{\rightarrow} \underset{\substack{\text{cyanopropyl}\\\text{radicals}}}{2(CH_3)_2\!\!-\!\!\underset{\overset{|}{CN}}{C}\!\cdot} + N_2$$

The radicals that are formed again recombine due to the solvent cage effect and so reducing the initiator efficiency. Photolysis gives better control over the production of radicals since the reaction can take place at relatively low temperatures and can be stopped by the removal of the light source.

Chemical reactions such as *redox reactions* may be used to produce radicals and they are particularly useful for the initiation of polymerization at low temperatures or for emulsion polymerization. The reaction between hydrogen peroxide and the ferrous ion produces ferric ions and hydroxyl radicals in solution

$$Fe^{2+} + HOOH \rightarrow Fe^{3+} + HO\cdot + OH^-$$

It will be shown later that emulsion polymerization is a process of considerable commercial importance.

2.2.2 *Formation of an active centre*

Once the free radicals have been formed the next stage of the initiation process is the formation of an active centre or chain carrier. If the free radical is denoted by $R\cdot$ and the monomer molecule as M then the reaction is simply

$$R\cdot + M \rightarrow RM_1\cdot$$

Very often the monomers are vinyl molecules of a general formula $CH_2{=}CHX$ where X can be the substituent Cl, C_6H_5, etc., and there are two possible reactions

$$
R\cdot + CH_2{=}CH \Big|_X
\begin{cases}
\nearrow & R{-}CH_2{-}CH\cdot \quad \Big|_X \qquad \text{I} \\
\searrow & R{-}CH{-}CH_2\cdot \quad \Big|_X \qquad \text{II}
\end{cases}
$$

The relative amounts of the two product radicals depends upon the difference between the activation energies of the two reactions. The activation energy for II is slightly higher than that for I because the X group which is usually large and bulky tends to hinder the approach of the $R\cdot$ radical. The lower activation energy for I leads to a much greater preponderance of the $R{-}CH_2{-}CHX\cdot$ radical.

2.2.3 *Propagation*

Chain propagation takes place by the rapid addition of monomer molecules to the growing chain

$$RM_1\cdot + M \rightarrow RM_2.$$

and this can be written, in general, ignoring the radical fragments as

$$M_i\cdot + M \rightarrow M_{i+1}\cdot$$

The average life-times of the growing chain are extremely short and several thousand additions can take place within a few seconds. This means that there may be a monomer addition every few milliseconds to each growing chain.

As with the reaction between the radical and the first monomer molecule described in the last section there is the problem of the way that the monomer adds on. There are two possibilities again for the way in which addition can take place

$$R—CH_2—CH\cdot + CH_2{=}CH$$

$$\begin{array}{c} R—CH_2—CH—CH_2—CH\cdot \quad (I) \\ \quad\quad | \quad\quad\quad\quad | \\ \quad\quad X \quad\quad\quad\quad X \end{array}$$

$$\begin{array}{c} R—CH_2—CH—CH—CH_2\cdot \quad (II) \\ \quad\quad | \quad\quad | \\ \quad\quad X \quad\quad X \end{array}$$

The first type of addition (I) is known as 'head-to-tail' addition and the second kind is called 'head-to-head'. For mainly steric reasons the addition reactions with vinyl polymers are predominantly the 'head-to-tail' variety giving a polymer which has alternating —CH_2— and —CHX— units as shown below.

$$\begin{array}{c} RCH_2—CHCH_2—CHCH_2—CHCH_2—\cdots\cdots\cdots \\ \quad\quad\quad | \quad\quad\quad\quad | \quad\quad\quad | \\ \quad\quad\quad X \quad\quad\quad X \quad\quad\quad X \end{array}$$

2.2.4 Termination

Free radicals are particularly reactive species and there are several ways in which the growing chains can react to form inert covalently bonded polymer molecules. The most important mechanisms of termination are when two growing chains interact with each other and become mutually terminated by one of two specific reactions. One of these reactions is *combination* where the two growing chains join together to form a single polymer molecule

$$\begin{array}{c} \text{www}CH_2—CH\cdot + \cdot HC—CH_2\text{www} \rightarrow \text{www}CH_2—CH—CH—CH_2\text{www} \\ \quad\quad | \quad\quad | \quad\quad\quad\quad | \quad\quad | \\ \quad\quad X \quad\quad X \quad\quad\quad\quad X \quad\quad X \end{array}$$

This combination reaction results in a 'head-to-head' linkage. Alternatively a hydrogen atom can be transferred from one chain to the other in a reaction known as *disproportionation*

$$\begin{array}{c} \text{www}CH_2—CH\cdot + \cdot HC—CH_2\text{www} \rightarrow \text{www}CH_2—CH_2 + HC{=}CH\text{www} \\ \quad\quad | \quad\quad | \quad\quad\quad\quad\quad | \quad\quad\quad | \\ \quad\quad X \quad\quad X \quad\quad\quad\quad\quad X \quad\quad\quad X \end{array}$$

The two growing chains become two polymer chains, one with a saturated group at the end and the other with an unsaturated group at its end. Also the chains have initiator fragments at only one end whereas polymer molecules formed by combination termination reactions have initiator fragments at both ends.

In general it is found that both types of termination take place in any particular system but to different extents. For instance it is found that polystyrene terminates principally by combination whereas poly(methyl methacrylate) terminates almost exclusively by disproportionation at high temperatures and by both mechanisms at lower temperatures.

Termination can also take place by reaction between a growing chain and a radical which has not reacted with any monomer molecules. It tends to occur particularly when high concentrations of initiator are used or when the concentration of monomer is depleted. It can also take place when the system increases in viscosity so that the growing chains lose their ability to diffuse around making it increasingly more difficult for two chain ends to meet each other.

The other main mechanism of termination is by *chain transfer* which can take place in a number of different ways. It is a process whereby a chain stops growing but the polymerization process does not stop because a new radical is generated. It is discussed in detail in Section 2.2.9.

2.2.5 *Kinetics*

The kinetics of addition polymerization contrast greatly with those for step-growth reactions since three distinct steps are involved. The *initiation* reaction takes place in two stages. First of all, the initiator molecules break down to form free radicals and then the radicals react with monomer molecules to form active centres. These two reactions have their own rate constants.

The initiator molecule I breaks down to form two radical fragments R·

$$I \xrightarrow{k_i} 2R \cdot$$

This reaction with rate constant k_i is slow and rate-determining and the rate of loss of initiator molecules is given by a rate equation for simple first order decay

$$\frac{-d[I]}{dt} = k_i[I] \tag{2.25}$$

where [I] is the concentration of initiator. Since two radicals are formed from one initiator molecule then the rate of formation of radicals is

$$\frac{d[R \cdot]}{dt} = 2k_i[I] \tag{2.26}$$

if the initiator is 100 percent efficient. The formation of an active centre

$$R\cdot + M \xrightarrow{k_r} RM_1\cdot$$

is a fast reaction and so the rate of active-centre formation is controlled by rate of formation of radicals (Equation 2.26). The amount of monomer consumed during the formation of active centres is negligible.

Propagation takes place by the addition of monomer molecules to the growing chain

$$M_i\cdot + M \xrightarrow{k_p} M_{i+1}\cdot$$

and it is assumed that the propagation rate constant k_p is independent of the length of the growing chain. Monomer is consumed by all growing chains of any length i and so the rate of consumption of monomer is given by

$$\frac{-d[M]}{dt} = k_p[M_1\cdot][M] + k_p[M_2\cdot][M] + \ldots + k_p[M_i\cdot][M] + \ldots$$

or

$$\frac{-d[M]}{dt} = k_p[M]([M_1\cdot] + [M_2\cdot] + \ldots + [M_i\cdot] + \ldots)$$

If the total concentration of all radical species is given by

$$\sum [M_i\cdot] = [M_1\cdot] + [M_2\cdot] + \ldots + [M_i\cdot] + \ldots$$

then

$$\frac{-d[M]}{dt} = k_p[M] \sum [M_i\cdot] \tag{2.27}$$

The *termination* reaction is a bimolecular process which can be either by combination of disproportionation.

$$M_i\cdot + M_j\cdot \begin{array}{c} \xrightarrow{k_{tc}} M_{i+j} \\ \xrightarrow{k_{td}} M_i + M_j \end{array}$$

The kinetic equation for either case is of a similar form and since two growing chains of any length are consumed by the reaction then

$$\frac{-d[M_i\cdot]}{dt} = 2k_{tc} \sum [M_i\cdot] \sum [M_j\cdot]$$

and

$$\frac{-d[M_i\cdot]}{dt} = 2k_{td} \sum [M_i\cdot] \sum [M_j\cdot]$$

Since the equations for the two mechanisms have the same form and both mechanisms often take place simultaneously then if $(k_{tc} + k_{td})$ is replaced by k_t.

$$\frac{-d[M_i\cdot]}{dt} = 2k_t\left(\sum [M_i\cdot]\right)^2 \qquad (2.28)$$

as chains of length i cannot be distinguished from those of length j. In the kinetic equations developed here the only termination mechanisms that will be considered are combination and disproportionation. If chain transfer occurs the analysis has to be modified.

In order to simplify the kinetic scheme a *steady-state approximation* has to be made. It is assumed that under steady-state conditions the net rate of production of radicals is zero. This means that in unit time the number of radicals produced by the initiation process must equal the number destroyed during the termination process. If this were not so and the total number of radicals increased during the reaction, the temperature would rise rapidly and there could even be an explosion since the propagation reactions are normally exothermic. In practice it is found that the steady-state assumption is usually valid for all but the first few seconds of most free radical addition polymerization reactions.

The rate of production of radicals is controlled by the breakdown of initiator molecules since this is more rapid than the formation of active centres. The rate of radical destruction is controlled by the termination reactions and so the steady-state assumption leads to

$$\frac{d[R\cdot]}{dt} = \frac{-d[M_i\cdot]}{dt}$$

and from Equations (2.26) and (2.28)

$$2k_i[I] = 2k_t\left(\sum [M_i\cdot]\right)^2$$

and hence the steady-state concentration of all growing chains is given by

$$\sum [M_i\cdot] = \left(\frac{k_i[I]}{k_t}\right)^{1/2} \qquad (2.29)$$

This means that $\sum [M_i\cdot]$ which is somewhat difficult to evaluate experimentally does not have to be measured for a particular reaction. In Equation (2.29) it is expressed in terms of parameters which are known or can be readily determined.

A small amount of monomer disappears with the formation of active centres but most of the monomer molecules are consumed during the propagation reaction for which the kinetic equation is

$$\frac{-d[M]}{dt} = k_p[M] \sum [M_i\cdot]$$ (2.27)

The steady-state concentration of active centres is given by Equation (2.29) and so

$$\frac{-d[M]}{dt} = k_p[M]\left(\frac{k_i[I]}{k_t}\right)^{1/2}$$ (2.30)

or

$$\frac{-d[M]}{dt} = \frac{k_p k_i^{1/2}}{k_t^{1/2}}[M][I]^{1/2}$$ (2.31)

The rate of consumption of monomer is usually called the *rate of polymerization* and is given by Equation (2.31) which applies only under steady-state conditions. Often the rate constant for initiation, k_i, is large and only a small proportion of the initiator breaks down into radicals. This means that [I] is effectively constant throughout the reaction and Equation (2.31) can be written as

$$\frac{-d[M]}{dt} = k_R[M]$$ (2.32)

which is simple first-order dependence on [M]. In this case k_R is a composite rate constant. Also the initiator may not be 100 percent efficient and some radicals may be lost by non-productive reactions (Section 2.2.1). In this case it may be necessary to replace k_i by (fk_i) where f is an initiator efficiency factor ($f \leqslant 1$).

The degree of polymerization of the polymer molecules produced during free radical addition polymerization can be calculated from the *kinetic chain length* $\bar{v}$ which is simply the number of times monomer molecules add on to the initiator radical during its life-time. It is given by

$$\bar{v} = \frac{\text{rate of addition of molecules to growing chains}}{\text{rate of formation of growing chains}}$$

and this is equal to

$$\bar{v} = \frac{\text{rate of propagation}}{\text{rate of initiation}} = \frac{-d[M]/dt}{d[R\cdot]/dt}$$

and so

$$\bar{v} = \frac{k_p[M]\sum[M_i\cdot]}{2k_i[I]}$$

Substituting for $\sum [M_i\cdot]$ from Equation (2.29) gives

$$\bar{\nu} = \frac{k_p[M]}{2(k_ik_t)^{1/2}[I]^{1/2}} \quad \text{at steady state} \tag{2.33}$$

The *number average degree of polymerization* $\bar{x}_n$, is simply related to the kinetic chain length through the mechanism of termination. If it is by combination then $\bar{x}_n = 2\bar{\nu}$ whereas if it is by disproportionation then $\bar{x}_n = \bar{\nu}$. In general both types of termination may take place and so

$$\bar{x}_n = a\bar{\nu}$$

where a is a number between 1 and 2 which depends upon the termination mechanism. The final expression for $\bar{x}_n$ is

$$\bar{x}_n = \frac{ak_p[M]}{2(k_ik_t)^{1/2}[I]^{1/2}} \tag{2.34}$$

The number average molar mass $\bar{M}_n$ is related to $\bar{x}_n$ and M_o, the molar mass of the monomer through the equation

$$\bar{M}_n = M_o\bar{x}_n = \frac{M_oak_p[M]}{2(k_ik_t)^{1/2}[I]^{1/2}}$$

or

$$\bar{M}_n = \frac{\beta[M]}{[I]^{1/2}}$$

where β is a constant for a given system at a particular temperature. The equation shows that $\bar{M}_n$ is proportional to the concentration of monomer and inversely proportional to the square root of the initiator concentration.

2.2.6 *Determination of individual rate constants*

In order to be able to predict the degree and rate of polymerization for a particular free radical addition polymerization system it is necessary to know the individual rate constants k_i, k_p and k_t. Also in order to determine $\bar{x}_n$, the mechanism of termination must be known. This can be done using radioactively labelled initiator molecules. The rate constant for the breakdown of initiator k_i can be determined from measurements upon the initiator alone. However, k_i, so determined, may be different from that in the presence of monomer molecules. So far we have developed two equations that can be used to determine the three rate constants. The equations are

$$\frac{-d[M]}{dt} = \frac{k_pk_i^{1/2}}{k_t^{1/2}}[M][I]^{1/2} \tag{2.31}$$

and

$$\bar{x}_n = \frac{ak_p[M]}{2(k_ik_t)^{1/2}[I]^{1/2}} \tag{2.34}$$

[M], $\bar{x}_n$ and [I] can be readily measured, but even if k_i is evaluated by another experiment it is only possible to calculate the ratio $k_p/k_t^{1/2}$ and not the two rate constants independently. A third equation containing either k_p or k_t is required. Such an expression can be obtained by considering the average life-time of the growing chains τ. This is given simply by

$$\tau = \frac{\text{concentration of growing chains at any time}}{\text{rate of disappearance of growing chains}}$$

and so

$$\tau = \frac{\sum [M_i\cdot]}{-d[M_i\cdot]/dt} \tag{2.36}$$

Using Equations (2.28) and (2.29) then

$$\tau = \left(\frac{k_i[I]}{k_t}\right)^{1/2} \Big/ \frac{2k_t k_i[I]}{k_t} \tag{2.37}$$

or

$$\tau = \frac{1}{2(k_ik_t[I])^{1/2}} \tag{2.38}$$

The three Equations (2.31), (2.34) and (2.38) contain three unknown parameters k_i, k_p and k_t which can therefore be evaluated if suitable measurements are made. All these measurements are normally done under steady-state conditions with the exception of the average radical lifetime τ which is most easily determined under non-steady-state conditions. The estimation of τ is most conveniently carried out in photo-initiated systems where measurements are taken with the illumination both switched alternately on and off and with the system receiving constant illumination.

2.2.7 *Distribution of molar mass*

In free radical addition polymerization the distribution of molar mass depends upon the mechanism of termination. The simplest case to consider is when the mechanism is through disproportionation and $\bar{x}_n$ is equal to the kinetic chain length $\bar{\nu}$. In this case the chain grows and terminates at the length to which it had grown. At this point it is necessary to define a new parameter α which is the probability that during an addition reaction a chain will propagate rather than terminate. It is analogous to the parameter p (the extent of reaction) in step-growth polymerization (Section 2.1.4). From the definition of α it follows that

$$\alpha = \frac{\text{rate of propagation}}{\text{rate of propagation} + \text{rate of all other reactions}}$$

and this can be used to calculate the probability, $P(i)$, of forming the polymer molecule M_i containing i monomer units.

$$\begin{array}{cccc} R\text{—}M\text{—}M\text{—}M\text{—} \ldots \text{—}M \\ 1 \quad 2 \quad 3 \quad (i-1) \end{array}$$

If the mechanism of termination is disproportionation then the degree of polymerization is the same as the kinetic chain length. The initial reaction between the radical and the first monomer molecule is fast and so only subsequent additions need be considered. The molecule M_i has been formed by $(i-1)$ addition reactions. The probability that one of these reactions has taken place is α and so using the analogy of step-growth polymerization the probability that $(i-1)$ successive addition reactions have taken place is $\alpha^{(i-1)}$. The probability that the last reaction is termination rather than propagation is $(1-\alpha)$ and so the probability that M_i has been formed is

$$P(i) = \alpha^{(i-1)}(1-\alpha) \tag{2.39}$$

Using similar arguments as for step-growth polymerization (Section 2.1.6) it is found that

$$\bar{M}_w/\bar{M}_n = (1+\alpha) \tag{2.40}$$

In order to obtain long molecules the rate of propagation needs to be much greater than the rate of all other reactions,

i.e. $\quad \alpha \to 1 \quad$ and $\quad \bar{M}_w/\bar{M}_n \to 2$

which is an identical result to that for step-growth polymerization.

A rather different result is obtained if the termination mechanism is combination. In this case the calculation is more complex since growing chains join together to give polymer molecules which are twice the kinetic chain length. It can be shown in this case that

$$\bar{M}_w/\bar{M}_n = (2+\alpha)/2 \tag{2.41}$$

and for long polymer molecules

$$\alpha \to 1 \quad \text{and so} \quad \bar{M}_w/\bar{M}_n \to 1.5$$

The two cases considered above are rather idealized and in practice the situation may be very much more complicated. For instance, the mechanism of termination may be a mixture of combination and disproportionation, branching reactions may occur and chain transfer can take place. All of these may affect the distribution of relative molecular mass.

2.2.8 Autoacceleration

It is often found that there can be serious deviations from steady-state kinetics during free radical addition polymerization especially towards the end of reactions using pure monomers or concentrated solutions. This can be most readily demonstrated by looking at the simplified equation for the rate of polymerization

$$\frac{-d[M]}{dt} = k_R[M] \tag{2.32}$$

If the concentration of initiator is effectively constant then it can be integrated if the monomer concentration is initially $[M]_0$ and becomes $[M]$ after time t, to give

$$\ln\left(\frac{[M]_0}{[M]}\right) = k_R t \tag{2.42}$$

A plot of $\ln([M]_0/[M])$ against t should be a straight line if the steady-state kinetics prevail. Fig. 2.5 shows some experimental data for the polymerization of methyl methacrylate solutions in benzene at various concentrations. The solution of lowest concentration obeys Equation (2.42), but as the methyl methacrylate concentration increases there is, at first, a significant and eventually a catastrophic upturn in the curves indicating an increase in k_R and considerable speeding up of the reaction. This phenomenon is known as *autoacceleration* or the *Trommsdorff–Norrish* effect or *gel effect*. It is due to an increase in viscosity of the reaction medium caused by the formation of polymer molecules and it tends to be more pronounced in concentrated solutions or with undiluted monomer. As the viscosity increases the active centres are unable to meet and terminate. The rate of termination is therefore dramatically reduced. The initiation and propagation reactions still take place at roughly the same rates and so the concentration of growing chains increases. The addition reactions are exothermic and so heat is evolved at an increasing rate which means that if energy dissipation is poor there may well be an explosion. The simplest way of avoiding the autoacceleration is to use dilute solutions or stop the reaction before chain diffusion becomes difficult.

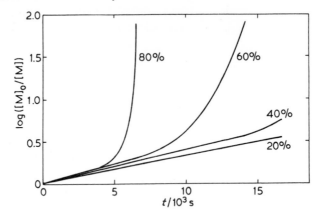

Fig. 2.5 *Variation of* log($[M]_0/[M]$) *with time for the polymerization of methyl methacrylate in benzene at 323 K at different concentrations.*

2.2.9 *Chain transfer*

In free-radical addition it is often found that growing chains are terminated prematurely in *chain transfer* reactions. The reactions can be represented as

$$M_i\cdot + TH \rightarrow M_iH + T\cdot$$

where TH is a general chain-transfer agent. In this reaction a hydrogen atom is normally abstracted from the transfer agent although other types of atoms can be abstracted in certain cases. The radical formed can go on to react with a monomer molecule.

$$T\cdot + M \rightarrow M_1\cdot$$

to form an active centre. If this reaction occurs rapidly the rate of polymerization is unaltered but since the growing chains are terminated prematurely the kinetic chain length and hence the degree of polymerization are both reduced.

A wide variety of substances have been found to act as chain transfer agents. Transfer to *polymer* does not lead to a reduction in the degree of polymerization, but it can cause branching. For example, polyethylene produced by free-radical polymerization is found to have some branches with five or six carbon atoms. It is thought that this is due to a 'back-biting' reaction

$$
\begin{array}{ccc}
& CH_2\!\!-\!\!CH_2 & CH_2\!\!-\!\!CH_2 \\
& \diagup \qquad \diagdown & \diagup \qquad \diagdown \\
\text{\textasciitilde}CH & CH_2 \rightarrow \text{\textasciitilde}C\dot{H} & CH_2 \\
| & | & | \\
H\!-\!-\!-\!-\!-\!-CH_2 & & CH_3
\end{array}
$$

The active centre is transferred from the end of the growing chain to a position along the back of the chain and chain growth proceeds from this position. Branched polyethylene cannot crystallize as easily as the linear form produced using special catalysts. The lower crystallinity causes it to be more flexible and tougher, but its melting temperature is lower than that of linear polyethylene.

Transfer can also take place to *monomer, initiator* or *solvent* molecules. In each case a hydrogen atom is usually, but not always, transferred as shown in the general reaction and the degree of polymerization is reduced. The kinetic equations in these cases are all of the same form:

$$\text{Monomer} \quad \frac{-d[M_i]}{dt} = k_{trM}[M] \sum [M_i\cdot] \tag{2.43}$$

$$\text{Initiator} \quad \frac{-d[M_i]}{dt} = k_{trI}[I] \sum [M_i\cdot] \tag{2.44}$$

$$\text{Solvent} \quad \frac{-d[M_i]}{dt} = k_{trS}[S] \sum [M_i\cdot] \tag{2.45}$$

where k_{trM}, k_{trI} and k_{trS} are the rate constants for the transfer reactions with monomer, initiator and solvent respectively. These transfer reactions modify the steady-state reaction kinetics outlined in Section 2.2.5. The kinetic chain length $\bar{\nu}$ was defined as

$$\bar{\nu} = \frac{\text{rate of propagation}}{\text{rate of initiation}}$$

but under steady-state conditions, rate of intiation = rate of termination and so

$$\bar{\nu} = \frac{\text{rate of propagation}}{\text{rate of termination}}$$

The transfer reactions can be considered as reactions which terminate the kinetic chains so it follows using Equations (2.27), (2.28), (2.43), (2.44) and (2.45) that

$$\bar{\nu} = \frac{k_p[M] \sum [M_i\cdot]}{2k_t \left(\sum [M_i\cdot] \right)^2 + \sum [M_i\cdot](k_{trM}[M] + k_{trI}[I] + k_{trS}[S])} \tag{2.46}$$

If it is assumed that the only normal termination mechanism is disproportionation then since $\bar{\nu} = \bar{x}_n$ for this mechanism

$$\frac{1}{\bar{x}_n} = \frac{2k_t \sum [M_i\cdot]}{k_p[M]} + C_M + C_I\frac{[I]}{[M]} + C_S\frac{[S]}{[M]} \tag{2.47}$$

TABLE 2.5 *Transfer constants for various*
substances with styrene. (Approximate values at
333 K)

Transfer Agent	Type	C
Styrene	Monomer	7×10^{-5}
Benzoyl peroxide	Initiator	0.05
Benzene	Solvent	2×10^{-6}
Toluene	Solvent	1.3×10^{-5}
Chloroform	Solvent	5×10^{-5}

where the C's are the ratios of k_{tr}/k_p and are called *transfer constants*. Using Equations (2.29) and (2.33) the first term in Equation (2.47) can be simplified to give

$$\frac{1}{\bar{x}_n} = \frac{1}{(\bar{x}_n)_o} + C_M + C_I\frac{[I]}{[M]} + C_S\frac{[S]}{[M]} \tag{2.48}$$

where $(\bar{x}_n)_o$ is the degree of polymerization obtained in the absence of chain transfer. Table 2.5 lists some values of C for various substances with styrene at 333 K. It can be seen that the benzoyl peroxide initiator ($C = 0.05$) is particularly prone to transfer. This is not unusual for organic peroxides and azo initiators, which are less vulnerable, are often used when transfer to initiator might give rise to problems.

2.2.10 *Inhibition and retardation*

Certain substances can be added to a polymerizing system which affect the rate of polymerization. *Retarders* are added to slow down the rate and *inhibitors* will completely stop polymerization. Both types of additive tend to act in the same way and differ only in their efficiency in rendering the growing chains inactive. The effect of the addition of a retarder or an inhibitor to a free-radical polymerization system is shown schematically in Fig. 2.6.

Nitrobenzene acts as a retarder for styrene through a chain-transfer reaction. The radicals yielded by this reaction are rather inactive and react only slowly with styrene monomer molecules. In this case both the rate and degree of polymerization are reduced. On the other hand benzoquinone inhibits the polymerization of styrene by scavenging every polymer radical formed and converting them into unreactive substances. The inhibitor itself is rendered inactive by the reaction and this leads to a phenomenon known as an *induction period* whereby inhibition only takes place for a limited period of time. After all the inhibitor molecules are used up polymerization proceeds in the normal way. Inhibitors are often added to

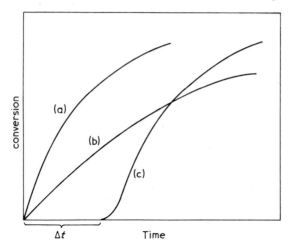

Fig. 2.6 *Effect of different additives upon free radical polymerization* (a) *Reaction without additives.* (b) *Reaction in the presence of a retarder.* (c) *Reaction in the presence of an inhibitor* (Δt *is the induction period*).

monomers such as styrene during transportation and storage otherwise they can undergo self-polymerization. The inhibitors have to be removed before any free radical polymerization experiments can be done in the laboratory and traces of inhibitor which are not removed can make measurements of rates of polymerization somewhat unreproducible.

2.2.11 *Effect of temperature*

The basic polymerization scheme outlined in Section 2.2.5 involves three basic reactions, initiation, propagation and termination, each of which has an individual rate constant. These constants depend upon temperature and can be expressed in terms of the appropriate Arrhenius expressions.

$$k_i = A_i \exp(-E_i/RT) \qquad (2.49)$$
$$k_p = A_p \exp(-E_p/RT) \qquad (2.50)$$
$$k_t = A_t \exp(-E_t/RT) \qquad (2.51)$$

where A_i, A_p and A_t are temperature-independent factors and E_i, E_p and E_t are the activation energies for the individual reactions. Approximate values of the activation energies and pre-exponential factors are given in Table 2.6 for different initiators and monomers. The values quoted in the literature can vary widely and so care must be taken if accurate calculations are to be made using the data in Table 2.6.

Since the rate constants k_i, k_p and k_t are temperature dependent it follows that the rate of polymerization, kinetic chain length and degree of

TABLE 2.6 *Arrhenius constants for several different monomers and initiators*

Initiator	$A_i/10^{13}\mathrm{s}^{-1}$	$E_i/\mathrm{kJ\ mol}^{-1}$
Azobisisobutyronitrile*	23	123
Benzoyl peroxide*	6.8	124
Monomer	$A_p/10^7\mathrm{dm}^3\mathrm{mol}^{-1}\mathrm{s}^{-1}$	$E_p/\mathrm{kJ\ mol}^{-1}$
Styrene	0.45	30
Methyl methacrylate	0.1	20
Vinyl chloride	0.3	15
Monomer	$A_t/10^9\mathrm{dm}^3\mathrm{mol}^{-1}\mathrm{s}^{-1}$	$E_t/\mathrm{kJ\ mol}^{-1}$
Styrene	0.06	8
Methyl methacrylate	0.1	5
Vinyl chloride	600	17

*Values determined for decomposition in benzene.
NB There is normally a great deal of scatter in determination of the Arrhenius constants for polymerization reactions. The ones quoted here can only be taken as approximate values.

polymerization must also vary with temperature. For example, the rate of polymerization is given by

$$\frac{-\mathrm{d}[M]}{\mathrm{d}t} = \frac{k_p k_i^{1/2}}{k_t^{1/2}}[M][I]^{1/2} \tag{2.31}$$

and so Equations (2.49)–(2.51) lead to

$$\frac{-\mathrm{d}[M]}{\mathrm{d}t} = \frac{A_p A_i^{1/2}}{A_t^{1/2}}\exp\left(\frac{E_t/2 - E_i/2 - E_p}{RT}\right)[M][I]^{1/2} \tag{2.52}$$

In this equation the only temperature-dependent factor is the exponential term

$$\exp\left(\frac{E_t/2 - E_i/2 - E_p}{RT}\right)$$

For free-radical polymerization the activation energies are generally such that this exponent is negative and so the rate of polymerization normally increases with temperature.

A different result is obtained for the degree of polymerization which is given by

$$\bar{x}_n = \frac{ak_p[M]}{2(k_i k_t)^{1/2}[I]^{1/2}} \tag{2.34}$$

which leads to

$$\bar{x}_n = \frac{aA_p}{2A_i^{1/2}A_t^{1/2}} \exp\left(\frac{E_i/2 + E_t/2 - E_p}{RT}\right)\frac{[M]}{[I]^{1/2}} \qquad (2.53)$$

In this case the exponential factor is often positive and so an increase in temperature will lead to a reduction in the degree of polymerization although the rate of polymerization continues to increase.

If the temperature is increased to a sufficiently high level the reverse of propagation, known as *depropagation* may occur. Instead of addition taking place

$$M_i \cdot + M \xrightarrow{k_p} M_{i+1} \cdot$$

the growing chain can loose a monomer unit

$$M_{i+1} \cdot \xrightarrow{k_{dp}} M_i \cdot + M$$

and the rate constant for this reaction is k_{dp}. The kinetic equation for depropagation is a simple first-order type

$$\frac{d[M]}{dt} = k_{dp} \sum [M_i \cdot] \qquad (2.54)$$

Combining this equation with Equation (2.30) leads to the overall rate of polymerization given by

$$\frac{-d[M]}{dt} = k_p[M] \sum [M_i \cdot] - k_{dp} \sum [M_i \cdot] \qquad (2.55)$$

if the radical concentration is at its steady-state value. Polymerization will cease if the temperature is raised sufficiently high that $d[M]/dt$ is equal to zero and Equation (2.55) becomes

$$k_p[M] = k_{dp} \qquad (2.56)$$

The variation of $k_p[M]$ and k_{dp} with temperature is shown for styrene in Fig. 2.7. The rise of k_{dp} with temperature is more rapid than that of $k_p[M]$ and it is eventually overtaken. The temperature, T_c, at which the two terms are equal is known as the *ceiling temperature* for the reaction. It can be raised by increasing the monomer concentration, but it has its highest value when pure monomer is used. Some ceiling temperatures are given in Table 2.7 for various pure olefin monomers.

TABLE 2.7 *Ceiling temperatures for different monomers*

Monomer	Ceiling temperature T_c
α-methyl styrene	334 K
Styrene	583 K
Methyl methacrylate	493 K

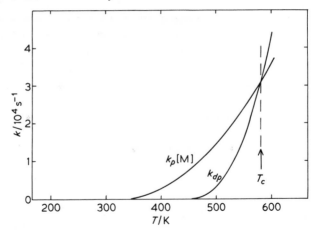

Fig. 2.7 *Variation of $k_p[M]$ and k_{dp} with temperature for the free radical polymerization of styrene. (After Dainton and Ivin, Quarterly Reviews, 12 (1958) 61).*

The values of T_c for styrene and methyl methacrylate are typical for monomers used in free-radical addition polymerization. The low value for α-methyl styrene ($\sim$60°C) is due to severe steric problems which arise during addition and reduce the value of k_p. A lot of fruitless experimental work was done with α-methyl styrene in an attempt to find suitable initiators until it was realized that the difficulties with polymerization were caused by depropagation reactions taking place.

2.2.12 *Homogeneous systems for free-radical polymerization*

Physical systems for polymerization can be conveniently divided into two main categories. One is *homogeneous* where the initiator, monomer and polymer are all mutually soluble and in a single phase and the other is *heterogeneous* where the three main components are not mutually soluble and the polymer forms as a discrete phase dispersed in a droplet or granular form.

Homogeneous polymerization systems are used mainly for laboratory preparations rather than for industrial synthesis. The choice of the method for preparation depends to a large extent upon the amount of polymer which is to be produced and the physical condition in which it is required. *Bulk polymerization* can be used for studies of polymerization kinetics and the production of small amounts of polymer. In this case no solvent is used and the initiator, monomer and polymer are mixed together in one phase. Care must be taken to remove heat from the system during polymerization otherwise serious problems can arise. The viscosity of the mixture increases as polymer is formed and the reactions are often stopped at about

10 percent conversion. The polymer can be extracted by the addition of a solvent which dissolves the initiator and monomer but causes the polymer to precipitate. Since long chains are formed at an early stage in the reaction high degrees of polymerization can be obtained by this method.

The method cannot be employed so easily to produce large amounts of polymer since autoacceleration (Section 2.2.8) can occur as soon as the conversion becomes appreciable. This gives rise to heat-transfer problems and 'hot-spots' can develop which may cause charring. A form of bulk polymerization is used to make optical grades of poly(methyl methacrylate). A low molar mass pre-polymer is made and this can be mixed with monomer and polymerized during casting to give a product with good optical clarity and relatively free from contamination. The use of a pre-polymer helps to reduce the heat evolution, but problems can be encountered with unreacted monomer becoming trapped in the polymer which can lead to unpleasant odours and may plasticize the material.

In *solution polymerization* many of the difficulties encountered with bulk systems are overcome. The addition of an inert solvent lowers the viscosity of the system which helps with heat transfer and reduces the likelihood of autoacceleration taking place. A useful way of removing heat is by refluxing whereby the solvent is allowed to boil, the vapour is cooled and condensed back into the reaction vessel. The boiling-point of the solvent is therefore an important consideration. The presence of a solvent may cause problems especially in an industrial situation. As the effective concentration of monomer is reduced the rate of polymerization (Equation 2.31) and the degree of polymerization (Equation 2.34) are lowered. Also, the removal of solvent and unreacted monomer from the final product can be expensive. Nevertheless, solution polymerization is often used industrially, particularly for the preparation of small amounts of specialized polymers.

2.2.13 *Heterogeneous polymerization systems*

From an industrial view-point bulk and solution polymerization give rise to significant problems of heat transfer in one case and the recovery of solvent in the other. A widely used industrial technique is *suspension polymerization* in which the monomer or a mixture of monomers is dispersed by vigorous stirring into an aqueous suspension of fine droplets typically $10\mu m$ to 10mm in diameter depending upon the conditions. The dispersing medium is rarely water alone, but often contains small amounts of stabilizer to stop coagulation of the droplets. With suspension polymerization the initiator is usually insoluble in water and soluble in the monomer. Thus each individual droplet acts as a small bulk polymerization system and the kinetics are similar to those for bulk polymerization. The aqueous suspending medium enables good heat dissipation, but autoacceleration can sometimes occur in the individual droplets leading to an increase in degree of polymerization. At low degrees of conversion the droplets can become viscous and tacky with a tendency to stick together and efficient stirring is essential to prevent coagulation. The product of the polymerization can be uniform spherical beads of polymer if the conditions are right, otherwise if coalescence takes place large lumps of polymer which are difficult to handle can be formed. Suspension polymerization is used for large-scale industrial free-radical polymerization of many olefinic monomers. Both homopolymers and copolymers of monomers such as methyl methacrylate, styrene, vinyl acetate and vinyl chloride can be readily made by this technique.

Another heterogeneous process of wide technological importance is *emulsion polymerization*. It differs from suspension polymerization in that the droplet size is much smaller, typically 0.05 to 5μm in diameter and the initiator is soluble in the aqueous phase rather than the monomer. The droplet particles are held in a stable colloidal suspension with the aid of surface-active agents (emulsifiers) and rapid agitation of the system produces a true emulsion. The mechanism of polymerization is thought to be quite complex and even now is not completely understood. However, the description of emulsion polymerization that is given below is consistent with the widely held views of the probable mechanism.

The most important characteristic of the process is the presence of surface-active agents which are often soap-type substances. When small quantities are added to water they migrate to the surface and lower the surface tension. If their concentration is increased above a level which saturates the surface the soap-like emulsifier molecules tend to aggregate within the body of the solution and form entities known as *micelles*. Soap molecules normally consist of short hydrocarbon chains of up to 30 carbon atoms and with a hydrophilic group at one end. The micelles are thought to contain of the order of 100 molecules and are typically 50 Å in diameter. They aggregate with the hydrocarbon chains pointing inwards and the hydrophilic groups in contact with the water molecules. When a monomer which is not soluble in water is added to the system two things tend to happen. A small quantity of the monomer enters the micelles because of the hydrophobic nature of both the monomer molecules and the hydrocarbon segments of the soap molecules. The presence of the monomer causes the micelles approximately to double in size. The majority of the monomer remains as droplets about a micron in size suspended in the aqueous phase with the aid of the emulsifier. A very small amount of monomer may be present as individual molecules dissolved in the aqueous phase.

Special water-soluble initiators are used in emulsion polymerization, such as redox (Section 2.2.1) or persulphate initiators. For example, potassium persulphate breaks up into ions in water and the persulphate ion breaks down into sulphate ion radicals at about 50°C

$$S_2O_8^{2-} \rightarrow 2SO_4^- \cdot$$

The ion radicals attack some of the monomer molecules in the aqueous phase to form soap-type radicals with both hydrophilic and hydrophobic ends. These radicals are attracted to the micelles and enter them initiating polymerization. As the monomer in the micelles is incorporated into polymer the supply of monomer is replenished by diffusion from the droplets. At this stage the rates of reaction are found to increase rapidly and the particles increase in size. As they grow larger they are stabilized by absorbing emulsifier obtained from the breakdown of inactive micelles. At a fairly early stage in the reaction all the available emulsifier becomes occupied in stabilizing the growing polymer particles and monomer droplets. When this happens the rate of polymerization becomes constant. These first two stages of the reaction designated I and II are shown schematically in Fig. 2.8. During stage II no micelles are present and particles of polymer of the order of 300 Å in diameter grow in size without increasing their number. Eventually in stage III the supply of monomer from the droplets becomes exhausted and the rate of polymerization drops as the remaining monomer absorbed in the polymer particle is used up.

The kinetics of emulsion polymerization are thought to be rather complex. However, it is possible to derive a relatively simple equation for the rate of polymerization during stages I and II. During this time the relative concentration of radicals to growing particles is such that each particle, at any time, only contains at the most one active radical. The first radical entering a micelle will initiate polymerization and the entry of a second radical causes immediate termination. The polymerization will only start again when a third radical enters the micelle. This means that on average each particle only contains an active radical for half of the total polymerization time. For free-radical addition the rate of polymerization is given by Equation (2.27) as

$$\frac{-d[M]}{dt} = k_p[M] \sum [M_i \cdot] \qquad (2.27)$$

where $\sum [M_i \cdot]$ is the concentration of growing chains. If each particle contains a growing chain for only half of the reaction time then if there are N particles per unit volume

$$\sum [M_i \cdot] = N/2 \qquad (2.57)$$

and so

$$\frac{-d[M]}{dt} = k_p \frac{[M]N}{2} \qquad (2.58)$$

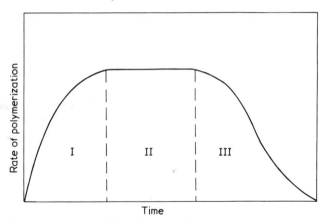

Fig. 2.8 *Schematic illustration of the change of the rate of polymerization with time during emulsion polymerization. (After Billingham and Jenkins).*

This equation which can usually be applied during stages I and II (Fig. 2.8), shows that the rate of polymerization is independent of both the initiation and termination rate constants. Emulsion polymerization is a useful and important method of synthesis for many polymers since it enables the molar mass of the polymer to be controlled without affecting the rate of polymerization. Reducing the concentration of radicals means that they enter the particles at a lower rate. This means that termination occurs less frequently and consequently the average length of the polymer chains is increased. On the other hand the rate of polymerization is virtually unaffected. Often high degrees of polymerization are obtained with emulsion-polymerization systems and sometimes chain-transfer agents are added to reduce the average chain length.

Emulsion polymerization is widely used industrially for the synthesis of many vinyl and diene polymers and for copolymerization. The polymer produced is in a latex of fine particles and the products can be used directly as bases for adhesives, paints, textile finishes etc. Otherwise the latices can be coagulated by the addition of chemicals to produce the polymer in a solid form.

2.3 Ionic polymerization

Ionic polymerizations are classified according to whether the polymeric ions are positively or negatively charged. If the ions are positively charged the polymerization is termed *cationic* and if they negatively charged it is known as *anionic* polymerization. It is normally possible to generalize about the mechanisms and kinetics of free-radical addition polymerization, but in the case of ionic initiated polymerization such generalizations cannot normally be made. Unlike free-radical systems the detailed mechanisms depend upon the type of initiator, monomer and solvent employed. When the chain carriers are ionic the addition reactions are often rapid and difficult to reproduce. Also high rates and degrees of polymerization can be obtained at low temperatures. Propagation rarely takes place through free ions. Normally there are counter-ions closely associated with the polymer ions during chain propagation. The nature of the solvent influences the strength of the association and the distance between the ions in the ion-pair. The presence of the counter-ion, in turn, can control the stereochemistry and the rate of monomer addition.

Ionic polymerization systems have been developed because some monomers which contain double bonds cannot be polymerized using free-radical initiators. Also ionic polymerization generally takes place at low temperatures and can offer better control of stereoregularity and relative molecular mass distribution. The chemical factors which control whether or not a particular monomer can be polymerized using free radical, cationic or anionic initiators are beyond the scope of this book. However, Table 2.8 gives an indication of what initiators can be used for different types of monomer. It can be seen that in general free radicals are less selective and can polymerize most types of monomers. The ionic initiators are only able to polymerize certain monomers which have specific types of substitutent groups.

2.3.1 *Cationic polymerization*

This type of polymerization takes place by the addition of monomer molecules to a positively charged growing chain, known as a carbonium ion. The reaction is initiated usually by the donation of a proton (H^+) to a monomer molecule by a strong acid. For example an ordinary mineral acid such as perchloric acid will initiate the polymerization of styrene in ethylene dichloride solution

$$HClO_4 + CH_2\!\!=\!\!CH \rightarrow CH_3\!\!-\!\!CH^+(ClO_4)^-$$

perchloric acid   ion pair

The carbonium ion and the perchlorate ion remain closely associated as an ion pair. The most important types of initiators for cationic polymerization are Lewis acids or Friedel–Crafts catalysts such as BF_3, $AlCl_3$ or $SnCl_4$. For Friedel–Crafts catalysts, proton transfer cannot take place without a source of H^+ ions. It is found BF_3 will catalyse the polymerization of styrene in carbon tetrachloride only when traces of water are present.

TABLE 2.8 *Ability of various monomers to polymerize by addition polymerization using different types of initiator.*

Monomer		Free radical	Anionic	Cationic
Ethylene	$CH_2{=}CH_2$	✓	x	✓
Styrene	$CH_2{=}CHC_6H_5$	✓	✓	✓
Vinyl halides	$CH_2{=}CHX$	✓	x	x
Dienes	$CH_2{=}CR{-}CH_2{=}CH_2$	✓	✓	✓
Vinyl ethers	$CH_2{=}CHOR$	x	x	✓
Dialkyl olefins	$CH_2{=}CR_1R_2$	x	x	✓

X = halogen
R = alkyl group

There is no reaction if water is carefully removed. The reaction is thought to take place in two stages.

$$BF_3 + H_2O \rightleftharpoons H^+ (BF_3OH)^-$$

with the water acting as a co-catalyst. The complex formed reacts with the monomer with the transfer of a proton.

$$H^+(BF_3OH)^- + CH_2{=}CH \rightarrow CH_3{-}CH^+(BF_3OH)^-$$

Propagation takes place through the repeated addition of monomer molecules to the ion pair in a head-to-tail manner. After each addition reaction the new ion pair differs only from the previous type by being one monomer unit longer.

e.g. $\text{\char"7F}CH_2CH^+(ClO_4)^- + CH_2{=}CH \rightarrow \text{\char"7F}CH_2CH^+(ClO_4)^-$

or $\text{\char"7F}CH_2CH^+(BF_3OH)^- + CH_2{=}CH \rightarrow \text{\char"7F}CH_2{-}CH^+(BF_3OH)^-$

The mechanism of addition depends upon the type of counter-ion, solvent and monomer used and also upon the reaction temperature. For example, when strong acid initiators such as BF_3 are used the reaction rates can be extremely rapid, but if weaker acids such as $SnCl_4$ are employed reactions may take several days to reach completion.

Termination is thought to take place by either a unimolecular rearrangement of the ion pair (spontaneous decomposition)

$$\text{\char"7F}CH_2CH^+(ClO_4)^- \rightarrow \text{\char"7F}CH{=}CH + H^+(ClO_4)^-$$

or through transfer to species such as monomer molecules.

$$\text{\tiny{www}}CH_2CH^+(ClO_4)^- + CH_2\!\!=\!\!CH \rightarrow \text{\tiny{www}}CH_2\!\!-\!\!CH_2 + CH_3CH^+(ClO_4)^-$$

It is thought that transfer reactions can also take place with polymer, solvent and even impurity molecules. The exact mechanisms of termination are not properly understood, but it is thought that the main reason only short polymer chains are obtained at room temperature is that these termination and transfer reactions predominate at higher temperatures.

2.3.2 Kinetics of cationic polymerization

Cationic polymerization is often rapid and heterogeneous and this tends to make the formulation of a general kinetic scheme somewhat difficult. However, it is possible to produce a kinetic scheme similar to that for free-radical addition (Section 2.2.5), but care must be taken in applying the scheme to any particular system.

The initiation reaction involves the breakdown of initiator, which is usually rapid, followed by the formation of an ion pair (active centre), which is normally rate determining. The reactions are then in general

$$H\!\!-\!\!X \rightleftharpoons H^+X^-$$

and $H^+X^- + M \xrightarrow{k_i} M^+X^-$

The rate of formation of the ion pairs is given by

$$\frac{d[M^+]}{dt} = k_i[HX][M] \tag{2.59}$$

where $[HX]$ is the concentration of initiator.

Propagation takes place through an addition reaction between the monomer molecule and the ion pair

$$M_i^+X^- + M \xrightarrow{k_p} M_{i+1}^+X^-$$

Using the analogy of free-radical addition (Equation 2.27) the rate of consumption of monomer or the rate of polymerization is given by

$$\frac{-d[M]}{dt} = k_p[M] \sum [M_i^+] \tag{2.60}$$

where $\sum [M_i^+]$ is the concentration of all the ion pairs or active centres. If termination is by spontaneous decomposition then the general reaction is

$$M_i^+X^- \xrightarrow{k_t} M_i + H^+X^-$$

The rate of loss of active centres is given by

$$\frac{-d[M_i^+]}{dt} = k_t \sum [M_i^+]$$ (2.61)

If the reaction is not too rapid and steady-state conditions prevail through most of the reaction then the rate of formation of active centres will be equal to the rate of their loss

i.e. $$\frac{d[M^+]}{dt} = \frac{-d[M_i^+]}{dt}$$ (2.62)

and so using Equations (2.59) and (2.61) the steady-state concentration of active centres is given by

$$\sum [M_i^+] = \frac{k_i}{k_t}[HX][M]$$ (2.63)

and the rate of polymerization becomes

$$\frac{-d[M]}{dt} = \frac{k_i k_p}{k_t}[HX][M]^2$$ (2.64)

which is second order with monomer concentration, in contrast to free-radical addition (Equation 2.31). If termination is by transfer to monomer it can be readily shown that under steady-state conditions

$$\frac{-d[M]}{dt} = \frac{k_i k_p}{k_{tr}}[HX][M]$$ (2.65)

where k_{tr} is the rate constant for the transfer reaction.

Since two growing chains are unable to terminate by combination the degree of polymerization will be the same as the kinetic chain length. The degree of polymerization is then given by

$$\bar{x}_n = \frac{\text{rate of polymerization}}{\text{rate of initiation}}$$

which leads to

$$\bar{x} = \frac{k_p}{k_t}[M]$$ (2.66)

when there is spontaneous decomposition of the active centre and

$$\bar{x}_n = k_p/k_{tr}$$ (2.67)

when termination is by transfer to monomer. In both cases $\bar{x}_n$ is independent of initiator concentration and when transfer to monomer takes place it is also independent of the monomer concentration.

An unusual feature of cationic addition polymerization is that in some systems both the degree and rate of polymerization *increase* as the temperature is *decreased*. The reason for this can be demonstrated by looking at the overall activation energy for the reaction. Each of the rate constants k_i, k_p, k_t, and k_{tr} can be represented by an Arrhenius expression of the type

$$k_i = A_i \exp(-E_i/RT) \quad \text{etc.}$$

as is the case for free radical addition (Section 2.2.11). The overall rate constant, k_R, for the polymerization reaction is given by

$$k_R = \frac{A_i A_p}{A_t} \exp\left(\frac{E_t - E_i - E_p}{RT}\right) \tag{2.68}$$

for termination by spontaneous decomposition or

$$k_R = \frac{A_i A_p}{A_{tr}} \exp\left(\frac{E_{tr} - E_i - E_p}{RT}\right) \tag{2.69}$$

when termination is by transfer. It is found that the propagation reaction has a rather low activation energy causing E_p to be much lower than E_i, E_t or E_{tr}. Consequently the overall activation energy terms in the exponentials in Equations (2.68) and (2.69) can be positive causing k_R to increase as the temperature is reduced.

The variation of the degree of polymerization, $\bar{x}_n$, with temperature is controlled by $\exp\{(E_t - E_p)/RT\}$ or $\exp\{(E_{tr} - E_p)/RT\}$ depending upon the mechanism of termination. However, E_t or E_{tr} are both normally greater than E_p and so the degree of polymerization will always decrease as the temperature of the reaction increases as is the case for free radical addition (Section 2.2.11).

2.3.3 *Practical systems for cationic polymerization*

The only cationic polymerization of major technological importance that is used at present is the low-temperature synthesis of butyl rubber by the polymerization of isobutylene in chlorinated solvents using $AlCl_3$ as a catalyst. Cationic polymerization is potentially an extremely useful method of polymer synthesis since with a given monomer it is possible to use a wide variety of initiators and the overall reactivity can be controlled by the choice of solvent. This type of polymerization is also important because of good control of stereoregularity that can be achieved in certain systems. However, there may be important practical limitations since often there is difficulty in producing polymers with high controlled degrees of polymerization. Also careful control of the reaction conditions is needed. It is often necessary to use low temperatures and inert atmospheres or high vacuum.

Also rigorous drying of solvents, monomers and initiators is usually required. The reactions are frequently rapid and violent and low-boiling-point solvents are sometimes needed in order to help to dissipate the heat evolved during polymerization by boiling of the solvent.

2.3.4 Anionic polymerization

Anionic polymerization involves the successive addition of monomers which contain double bonds to a negatively charged species known as a carbanion. It can be seen from Table 2.8 that this type of polymerization will take place only using monomers which contain strong electronegative groups such as $-C_6H_5$, $-CN$, etc. which help to form a stable carbanion. Initiation can take place in one of two ways. The first involves addition of a negative ion or group to the monomer molecule. Typical initiators of this type are alkali metal amides such as KNH_2 where the NH_2^- ion adds to the monomer styrene:

$$NH_2^- + CH_2{=}CH \rightarrow NH_2{-}CH_2{-}CH^-$$

This reaction takes place in liquid ammonia and in this solvent the active centres are present in the form of free ions. On the other hand, styrene initiated by butyl lithium in benzene polymerizes by addition to ion-pairs. The second type of initiation that can occur is by the transfer of an electron from an electron donor to a monomer molecule to form a radical anion. A simple example of this is the initiation of the polymerization of butadiene by sodium metal

$$CH_2{=}CH{-}CH{=}CH_2 + Na \rightarrow (CH_2{=}CH{-}CH{\dot{-}}CH_2)^- \ Na^+$$

The negatively charged active centre contains an extra electron which is unpaired and so the anion is often called an anion radical.

Propagation takes place through addition of monomer molecules to active centres which are either free ions or ion pairs:

(i) free ion propagation

$$\text{\textasciitilde}CH_2{-}\underset{X}{CH}^- + CH_2{=}\underset{X}{CH} \rightarrow \text{\textasciitilde}CH_2{-}\underset{X}{CH}^-$$

(ii) ion pair propagation

$$\text{\textasciitilde}CH_2{-}\underset{X}{CH}^-A^+ + CH_2{=}\underset{X}{CH} \rightarrow \text{\textasciitilde}CH_2{-}\underset{X}{CH}^-A^+$$

in this case A^+ is the counter ion.

The most striking aspect of anionic polymerization is that termination sometimes may not occur unless impurities are present. This can lead to species known as *'living' polymers*. The detailed mechanisms and kinetics of anionic polymerization are best considered for individual systems since each reaction is controlled by the nature of the initiator, monomer and solvent and the presence of impurities.

2.3.5 Polymerization of styrene in liquid NH_3 initiated by KNH_2

The polymerization of styrene in liquid ammonia, initiated by potassium amide was one of the first anionic polymerization reactions studied in detail. It is of particular interest because propagation takes place through free ions rather than by ion-pairs because of the high dielectric constant of liquid ammonia. The reaction takes place at 240 K ($-33°C$). Initiation consists of two processes. The first is the breakdown of initiator into ions

$$KNH_2 \rightleftharpoons K^+ + NH_2^-$$

The rate-determining step is the addition of a monomer molecule to the amide anion to form an active centre

$$NH_2^- + CH_2{=}CH \xrightarrow{k_i} NH_2CH_2{-}CH^-$$

The rate of production of active centres is given by

$$\frac{d[M^-]}{dt} = k_i[I][M] \tag{2.70}$$

where [I] is the concentration of initiator. Propagation is by the normal addition of monomer to the free carbanions

$$\text{\textasciitilde\textasciitilde}CH_2{-}CH^- + CH_2{=}CH \xrightarrow{k_p} \text{\textasciitilde\textasciitilde}CH_2{-}CH^-$$

and the rate of polymerization is given by

$$\frac{-d[M]}{dt} = k_p[M] \sum [M_i^-] \tag{2.71}$$

Termination takes place by the transfer of a proton from a solvent molecule.

$$\text{\textasciitilde\textasciitilde}CH_2{-}CH^- + NH_3 \xrightarrow{k_{tr}} \text{\textasciitilde\textasciitilde}CH_2{-}CH_2 + NH_2^-$$

This gives a polymer chain with a saturated end-group and regenerates a NH_2^- ion which can go on to initiate another polymer chain. The rate of termination is given by;

$$\frac{-d[M_i^-]}{dt} = k_{tr}[NH_3]\sum [M_i^-] \qquad (2.72)$$

If it is assumed that $\sum [M_i^-]$ is constant under steady-state conditions then it follows that since

$$\frac{d[M^-]}{dt} = \frac{-d[M_i^-]}{dt}$$

then from Equations (2.70) and (2.72)

$$\sum [M_i^-] = \frac{k_i}{k_{tr}}\frac{[I][M]}{[NH_3]} \qquad (2.73)$$

The rate of polymerization is then

$$\frac{-d[M]}{dt} = \frac{k_i k_p}{k_{tr}}\frac{[I][M]^2}{[NH_3]} \qquad (2.74)$$

As for free radical addition the degree of polymerization is given by

$$\bar{x}_n = \frac{\text{rate of polymerization}}{\text{rate of initiation}}$$

and so using Equations (2.70) and (2.74)

$$\bar{x}_n = \frac{k_p}{k_{tr}}\frac{[M]}{[NH_3]} \qquad (2.75)$$

Equations (2.74) and (2.75) are unusual in that they contain the concentration of solvent $[NH_3]$ since the termination mechanism is transfer to solvent. This transfer reaction is highly competitive with propagation and so k_p/k_{tr} is not large and the reaction produces only short polymer chains. On the other hand, the effect of changing the reaction temperature is similar to that for free-radical polymerization. Reducing the temperature increases the degree of polymerization but reduces the rate of polymerization.

2.3.6 'Living' polymers

In certain anionic polymerization systems there can be a total absence of any formal termination reaction if all impurities are excluded. The exact details of each system depend upon the nature and type of initiator, monomer and solvent that are used. Propagation will continue until all the monomer is consumed and the carbanions remain active and are not terminated. If further monomer molecules of the same or even a different type are added propagation will restart and continue until the extra monomer is consumed. These active carbanions are termed *'living' polymers*. They are best demonstrated for particular systems used for the polymerization of styrene.

Sodium naphthalide will initiate the polymerization of styrene in an inert solvent such as tetrahydrofuran. This was one of the first 'living' polymer systems studied. The initiator is made first of all by adding sodium to a solution of naphthalene in tetrahydrofuran.

$$Na + \text{[naphthalene]} \rightarrow [\text{naphthalide}]^{-}\ Na^{+}$$

The solution of sodium naphthalide is green in colour because of the presence of the naphthalide anion radical. The ions in the initiator are thought to be closely associated in an ion pair. The initiator is unstable and will readily react with any impurities such as water, oxygen and carbon dioxide. The usefulness of the anion radical lies in its ability to transfer its extra electron to a monomer such as styrene.

$$[\text{naphthalide}]^{-}\ Na^{+} + CH_2{=}CH\overset{|}{\underset{C_6H_5}{}} \rightleftharpoons [CH_2{-}\dot{C}H]^{-}\ Na^{+} + \text{[naphthalene]}$$

The styryl anion radical that is formed gives a red colouration. However, the reaction does not stop at this stage and the styryl anion radicals can combine together to form a dianion

$$[CH_2{-}\dot{C}H]^{-}\ Na^{+} + [CH_2{-}\dot{C}H]^{-}\ Na^{+}$$

$$\rightarrow Na^{+}\ ^{-}[CH{-}CH_2{-}CH_2{-}CH]^{-}\ Na^{+}$$

This reaction is very rapid and the dianion formed is capable of propagation through addition reactions at both ends

$$Na^{+}\ ^{-}[CH{-}CH_2{-}CH_2{-}CH]^{-}\ Na^{+} + nCH_2{=}CH$$

$$\rightarrow Na^{+}\ ^{-}[CH{-}CH_2{\sim\!\sim}CH_2{-}CH]^{-}\ Na^{+}$$

If purified materials are used and the reaction is carried out under high vacuum there is no termination reaction. Since the formation of the dianions is very rapid most polymer chains are initiated at the same time. Since they all grow at approximately the same rate the polymer chains formed all have about the same degree of polymerization and so the polymer is virtually *monodisperse*. The reaction can be 'killed' at any time by the addition of impurities. A typical value of $\bar{M}_w/\bar{M}_n$ would be about 1.1. The degree of polymerization can also be calculated simply. If the initial concentration of sodium naphthalide is $[I]_0$ and the initial concentration of monomer is $[M]_0$ then the concentration of active centres is $[I]_0/2$ since they are in the form of dianions. In unit volume of the system $[M]_0$ moles are distributed between $[I]_0/2$ moles of polymer and so the number average degree of polymerization, $\bar{x}_n$, is given by

$$\bar{x}_n = \frac{2[M]_0}{[I]} \tag{2.76}$$

Many other 'living' polymer systems have since been studied. Often the chains are initiated at different times during the reaction and so the polymers may have a higher degree of polydispersivity with $\bar{M}_w/\bar{M}_n > 1.1$. Even so an equation of the form of Equation (2.76) will still hold. If the 'living' polymers are present as single anions the equation becomes

$$\bar{x}_n = \frac{[M]_0}{[I]_0} \tag{2.77}$$

'Living' polymers are becoming increasingly important in the preparation of monodisperse polymers and specialized types of copolymers. For example block copolymers can be readily made allowing one type of monomer to be used up and then changing the monomer and repeating the process as required.

2.4 Copolymerization

It is common practice in materials science to improve the properties of one material by mixing it in some way with another material. Metal alloys have mechanical properties which are vastly superior to those of pure metals. In polymer science this is rather more difficult to do as few polymers can be mixed or blended satisfactorily and the properties of the blend are often inferior to those of the pure polymers. An alternative approach which can be adopted with polymers is to synthesize chains which have more than one type of monomer unit. *Copolymers* which are made by this method can have qualities which may be better than those of the parent homopolymers, but the properties of the copolymer are usually characterized by the detailed arrangements of the monomer units on the polymer chain.

2.4.1 *Classification of copolymers*

The simplest types of copolymer contain only two types of monomer unit. There are several ways in which they can be arranged in the macromolecules. If the two types of monomer are A and B leading to homopolymers of the type —A—A—A—A—A—A— or —B—B—B—B—B— then the copolymers types are as outlined below.

    Random copolymers are formed when there is a random arrangement of A and B units along the molecule

    —A—B—A—A—B—A—B—B—B—A—A—B—

Truely random copolymers are only found under specific polymerization conditions.

    In *alternating copolymers* the A and B units are placed alternately along the polymer chain.

    —A—B—A—B—A—B—A—B—A—B—A—B—

Of course, this type of copolymer is formed during condensation polymerization when two different types of monomer such as a diacid and diol are used (see Section 2.1) but in this case the repeat unit is usually taken as —A—B— and the alternating copolymer treated as a homopolymer with this repeat unit.

    Block copolymers contain long sequences of the monomer in a linear copolymer of the form

—A—A—A—A—A—A—B—B—B—B—B—B—A—A—A—A—A—

Sometimes the block copolymer is made up of just two or three long blocks of each type of unit giving a material with interesting and unusual properties.

    The last main type of copolymer that is usually considered are *graft copolymers*. They consist of a main homopolymer chain with branches of another type of homopolymer

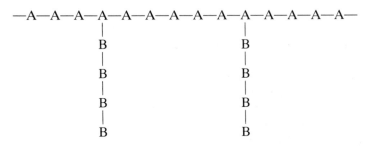

There are other types of special copolymers which will not be considered here. Stereoblock copolymers can be formed using the same type of monomer, but with blocks of different tacticity. In general it is found that block and graft copolymers have properties of both the homopolymers

whereas alternating and random copolymers possess properties somewhere between those of the two homopolymers.

2.4.2 *Copolymer composition equation*

It is clear that copolymerization is a much more complex process than polymerization using a single monomer. For example, in addition copolymerization using two monomers the tendency of each type of monomer to add to the growing chain may be different. This can lead to a variation of copolymer composition during the reaction even when equimolar amounts of the two types of monomer are used initially. This phenomenon is known as *composition drift* and is a common feature in copolymerization reactions.

It is possible to construct a kinetic scheme for a general addition copolymerization reaction between two types of monomer, A and B and this can go some way towards explaining the effect outlined above. As chain growth takes place there are in general two types of monomer unit at the active centre:

$$\text{\textasciitilde\textasciitilde}A^* \quad \text{and} \quad \text{\textasciitilde\textasciitilde}B^*$$

where the asterisk refers to a single unpaired electron or a positive or negative charge depending upon the type of active centre present. In general an A or B monomer unit can add on to the growing chain and so there will be four possible addition reactions that can occur, each with its own rate constant.

$$\text{\textasciitilde\textasciitilde}A^* + A \xrightarrow{k_{11}} \text{\textasciitilde\textasciitilde}AA^*$$

$$\text{\textasciitilde\textasciitilde}A^* + B \xrightarrow{k_{12}} \text{\textasciitilde\textasciitilde}AB^*$$

$$\text{\textasciitilde\textasciitilde}B^* + B \xrightarrow{k_{22}} \text{\textasciitilde\textasciitilde}BB^*$$

$$\text{\textasciitilde\textasciitilde}B^* + A \xrightarrow{k_{21}} \text{\textasciitilde\textasciitilde}BA^*$$

The reactions with rate constants k_{11} and k_{22} are known as self-propagation and those with constants k_{12} and k_{21} are called cross-propagation. If it is assumed that the reactivity of the growing chain is independent of the chain length and dependent only upon the nature of the terminal group then the rate of consumption of monomer A is given by

$$\frac{-d[A]}{dt} = k_{11}[A] \sum [A^*] + k_{21}[A] \sum [B^*] \tag{2.78}$$

where $\sum[A^*]$ and $\sum[B^*]$ are the concentrations of all the active centres terminating in A and B units. Similarly the rate of consumption of B is given by

$$\frac{-d[B]}{dt} = k_{22}[B] \sum [B^*] + k_{12}[B] \sum [A^*] \tag{2.79}$$

At any instant of time during the reaction the ratio of the amounts of each type of monomer being incorporated into the copolymer chain is found by dividing Equations (2.78) and (2.79) to be

$$\frac{d[A]}{d[B]} = \frac{[A]}{[B]}\left[\frac{k_{11}\sum[A^*]/\sum[B^*] + k_{21}}{k_{12}\sum[A^*]/\sum[B^*] + k_{22}}\right] \qquad (2.80)$$

The ratio $\sum[A^*]/\sum[B^*]$ can be evaluated by making a steady-state approximation. In the case of copolymerization the assumption is that the concentration of each of the two types of active centre remains constant

i.e. $\quad \dfrac{d\sum[A^*]}{dt} = 0$

and $\quad \dfrac{d\sum[B^*]}{dt} = 0$ $\qquad$ at steady state

By far the most important reactions in which they are created and destroyed are the cross-propagation reactions. If only the $\sim\!\!A^*$ chains are considered then the rate of creation is given by

$$\frac{d\sum[A^*]}{dt} = k_{21}[A]\sum[B^*] \qquad (2.81)$$

and their rate of destruction is given by

$$\frac{-d\sum[A^*]}{dt} = k_{12}[B]\sum[A^*] \qquad (2.82)$$

At steady-state the rate of creation equals the rate of destruction and so

$$k_{21}[A]\sum[B^*] = k_{12}[B]\sum[A^*]$$

or $\quad \dfrac{\sum[A^*]}{\sum[B^*]} = \dfrac{k_{21}[A]}{k_{12}[B]} \qquad (2.83)$

Substituting Equation (2.83) into Equation (2.80) and simplifying leads to

$$\frac{d[A]}{d[B]} = \frac{[A]}{[B]}\left[\frac{k_{11}[A]/k_{12} + [B]}{[A] + k_{22}[B]/k_{21}}\right] \qquad (2.84)$$

If the reactivity ratios of the two types of monomer are defined as

$$r_1 = k_{11}/k_{12}$$
$$r_2 = k_{22}/k_{21}$$

then

$$\frac{d[A]}{d[B]} = \frac{[A]}{[B]} \left[\frac{r_1[A] + [B]}{[A] + r_2[B]} \right] \tag{2.85}$$

This equation is known as the *copolymer equation* and it gives the composition of the copolymer being formed *at any instant* during the reaction. It does not necessarily describe the structure of the copolymer produced at the end of the reaction since [A] and [B] can vary during the course of the reaction.

2.4.3 *Reactivity ratios and copolymer structure*

The reactivity ratios, r, are the ratios of the reactivity of each species with its own monomer to its reactivity with the other type of monomer. If $r_1 > 1$ then the monomer A tends to polymerize whereas if $r_1 < 1$ copolymerization is preferred. Similarly r_2 describes the behaviour of monomer B. In order to understand a copolymerization reaction completely it is necessary to obtain reliable values of r. This can be done by taking monomer mixtures with various values of [A]/[B] and analysing the copolymer produced at low conversion. If X_1 and X_2 are defined as the mole fractions of the monomers A and B being incorporated into the growing chain at any instant of time, and x_1 and x_2 the corresponding mole fractions in the feed then it follows that

$$\frac{d[A]}{d[B]} = \frac{X_1}{X_2} = X$$

and

$$\frac{[A]}{[B]} = \frac{x_1}{x_2} = x$$

and so Equation (2.85) becomes

$$X = x \left(\frac{xr_1 + 1}{x + r_2} \right)$$

or

$$\frac{x(1 - X)}{X} = r_2 - \frac{x^2}{X} r_1 \tag{2.86}$$

which is another form of the copolymer equation. A plot of $x(1 - X)/X$ against x^2/X should give a straight line of slope $-r_1$ and intercept r_2. Hence, r_1 and r_2 can be determined if suitable measurements are made.

TABLE 2.9 *Some typical values of reactivity ratios for free-radical copolymerization at 60°C.*

M_1	M_2	r_1	r_2
Styrene	Methyl methacrylate	0.5	0.5
Styrene	Acrylonitrile	0.4	0.05
Styrene	Vinyl acetate	55	0.01
Styrene	1,3-Butadiene	0.8	1.4
Styrene	Vinyl chloride	17	0.02
Methyl methacrylate	Vinyl acetate	20	0.015
Methyl methacrylate	Vinyl chloride	12	~0
Methyl methacrylate	Acrylonitrile	1.2	0.15

Table 2.9 lists values of reactivity ratios for several pairs of monomers. It can be seen that r_1 and r_2 vary widely and the structure of the polymer produced will be controlled both by the composition of the feed and r_1 and r_2. This effect can most readily be seen by taking specific examples of reactivity ratios.

— EXAMPLES —

(i) $r_1 \simeq r_2 \simeq 1$

This is a special case which is only found under particular conditions. It means that there is no preference for either type of monomer to add onto a particular type of active centre. The copolymer produced will therefore have entirely random distribution of monomer units. When the reactivity ratios are of the order of unity it means that $k_{11} \simeq k_{12}$ and $k_{22} \simeq k_{21}$. Copolymers which have completely random sequences of monomer units are known as *ideal copolymers*.

(ii) $r_1 \simeq r_2 \simeq 0$

In this case k_{11} and k_{22} are very small and it means that there is no tendency for the monomers to homopolymerize, but they may form a copolymer. In the copolymer the two types of monomer units must therefore alternate along the polymer chain leading to a completely alternating copolymer.

(iii) $r_1 \gg$ and $r_2 \ll 1$

In this case the copolymer Equation (2.85) becomes

$$\frac{d[A]}{d[B]} = 1 + r_1\frac{[A]}{[B]}$$

Since $r_1 \gg 1$ then the monomer A is incorporated in the copolymer at a much greater rate than B and so the polymer produced has long sequences of A interspersed with a few B units. However, things can change towards the end of the reaction since [A] reduces at a much greater rate than [B] and so longer sequences of B may be found in the latter stages of the reaction. Composition drift is always found when $r_1 \neq r_2$.

Normally reactivity ratios lie between 0 and 1 (Table 2.9) and so there is usually a tendency toward alternation in most copolymerization reactions. It is found that for the same pair of monomer molecules the reactivity ratios can differ greatly depending upon the nature of the chain carrier used (i.e. free radical, cationic or anionic). Obviously the rate constants k_{11}, k_{12}, k_{22} and k_{21} will be affected in different ways by the nature of the active centre and it is found that the relative reactivity of different monomers can be correlated with resonance stability, polarity and steric effects. Such correlations are beyond the scope of this book and the reader is directed towards more advanced texts.

2.4.4 *The Q–e scheme*

A semi-empirical method has been devised to predict the reactivity ratios for a pair of monomers. It eliminates the need to determine the reactivity ratios experimentally for each monomer in a particular free-radical copolymerization reaction. The method is known as the *Q–e scheme* and each monomer is assigned a particular value of Q and e relative to styrene which is given arbitrary reference values of $Q = 1.0$ and $e = -0.8$. The reactivity ratios are given for the two monomers by

$$r_1 = (Q_1/Q_2) \exp[-e_1(e_1 - e_2)]$$ (2.87)
$$r_2 = (Q_2/Q_1) \exp[-e_2(e_2 - e_1)]$$ (2.88)

where Q_1 and Q_2 are measures of the reactivity of monomers A and B and the constants e_1 and e_2 depend upon the polarity of the two monomers. Several representative values of Q and e are given in Table 2.10. It should be pointed out that these values are approximate and provide only a rough guide to the reactivity ratios of pairs of monomers. Nevertheless, the Q–e scheme is extremely useful in predicting the reactivity ratios when direct experimental data are unavailable.

2.4.5 *Block copolymers*

Block copolymers are polymers which contain long sequences of two or more types of monomer unit. As far as mechanical properties are concerned some of the most interesting types of block copolymers are those consisting of two or three sequences only such as A_mB_n, $A_mB_nA_p$ or even $A_mB_nC_p$ where A, B and C are different types of monomer and m, n and p are the number of units in each sequence.

The most convenient way of producing block copolymers is through the

TABLE 2.10 *Values of Q and e for various monomers undergoing free radical addition copolymerization. (Data taken from the 'Polymer Handbook' edited by Brandrup and Immergut, reproduced with permission).*

Monomer	Q	e
Styrene (base)	1.0	−0.80
Methyl methacrylate	0.74	0.40
Ethylene	0.015	−0.20
Vinyl acetate	0.026	−0.22
Acrylonitrile	0.60	1.20
1,3-Butadiene	2.39	−1.05

use of living polymers (Section 2.3.6) in anionic systems. The method consists of, first of all, preparing a living homopolymer and preventing termination by rigorous exclusion of all impurities such as oxygen or water. If a second monomer is then added a block copolymer can be formed. It is not always possible to form block copolymers for any pair of monomers in this way and even the order of monomer addition can have an important effect. For example, a styrene–methyl methacrylate block copolymer can be made by the addition of methyl methacrylate to living polystyrene, but it cannot be made by adding styrene to a living polymer of poly(methyl methacrylate). Living dianions which are capable of propagating from both ends are useful in preparation of the $A_mB_nA_p$ type triblock copolymers. Other types of triblock copolymers can be made from three different monomers such as styrene, methyl methacrylate and acrylonitrile.

If living polymers are unavailable block copolymers can be prepared in other ways. It is necessary, first of all, to make a homopolymer and then use it as the starting material for the block copolymer. If the homopolymer can be induced to undergo chain scission by the use of high-energy radiation or even mastication or high-speed stirring, radicals will be formed at the broken chain ends. When another monomer is present the molecules can add onto the broken chains to give A_mB_n type block copolymers. Alternatively, homopolymers can be formed which have end-groups amenable to removal by, for example, high-energy radiation or ultra-violet light. If this is done in the presence of a second monomer and free radicals are formed at the chain ends block copolymers may again be formed. Block copolymers can be prepared with interesting morphologies and unusual mechanical properties. In general they are found to behave rather like blends or mixtures of homopolymers especially when the blocks are relatively long.

2.4.6 *Graft copolymers*

Polymers which consist of a homopolymer backbone and have branches of another type of polymer are known as graft copolymers. Normally the backbone polymer is made first of all and then an active site is created on the chain. The most widely used method of providing these sites is *irradiation.* This can be done using ultraviolet light or ionizing radiation and in all cases free radicals are involved through hydrogen abstraction from the polymer chain. Usually the backbone polymer and second monomer are irradiated together and monomer addition takes place at the active sites created by the radiation. A major difficulty with this method is homopolymerization of the second monomer can occur as well as grafting. This can be overcome to a certain extent by pre-irradiating the backbone polymer in the absence of air and then adding the monomer to form the branches. However, the most extensively used method is mutual irradiation of the polymer and monomer. The types of monomer and polymer are chosen carefully so as to avoid homopolymerization. The best results are obtained by choosing a polymer which is very sensitive to radiation and forms a high concentration of radicals along the backbone and a monomer which is relatively insensitive.

It is possible to produce graft copolymers by reacting the backbone monomer with an initiator (usually free radical) in the presence of a second monomer. The reaction again normally proceeds through the formation of an active site by hydrogen abstraction from the polymer backbone which is followed by successive monomer addition at the active sites. This method is called *transfer grafting* since it takes advantage of the chain-transfer reactions which are often encountered in addition polymerization. The problem is that as well as producing a graft copolymer the product usually also contains some homopolymer and unreacted backbone chains. It is found that the efficiency of transfer grafting is a function of a number of variables, the most important of which is the type of initiator employed. For instance, benzoyl peroxide is a good initiator for grafting methyl methacrylate onto polystyrene whereas azobisisobutyronitrile shows little tendency to cause grafting with this system. Grafting branches onto a polymer chain can lead to improvements in certain properties of the polymer. For example, many synthetic fibres are resistant to dyeing and this can be overcome by grafting suitable branches onto the backbone molecules.

2.5 Stereoregular polymerization

So far only the overall chemical structure of the polymer molecules produced in polymerization reactions has been considered. The physical properties of polymers are controlled principally by the basic chemical composition of the material, but they are also strongly dependent upon the detailed microstructure of the molecules. This microstructure is usually considered in terms of structural or configurational isomerism which describes the detailed ways in which the atoms and groups in the polymer molecules are arranged. It is determined by the mechanisms and conditions of the polymerization processes.

2.5.1 *Structural isomerism*

There are several types of structural isomers which can be obtained for polymers containing an identical number and type of atoms in the repeat unit. For example, polystyrene and poly (*p*-xylylene) are isomers but have different physical properties.

$$(-CH_2-CH-)_n \qquad (-CH_2-CH_2-\langle O \rangle-)_n$$

  polystyrene   poly(*p*-xylylene)

There are many other examples of similar isomers and they are discussed in the section on the glass transition (Section 4.4.3).

Even when polymers have identical repeat units the molecules can differ in architecture leading to further types of isomerism. *Branching* can have a profound effect upon the properties of the polymer. It is normally caused during addition polymerization by transfer reactions with polymer molecules (Section 2.2.9). The tendency for the transfer to take place depends principally upon the reactivity of the active centre. Free radicals are more reactive than ions and branching is more common in polymers produced by free radical addition than by ionic polymerization. In general it is found that the tendency for branching depends upon the type of polymerization system as follows:

  free radical > cationic > anionic
  (common)   (rare)

Branching greatly affects the ability of a polymer molecule to crystallize. Linear polyethylene is more crystalline than the branched variety and is found to be more dense, stiffer but less tough than branched polyethylene.

Another type of isomerism that can occur is that caused by different *orientations* of monomer addition at the active centre during vinyl polymerization. This problem has been discussed in the case of free-radical addition with respect to active centre formation (Section 2.2.2) and the propagation reaction (Section 2.2.3). The normal orientations that are found with vinyl monomers are the head-to-tail configurations

$$\text{\small ᴡᴡCH}_2\text{—CH—CH}_2\text{—CH—CH}_2\text{—CH—CH}_2\text{—CHᴡᴡ}$$
$$\qquad\quad |\qquad\quad |\qquad\quad |\qquad\quad |$$
$$\qquad\quad X\qquad\quad X\qquad\quad X\qquad\quad X$$

Occasionally a head-to-head configuration is found

$$\text{\small ᴡᴡCH}_2\text{—CH—CH—CH}_2\text{—CH}_2\text{—CH—CH—CH}_2\text{ᴡᴡ}$$
$$\qquad\quad |\quad\ |\qquad\qquad\quad |\quad\ |$$
$$\qquad\quad X\quad X\qquad\qquad\quad X\quad X$$

but this is not favoured because of steric effects. Of course, head-to-head orientations are created if termination is by combination. A mixture of orientations will lead to an irregular polymer chain and so may affect the ability of the polymer molecule to crystallize which in turn could affect the physical properties of the polymer.

2.5.2 *Configurational isomerism*

There are two distinct stereochemical arrangements of polymers prepared from a general vinyl monomer of the type CH_2=CHX. In one type the carbon atoms containing the X group have the same configuration. A polymer for which this is the case is said to be *isotactic.* In the other type alternate carbon atoms have the same configuration and this type of polymer is termed *syndiotactic.* Polymers from this type of monomer in which the X groups are randomly placed are known as *atactic* polymers. These three stereochemical configurations are shown schematically in Fig. 2.9. It must be stressed that it is impossible to change the tacticity of a polymer by rotation about the C—C bonds. The conditions under which polymers with different tacticities can be synthesized will be discussed later.

Similar stereochemical effects can be obtained for other addition polymers made from monomers which contain double bonds. Monomers of the type CH_2=CXY and XCH=CHX can both lead to polymers with isotactic, syndiotactic or atactic forms. However, addition polymers from the monomer CH_2=CX$_2$ are always stereoregular since the polymer contains no assymetric carbon atoms. The situation is rather more complicated with polymers prepared from XCH=CHY, where X and Y are different groups. In this case it is possible to have two isotactic forms and one syndiotactic form depending upon the configuration of the planar monomer

$$\begin{array}{ccc}
\text{X}\diagdown\quad\diagup\text{H} & & \text{X}\diagdown\quad\diagup\text{Y} \\
\quad\text{C}{=}\text{C} & \text{or} & \quad\text{C}{=}\text{C} \\
\text{H}\diagup\quad\diagdown\text{Y} & & \text{H}\diagup\quad\diagdown\text{H} \\
\textit{trans} & & \textit{cis}
\end{array}$$

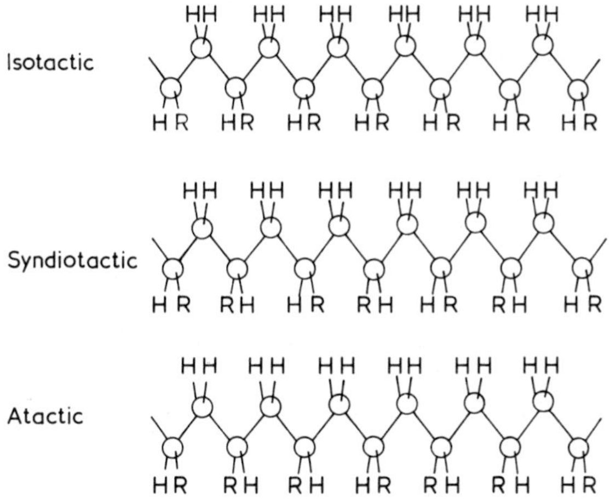

Fig. 2.9 *The different stereochemical configurations of polymers formed from the general vinyl monomer* CH$_2$=CHR.

and the mechanism of propagation. There are many other examples of more complex polymers with different stereochemical forms and details can be found in more specialized texts.

2.5.3 *Homogeneous vinyl polymerization*

It will be shown later that a high degree of control of stereoregularity and branching can be obtained using heterogeneous catalysts. However, it has been known for many years that with conventional homogeneous free radical or ionic initiated addition reactions a certain amount of control can be exercised over the stereoregularity of the polymer produced by varying the type of initiator and the reaction conditions. With simple vinyl monomers of the type CH$_2$=CHX the three basic types of polymer that can be produced are isotactic, syndiotactic and atactic. But in general using homogeneous systems the tacticity is mixed in any polymer produced and it is only possible to talk in terms of homogeneous reactions having tendencies to give a particular type of tacticity.

The tacticity of the polymer produced is controlled by the way in which addition takes place which in turn depends upon interaction between the terminal chain carbon atom and the approaching monomer molecule. For *free radical* addition polymerization reactions it is presumed that the terminal carbon atom has sp^2 hybridization and so has planar bonds. If this is the case then there are two ways in which it can be approached by the monomer molecule:

In case I all X side groups are on the same side and so the placement is isotactic whereas in case II the X group is on opposite sides leading to syndiotactic placement. If there is a succession of identical placements and no free rotation about the terminal C—C bond on the polymer chain a monotactic polymer will be produced. In the case of free radical addition polymerization atactic polymer is produced at high temperatures. On the other hand, it is found that if X is relatively large and polar then there is some tendency towards syndiotactic placements as the reaction temperature is decreased. It appears that when this is so the X groups tend to repel each other and addition is easier when they are on opposite sides (II). For syndiotactic polymer to be produced no rotation about the terminal C—C bond must take place. Such a process would involve an activation energy which would be relatively high if X were large and polar and so it would be expected that the rotation would be easier at higher temperatures leading to a lower degree of stereoregularity as the temperature is increased as observed in practice.

It has long been recognized that addition polymerization with ionic initiators can lead to some control of tacticity and a stereoregular structure. The main observations are as follows.

(i) A greater degree of stereoregularity is achieved in reactions at lower temperatures.

(ii) The type of initiator can affect the stereoregularity.

(iii) The degree and nature (isotactic or syndiotactic) of the stereoregularity is affected by the polarity of the solvent.

The mechanisms of homogeneous ionic stereoregular addition polymerization are not properly understood and still a matter of some debate. The most important difference between ionic and free-radical addition is the presence of the counter-ion (Section 2.3) at the active centre. In polar solvents the counter-ions are not very closely associated with the active centre and so the ions at the chain end are effectively free. If this is the case syndiotactic addition is often preferred for the same reasons as for free radical addition. In non-polar solvents the counter-ions are more tightly bound to the chain end and the tightness depends upon the polarity of the solvent and the nature of the counter-ion. In this case the counter-ions can affect the mechanism of addition leading to a tendency for isotactic placements but again these mechanisms are not fully understood.

2.5.4 *Homogeneous diene polymerization*

A variety of different products can be obtained by the addition polymerization of conjugated dienes. Reactions of particular interest are those of 1,3-butadiene, isoprene and chloroprene

$$CH_2{=}CH{-}CH{=}CH_2 \qquad CH_2{=}\underset{\underset{CH_3}{|}}{C}{-}CH{=}CH_2 \qquad CH_2{=}\underset{\underset{Cl}{|}}{C}{-}CH{=}CH_2$$

1,3-butadiene   isoprene   chloroprene

The addition of an initiator radical to a 1,3-diene leads to a radical which can have two equivalent resonance forms as in the case of 1,3-butadiene

$$R{\cdot} + CH_2{=}CH{-}CH{=}CH_2 \begin{cases} \nearrow R{-}CH_2{-}\dot{C}H{-}CH{=}CH_2 \qquad I \\ \searrow R{-}CH_2{-}CH{=}CH{-}\dot{C}H_2 \qquad II \end{cases}$$

In the case of I, if propagation takes place by addition at carbon 2 and subsequent propagation is through a radical of the same form (I) then the polymerization is called 1,2-addition.

$$\underset{\underset{CH{=}CH_2}{|}}{[{-}CH_2{-}CH{-}]} \qquad \text{1,2-addition}$$

On the other hand if attachment of successive monomer units takes place at carbon 4 through a radical of the form II then this is referred to as 1,4-addition. In this case, however, two stereoisomeric configurations known as *cis* and *trans* are possible

$$\begin{array}{cc} [{-}CH_2 \qquad CH_2{-}] \\ \diagdown \qquad \diagup \\ C{=}C \\ \diagup \qquad \diagdown \\ H \qquad\qquad H \end{array} \qquad\qquad \begin{array}{cc} [{-}CH_2 \qquad H \\ \diagdown \qquad \diagup \\ C{=}C \\ \diagup \qquad \diagdown \\ H \qquad\qquad CH_2{-}] \end{array}$$

cis 1,4-addition   *trans* 1,4-addition

The occurrence of the three possible structures depends upon the activation energies for the different reactions. Using free-radical initiators it is generally found that about 20 percent of the polymer has the 1,2 structure, whereas most of the polymer has the *trans*-1,4 structure. If the reaction temperature is increased the amount of the 1,2 structure remains almost constant whereas the proportion of *cis*-1,4 increases.

The situation is more complicated for substituted dienes such as isoprene and chloroprene since the results of 1,2 and 3,4 addition are now different. In the case of isoprene we have

$$
\begin{array}{c c}
\quad CH_3 & \quad H \\
\quad | & \quad | \\
[-CH_2-C- \quad] & [-CH_2-C- \quad] \\
\quad | & \quad | \\
\quad CH=CH_2 & \quad CH_3-C=CH_2 \\
\text{1,2-addition} & \text{3,4-addition}
\end{array}
$$

The 1,4 structure can still have *cis* and *trans* forms as for isoprene

$$
\begin{array}{c c}
[-CH_2 \quad\quad CH_2-] & [-CH_2 \quad\quad H \\
\quad \diagdown \quad / & \quad \diagdown \quad / \\
\quad C=C & \quad C=C \\
\quad / \quad \diagdown & \quad / \quad \diagdown \\
CH_3 \quad\quad H & CH_3 \quad\quad CH_2-] \\
\textit{cis} & \textit{trans}
\end{array}
$$

There is also the added possibility of head-to-head rather than head-to-tail arrangements. For the free-radical polymerization of both isoprene and chloroprene it is found that the *trans*-1,4 structure predominates. The tendency for most simple diene monomers to form the *trans* structures may be because it is thought that the monomers exist mainly in transoid conformations and that this stereochemistry is retained during polymerization.

It is possible to polymerize conjugated dienes into polymers with more *cis*-1,4 segments using ionic initiators. In particular it is found anionic initiators such as lithium or organolithium compounds in non-polar solvents will polymerize isoprene to produce a product which is very similar in structure to natural rubber (*cis*-1,4-polyisoprene). It is thought that the Li^+ ion holds the isoprene molecule in a cisoid conformation through the formation of a π-complex type of structure.

$$
\begin{array}{c}
H \quad\quad\quad CH_3 \\
\diagdown \quad\quad / \\
C-C \\
/\!/ \quad\quad \backslash\!\backslash \\
H_2C \quad\quad\quad CH_2 \\
\quad \cdots \quad\quad \cdots \\
\quad\quad Li^+
\end{array}
$$

This tendency to form a *cis*-1,4 polymer is reduced if higher alkali metals are used. Also it is found that the effect is more marked for 1,3-isoprene than for 1,3-butadiene. It is thought that this is because the isoprene will adopt a cisoid conformation more readily because of steric effects due to the methyl group. For example, the polymerization of 1,3-butadiene in pentane using Li as an initiator produces a polymer with only about 35 percent *cis*-1,4 segments whereas the same system with 1,3-isoprene gives a polymer which is almost entirely *cis*-1,4-polyisoprene.

2.5.5 *Ziegler–Natta catalysts*

During experiments in the 1950s on the polymerization of ethylene using organometallic catalysts in inert atmospheres, Ziegler discovered that it was possible to make a linear form of polyethylene which was highly crystalline in contrast to the low crystallinity branched form prepared by conventional high-pressure polymerization (Section 2.2.9). Subsequently Natta showed that similar catalysts could be used to synthesize α-olefins such as 1-butene or propylene which could not be polymerized using conventional free radical or ionic initiators. The polymers produced using these catalysts can also possess a high degree of stereoregularity and Ziegler–Natta catalysts are now extremely important as commercial catalysts for the production of linear polyethylene and isotactic polypropylene, with their use increasing year by year.

There is a whole variety of combinations of many compounds which fall into the category of Ziegler–Natta catalysts. The catalysts are made up principally of a combination of hydrides or alkyls of metals from groups I–III and salts of transition metals from groups IV–VIII. A typical catalyst combination for the polymerization of propylene is triethyl aluminium, $Al(C_2H_2)_3$, and titanium chloride, $TiCl_3$. Many different catalysts have been used, but even apparently similar systems can give quite different results. For example, the stereoregularity can be affected by the crystal form of $TiCl_3$ when it is employed as a cocatalyst with $Al(C_2H_5)_3$. The α form of $TiCl_3$ gives highly stereoregular polypropylene whereas if the β form is used the stereoregularity of the polymer is reduced. This observation gives a clue as to the mechanism of the action of the catalysts and shows that the polymerization reaction is heterogeneous. It is thought that crystals of the two forms have different surface properties and that the crystal structure of α-$TiCl_3$ leads to surface sites at which there can be better control over the stereoregularity of addition than with β-$TiCl_3$.

The mechanisms and kinetics of Ziegler–Natta polymerization are known to be complex and are not yet fully understood. It is clear that the reaction takes place at the surface of the solid catalyst and that there is some type of complex formed at the surface. It is thought that the stereoregulation occurs from the way in which the monomer molecules are oriented during addition to the growing chain at the catalyst surface, although there is still considerable argument over the exact way in which this occurs. The suggested mechanisms of the action of Ziegler–Natta catalysts have been reviewed extensively in the literature and the reader is directed to more advanced texts for further details.

Ziegler–Natta catalysts are now of great importance for the production of polymers on both the laboratory and commercial scales. For example, linear polyethylene has greater mechanical strength and stiffness and a higher melting point than the high-pressure branched variety. Also it has facilitated the synthesis of polymers which could not be otherwise made and enabled highly stereoregular forms of many polymers to be produced.

2.5.6 *Solid-state polymerization*

It has been known for many years that there are numerous monomers capable of undergoing polymerization in the solid state, but it has been rather difficult to decide whether the reactive species in each case are free radicals or ions. Styrene and acrylonitrile will polymerize more rapidly in the solid state than as liquids just above their melting temperatures when they are irradiated with ionizing radiation. On the other hand, methyl methacrylate and vinyl chloride will not polymerize as solids. In recent years there have been important developments in using solid-state polymerization reactions to prepare highly stereoregular polymers. One of the first methods involved the inclusion of monomer molecules in complexes with urea or thiourea and polymerizing the monomer with ionizing radiation. It is possible to synthesize *trans*-1,4 butadiene by photopolymerization if butadiene is prepared in a crystal complex as shown schematically in Fig. 2.10. It is possible to perform similar reactions with other dienes and it is even possible to make a stereoregular crystalline form of vinyl chloride by a similar reaction. When the monomer molecules are contained in a crystal lattice, the lattice acts as a template and the orientation of the molecules is fixed so that polymerization can only take place in one way.

There are certain monomers which do not polymerize in the liquid state but will undergo solid-state polymerization. One such monomer that has now been extensively studied is trioxane where single crystals of the monomer will polymerize to give polymer crystals when they are irradiated.

$$ n \underset{\text{trioxane}}{\left[\begin{array}{c} \text{CH}_2 \\ \diagup \quad \diagdown \\ \text{O} \qquad \text{O} \\ | \qquad | \\ \text{CH}_2 \quad \text{CH}_2 \\ \diagdown \quad \diagup \\ \text{O} \end{array} \right]} \xrightarrow{\text{radiation}} \underset{\text{polyoxymethylene}}{(-\text{CH}_2-\text{O}-)_n} $$

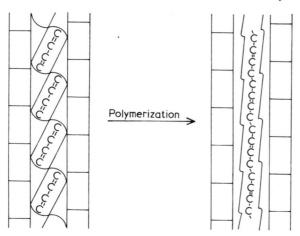

Fig. 2.10 *Schematic illustration of the polymerization of a diene monomer in the canals of a urea or thiourea complex. (After Brown and White, J. Am. Chem. Soc.* 82 (1960) 5671).

The polymer crystals that are formed have good orientation of the polymer chain axes, but are not in fact true single crystals. They contain many small single crystal microfibrils which are slightly misaligned with respect to each other. It is thought that this is because there are significant changes in the atomic packing during polymerization and different possible reaction paths.

In the late 1960s Wegner discovered that it was possible to polymerize certain diacetylene monomer single crystals to give the polymer single crystals which were virtually defect free. He found that such reactions could be activated either thermally or by using ionizing radiations and this type of reaction has been termed *topochemical*. The reaction for a substituted diacetylene monomer is illustrated schematically in Fig. 2.11 and some typical substituent R groups are also given. The reaction takes place by the direct transition from monomer molecules to polymer without any destruction of the crystal lattice and so maintaining the three-dimensional order. There is usually a slight rotation of the monomer molecules, but the amount of atomic movement is usually small. It is thought that the reactive species are carbenes rather than radicals. The polymers produced are completely crystalline with no amorphous content and the polymer molecules are unbranched, stereoregular and lie in the polymer crystals in a chain-extended conformation. They usually have very high degrees of polymerization and when fully polymerized are insoluble in all common solvents. Examples of polydiacetylene single crystals are shown in Fig. 4.12.

There is now a great deal of research interest in polydiacetylene single

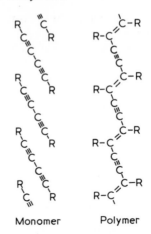

Monomer   Polymer

Fig. 2.11 *Schematic representation of the solid-state polymerization of a substituted diacetylene monomer. Typical R groups are* —CH_2—O—SO_2—C_6H_4—CH_3, —CH_2—O—CHNH—C_6H_5 *etc.*

crystals produced by solid-state polymerization, although they do not yet have any commercial applications. They have interesting and unusual optical, electronic and mechanical properties and their availability has greatly increased our knowledge of the fundamental physical properties of crystalline polymers.

Further reading

Billingham, N.C. and Jenkins, A.D. (1972), in *Polymer Science*, (Ed. A.D. Jenkins), North-Holland, London, Chap. 1.
Cowie, J.M.G. (1973), *Polymers: Chemistry and Physics of Modern Materials*, International Textbook Company, Aylesbury, U.K.
Flory, P.J. (1953), *Principles of Polymer Chemistry*, Cornell University Press, Ithaca, N.Y.
Lenz, R.W. (1967), *Organic Chemistry of High Polymers*, Interscience Publishers Inc., New York.
Margerison, D. and East, G.C. (1967), *An Introduction to Polymer Chemistry*, Pergamon Press, London.
Odian, G. (1970), *Principles of Polymerization*, McGraw-Hill Inc., New York.
Stevens, M.P. (1975), *Polymer Chemistry-An Introduction*, Addison Wesley Inc., Reading, Mass.

Problems

2.1 Write down the reaction scheme for the polymerization of an ω-amino carboxylic acid.

The following data were obtained during the condensation of 12-hydroxy stearic acid at 433.5 K in the molten state by sampling the reaction mixture at various times. [COOH] was determined for each sample by titrating with ethanolic sodium hydroxide.

t(h)	[COOH] (mol dm^{-3})
0	3.10
0.5	1.30
1.0	0.83
1.5	0.61
2.0	0.48
2.5	0.40
3.0	0.34

Determine the rate constant for the reaction under these conditions and the order of the reaction. Also suggest whether or not a catalyst was used.

What would be the extent of the reaction after 1 hour and 5 hours?

2.2 Neglecting the effect of end groups, calculate the percentage conversion into polymer to obtain a number average molar mass of 24 000 for the monomer

$$HO(CH_2)_{14}COOH$$

2.3 A condensation polymerization reaction takes place between 1.2 moles of a dicarboxylic acid, 0.4 moles of glycerol (a triol) and 0.6 moles of ethylene glycol (a diol).

Calculate the critical extents of reaction for gelation using the Statistical Theory of Flory and the modified Carothers equation.

Comment on the observation that the measured value of the critical extent of reaction is 0.866.

2.4 1 kg of a polyester of number average molar mass of 10 000 is mixed with 1 kg of another polyester of number average molar mass of 30 000. The mixture is then heated to a temperature at which it undergoes an ester interchange reaction.

Assuming that the two original polymer samples and the polymer produced by the ester interchange reaction have the ideal distributions of molar mass calculate $\bar{M}_n$ and $\bar{M}_w$ for the mixture before and after the ester interchange reaction.

2.5 Write down the reaction scheme for the polymerization of acrylonitrile (CH$_2$=CHCN) using azobisisobutyronitrile as an initiator. Assume that termination takes place by combination.

Using radioactive AZBN as an initiator, a sample of styrene is polymerized to a number average degree of polymerization of 10 000. The AZBN has a radioactivity of 6×10^9 counts s^{-1} mol^{-1} in a liquid-scintillation counter. If 0.001 kg of the polystyrene has a radioactivity of 6×10^3 counts s^{-1} determine whether the termination mechanism is combination or disproportionation.

2.6 Calculate the half-life of benzoyl peroxide in benzene at 333 K given that the rate constant for initiation of polymerization (k_i) by this initiation under these conditions is $3.4 \times 10^{-6}s^{-1}$. What would be the change in initiator concentration after 1 hour? Comment on the values obtained.

2.7 Calculate the initial rate of polymerization of acrylonitrile in homogeneous solution in benzene under the following conditions:

0.2 M-acrylonitrile
Temperature 333 K
0.05 M initiator

given the following rate constants at 333 K.

Initiation $k_i = 1.5 \times 10^{-5}s^{-1}$
propagation $k_p = 2.0 \times 10^3 dm^3 mol^{-1}s^{-1}$
termination $k_t = 7.8 \times 10^8 dm^3 mol^{-1}s^{-1}$

NB A M solution is one which has a concentration of one mole in one $(decimetre)^3$.

2.8 In a free radical addition polymerization reaction what would be the effect of
(a) increasing $[M]_0$ 4 times at constant $[I]_0$ or
(b) increasing $[I]_0$ 4 times at constant $[M]_0$ upon

1. The total radical concentration at steady state
2. The rate of polymerization
3. The number average degree of polymerization

2.9 Assuming that in free-radical addition polymerization the rate constants can be replaced by the appropriate Arrhenius expressions:

$$k_i = A_i \exp(-E_i/RT)$$
$$k_p = A_p \exp(-E_p/RT)$$
$$k_t = A_t \exp(-E_t/RT)$$

Calculate the change in rate of polymerization and degree of polymerization by increasing the temperature of the polymerization of styrene in benzene initiated by AZBN from 60°C to 70°C given the following information.
    For styrene

$A_p = 2.2 \times 10^7 mol^{-1}dm^3 s^{-1}$
$A_t = 2.6 \times 10^9 mol^{-1}dm^3 s^{-1}$
$E_p = 34 \text{ kJ mol}^{-1}$
$E_t = 10 \text{ kJmol}^{-1}$

and for AZBN

$A_i = 5.6 \times 10^{14}s^{-1}$
$E_i = 126 \text{ kJ mol}^{-1}$

Assume that the concentration of monomer and initiator and the mechanism of termination remain unchanged when the temperature is increased.
    If the rate constant for the depropagation reaction k_{dp} can also be given by an Arrhenius equation calculate the ceiling temperature for the reaction given

$$A_{dp} = 10^{13} s^{-1}$$
$$E_{dp} = 122 \text{ kJ mol}^{-1}$$
and $\quad [M] = 1 \text{ mol dm}^{-3}$

(The molar gas constant R is 8.314 J mol^{-1} K^{-1})

2.10 Calculate the ultimate number average degree of polymerization for the anionic polymerization of styrene in benzene at 25°C initiated by butyl lithium. The initial concentration of styrene was 10^3 mol m^{-3} and that of butyl lithium was 0.5 mol m^{-3}. Account for the observation that even though this is a 'living' polymer system a polydisperse sample is produced.

2.11 Wooding and Higginson (*J. Chem. Soc.* (1952) 760) studied the polymerization of styrene in liquid ammonia using potassium amide as initiator. They found that the individual rate constants k_i, k_p and k_{tr} could be given by Arrhenius expressions.

Calculate the changes in rate of polymerization and degree of polymerization by changing the temperature of polymerization from -33.5°C to -70°C given the following information

$$E_i = 54.6 \text{ kJ mol}^{-1}$$
$$E_p = 25.2 \text{ kJ mol}^{-1}$$
$$E_{tr} = 42.0 \text{ kJ mol}^{-1}$$

3 Characterization

3.1 Macromolecules in solution

Many techniques of polymer characterization are performed when the polymer molecules are in solution. This is because the thermodynamic properties of polymer solutions can be readily measured and the results interpreted in terms of the size and structure of the macromolecules and so enabling characterization of the polymer. Because of the importance of polymer solutions it is necessary, first of all, to consider the conformations that polymer molecules adopt in solutions and then go on to look at the various theories which have been developed to explain the thermodynamic properties of polymer solutions.

3.1.1 *Conformations of hydrocarbons*

Polymer molecules tend to adopt a range of different conformations in solution and these conformations are continually changing. It is important at this stage to define carefully what is meant by the term conformation. Many textbooks use different definitions of this term and sometimes confuse it with the term configuration which was used earlier in Section 2.5.2. Both terms refer to isomerism but configurational isomerism is concerned with arrangements of atoms fixed by chemical bonding which cannot be changed without breaking and reforming bonds. An example of this is given in Section 2.5.2 for the *cis* and *trans* forms of the monomer XCH=CHY. This *cis–trans* configurational isomerism can also be found in polymer molecules which contain double bonds in the polymer chain such 1,4-polyisoprene (Section 2.5.4).

The term conformation is retained for different three-dimensional spatial arrangements of molecules which are altered by rotation about single bonds and do not involve any bond breaking to change them. In saturated hydrocarbon polymers the carbon atoms are sp^3 hybridized and the angles between the C—C bonds are fixed normally at 109.5°. If n-butane is used as an example it is possible to consider the various conformations that a segment of polymer chain can adopt. Fig. 3.1 gives a schematic diagram of the n-butane molecule and it can be seen that there are two extreme conformations the molecule can adopt and they are again termed *cis* and *trans*. The —CH_3 groups at the two ends of the n-butane molecule tend to be rather bulky. In the *cis* conformation the distance

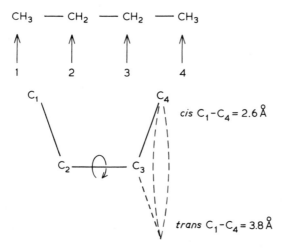

Fig. 3.1 *The different conformations of the n-butane molecule.*

between the C_1 and C_4 atoms is 2.6Å in contrast to 3.8Å in the *trans* conformation. The result of this is that the two —CH_3 groups tend to interfere with each other in the *cis* conformation and so the molecule prefers to adopt the *trans* conformation. This effect is called *steric hindrance* and is extremely important in controlling the conformations of longer-chain molecules.

It is possible to characterize the conformations of n-butane from the angle ϕ through which the C_3–C_4 bond is rotated about the C_2–C_3 bond from the *trans* conformation ($\phi = 0$ for *trans*, $\phi = 180°$ for *cis*). Fig. 3.2 shows different projections of the n-butane molecule along the C_2–C_3 bond. These types of projections are known as Newman projections and give an extra insight into the possible conformations that can be adopted. The *trans* conformation is known as a staggered state with the —H and —CH_3 atoms and groups all at 60° to each other. This is a low energy position and it can be seen that there also must be two other similar staggered positions known as *gauche* which should also be of low energy. It is possible to plot out the relative energy of the different positions as a function of ϕ graphically as shown in Fig. 3.3. The energy minimum is when $\phi = 0$ in the *trans* conformation. It can be seen that there are also two other minima, at an energy of about 3.3 kJ mol^{-1} higher than the *trans* position, corresponding to the two *gauche* positions. This means that the *trans* position will be most favoured conformation at normal temperatures, but *gauche* conformations will also be found. Any other conformations will be highly unlikely.

Although these considerations are only for n-butane it is possible to extend them with certain reservations to longer molecules such as polyethylene. In general, each successive linkage will have three possible

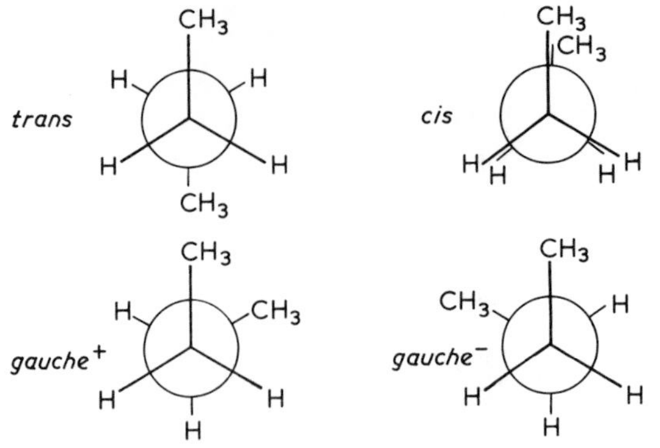

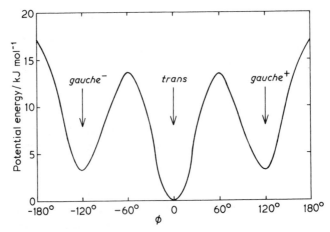

Fig. 3.2 *Newman projections of the different conformations of the n-butane molecule.*

low-energy positions, one *trans* and two *gauche*. This will lead to an enormous number of possible conformations for a long polymer molecule with a large number of linkages. However, normally not all of the theoretical conformations are possible because of steric effects. This can be demonstrated by considering n-pentane which is similar to n-butane but contains an extra —CH_2— unit. The n-pentane molecule is shown schematically in Fig. 3.4. If the middle atom is fixed then it can be readily seen that there are only four distinct conformations that the molecule can adopt. Three of these are relatively low energy positions, but in the *gauche*$^+$–*gauche*$^+$ conformation the —CH_3 groups at the ends of the molecule interfere with each other and so the conformation has high energy and so is not favoured. This effect is called *pentane interference* and can occur between every fifth C atom on a hydrocarbon polymer chain.

Fig. 3.3 *Energy of the different conformations of the n-butane molecule as a function of φ.*

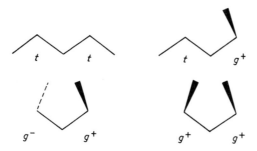

Fig. 3.4 *Conformations of the n-pentane molecule. Pentane interference occurs in the gauche⁺–gauche⁺ conformation.*

There are also other conformations where there is interference between segments of the molecule and so in general not all combinations of *trans* and *gauche* positions are possible along a given chain.

3.1.2 *Freely-jointed chains*

It is possible to gain an idea of the size and spatial arrangements of a long polymer chain by, first of all, determining the dimensions of a freely-jointed chain and then modifying the calculations to take into account the differences between the freely-jointed and polymer chains. A *freely-jointed chain* is a chain of n links of length l which has no fixed bond angles and there is no restriction of rotation at any bond. The links are in fact vectors and so there are no steric restrictions on the conformations the chain can adopt. Although this is an unrealistic model it is possible to calculate important parameters rather easily and simple modifications can be made for the case of a real polymer chain.

At this stage it is necessary to define some terms which are used to characterize chain dimensions. The total length of a chain is called the *contour length*. For a freely-jointed chain the contour length is nl. If the chain has fixed bond angles of Θ then it can be shown by a simple geometric construction that the contour length is reduced to $nl \sin(\Theta/2)$. An important parameter that is often used to describe the dimensions of a chain that is not extended is the *displacement length* (r) which is the end-to-end separation of a coiled chain as shown in Fig. 3.5. It is a parameter which can be used to define the dimensions of a polymer chain in solution and can be readily calculated by using standard relationships from statistical physics. For a polymer chain in solution r will be continually changing and since there are often many chains of different lengths we normally use an average value. The one that is usually quoted and is related to experimental parameters is the root mean square (RMS) displacement length $\langle r^2 \rangle^{1/2}$. If we have N_1 molecules with displacement

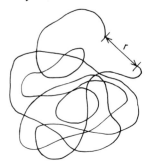

Fig. 3.5 *Schematic representation of a coiled polymer chain showing the displacement length, r.*

lengths of r_1 and N_2 molecules with displacement length r_2 and in general N_i molecules with displacement length r_i then the RMS displacement length is given by

$$\langle \overline{r^2} \rangle^{1/2} = \left\{ \frac{N_1 r_1^2 + N_2 r_2^2 + \ldots + N_i r_i^2 + \ldots}{N_1 + N_2 + \ldots + N_i + \ldots} \right\}^{1/2} \tag{3.1}$$

An equivalent measure of the dimensions of a simple chain is the RMS *radius of gyration* $\langle \overline{s^2} \rangle^{1/2}$ which is the RMS distance of the elements of the chain from its centre of gravity. It can be shown that $\langle s^2 \rangle^{1/2}$ is related simply to $\langle r^2 \rangle^{1/2}$ through the equation

$$\langle \overline{r^2} \rangle^{1/2} = \langle \overline{6s^2} \rangle^{1/2} \tag{3.2}$$

A polymer chain is modelled as a freely-jointed chain because the dimensions of such a chain can be readily determined using the results of the well-known *random-walk* calculation in three-dimensions. This was developed originally to describe the path of the particles in an ideal gas. This is often referred to in two dimensions as a *drunkard's walk*. For the freely-jointed chain a three-dimensional rectangular co-ordinate system is normally used as shown in Fig. 3.6. The chain has one end fixed at the origin 0 and the other end is at a distance r from 0. It is normally assumed that the distribution of the end-to-end distances of the chain will be approximately *Gaussian*. The probability per unit volume of finding the end of the chain at a particular point with co-ordinates (x,y,z) is given by

$$W(x,y,z) = (\beta/\pi^{1/2})^3 \exp[-\beta^2(x^2 + y^2 + z^2)] \tag{3.3}$$

where β is given $(3/(2nl^2))^{1/2}$ for a chain of n links of length l. Since r^2 is equal to $(x^2 + y^2 + z^2)$ then the probability per unit volume of finding the chain end at (x,y,z) can be written in terms of r as

$$W(x,y,z) = (\beta/\pi^{1/2})^3 \exp(-\beta^2 r^2) \tag{3.4}$$

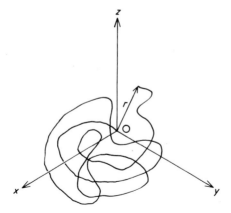

Fig. 3.6 *Rectangular coordinate system used to describe a freely-jointed polymer chain of displacement length r with one end fixed at the origin O.*

It must be stressed that this function, which is plotted in Fig. 3.7(a) is the probability per volume of the chain end being at (x,y,z) and has a maximum at the origin. We are normally interested in the probability of the chain end being at a distance r in any direction from the origin. This can be readily determined by calculating the probability of finding the chain end in a spherical shell of thickness dr and radius r and centred at the origin. The volume dV of such a shell is then

$$dV = 4\pi r^2 dr \qquad (3.5)$$

If $W(r)dr$ is the probability of finding the chain end in this shell then

$$W(r)dr = W(x,y,z)4\pi r^2 dr \qquad (3.6)$$

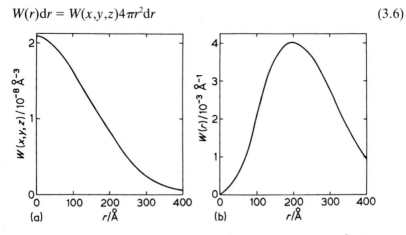

Fig. 3.7 (a) *Plot of Equation (3.4) for a chain of 10^4 links each of length 2.5 Å. (b) Plot of Equation (3.7) for a chain of 10^4 links each of length 2.5 Å.*

and so it follows that

$$W(r)dr = (\beta/\pi^{1/2})^3 4\pi r^2 \exp(-\beta^2 r^2)dr \tag{3.7}$$

This function is plotted in Fig. 3.7(b) and it can be shown by simple differentiation (see problems) that it has a maximum value at $r = 1/\beta$. One problem with the Gaussian distribution is that it predicts that $W(r)dr$ is still finite at $r > nl$. This is clearly not correct as it means that the displacement length of the chain could be greater than its contour length! In fact the Gaussian approximation although good for coiled chains breaks down as the displacement length approaches nl and better agreement is obtained using more sophisticated functions. However, such functions are beyond the scope of this book and it suffices to say that for polymer chains in dilute solution the Gaussian distribution can be used to predict the chain dimensions with very little error.

It is quite simple to calculate the RMS displacement length $\langle r^2 \rangle^{1/2}$ from Equation (3.7) since $\langle r^2 \rangle$ can be given in an integral form as

$$\langle r^2 \rangle = \int_0^\infty r^2 W(r)dr \tag{3.8}$$

Combining Equations (3.7) and (3.8) and integrating gives the result (see problems)

$$\langle r^2 \rangle = 3/2\beta^2 \tag{3.9}$$

Since β^2 is given by $(3/2nl^2)$ then the RMS displacement length is

$$\langle r^2 \rangle^{1/2} = n^{1/2}l \tag{3.10}$$

This rather simple result is very important and it shows to what extent a polymer chain will be coiled in a dilute solution. For example a freely-jointed chain of 10 000 segments of 2 Å long will have a fully extended contour length of 20 000 Å whereas the value of $\langle r^2 \rangle^{1/2}$ for such a chain will only be 200 Å. In practice there are several factors which cause $\langle r^2 \rangle^{1/2}$ to be somewhat greater than the value given for a freely-jointed chain and these are discussed below.

3.1.3 *Extension to real chains*

In the preceding section it was shown that the mean square displacement length of a freely-jointed chain $\langle r^2 \rangle_{fj}$ is given by

$$\langle r^2 \rangle_{fj} = nl^2 \tag{3.11}$$

This equation must, of course, be modified for polymer chains as the bond angles (Section 3.1.1) are fixed in real macromolecules. If the polymer molecule has a backbone with only one type of atom and bond and the bond angle is Θ then Equation (3.11) can be modified to take into account the fixed bond angles and the mean square displacement length becomes

$$\langle \overline{r^2} \rangle_{\text{fa}} = nl^2(1 - \cos \Theta/(1 + \cos \Theta) \tag{3.12}$$

Normally $\Theta > 90°$ and so $\cos \Theta$ is negative and, therefore, the effect of a chain having fixed bond angles is to expand the coil. For a simple molecule such as the polyethylene chain $\Theta \sim 109.5°$ and so $\cos \Theta$ is $-1/3$ and Equation (3.12) becomes

$$\langle \overline{r^2} \rangle_{\text{fa}} = 2nl^2 \tag{3.13}$$

This means that the polyethylene chain will be expanded significantly compared with a freely-jointed chain. The freely-jointed chain is assumed to be made up of a series of vectors with continually variable bond angles and no preferred conformations. For a real polymer chain this is not the case and allowance for this can be made by taking into account the effect of restricted rotation. This is due to the preference of the segments of the molecule to adopt low energy *trans* and *gauche* conformations. Equation (3.12) can be further modified to allow for this effect and becomes

$$\langle \overline{r^2} \rangle_0 = nl^2\frac{(1 - \cos \Theta)}{(1 + \cos \Theta)} \cdot \frac{(1 + \overline{\cos \phi})}{(1 - \overline{\cos \phi})} \tag{3.14}$$

and $\langle \overline{r^2} \rangle_0$ is now known as the mean square of the *unperturbed dimension* of the polymer chain. The term $\overline{\cos \phi}$ is the average value of $\cos \phi$ where ϕ is the internal angle of rotation of the bond (Fig. 3.3). It turns out that if the *trans* and *gauche* conformations have the same energy then $\overline{\cos \phi}$ is equal to zero and Equations (3.14) and (3.12) become identical. However, if the behaviour is as shown in Fig. 3.3 and values of $|\phi| < 90°$ are favoured then $\overline{\cos \phi}$ will be positive and the coil further expanded. If side groups are present on the chain they will have a profound effect upon the ease of bond rotation and hence the dimensions of the coiled chain. Such steric interactions are very difficult, if not impossible, to calculate and so Equation (3.14) is normally written in the form

$$\langle \overline{r^2} \rangle_0 = \sigma^2 nl^2(1 - \cos \Theta)/(1 + \cos \Theta) \tag{3.15}$$

where σ^2 is a steric parameter which has the value of $(1 + \overline{\cos \phi})/(1 - \overline{\cos \phi})$ for simple chains. It is rather difficult to calculate σ, but it can be obtained from measurements on dilute polymer solutions which will be discussed later. The unperturbed dimension $\langle \overline{r^2} \rangle_0$ will be characteristic for a given polymer chain. It depends only on the geometry and chemical constitution of the polymer molecule and not upon other factors such as interactions with solvent molecules. An idea of the stiffness of a polymer chain can be gained from either the value of σ or the *characteristic ratio*, $\langle r^2 \rangle_0/nl^2$, which can both be obtained from measurements in a Θ solvent (see Section 3.1.7). Values of σ and the characteristic ratio for particular polymer molecules are given in Table 3.1. It is found that σ is typically between 1.5 and 2.5 and the combination of steric effects and fixed bond angles increase the RMS displacement lengths of polymer molecules by a factor of about 2 or 3 over those of freely-jointed chains.

The only effects that have been considered so far in determining the differences between the dimensions of a freely-jointed chain and a polymer chain are essentially short-range interactions between neighbouring groups or atoms on the chain. However, interactions between more widely spaced segments of the chain can also occur. The freely-jointed chain is assumed to be a series of connected vectors which can, of course, occupy the same area of space. In reality the segments of a polymer chain have a finite diameter and so will tend to interfere with each other which will have the effect of expanding the coil. Pentane interference (Section 3.1.1) is a simple example of this although it can happen between more widely spaced segments as well. These long-range interactions tend to expand the polymer chain from its unperturbed dimension and the extent of this perturbation is usually defined in terms of an expansion factor α where the RMS displacement length of the expanded coil is given by

$$\langle r^2 \rangle^{1/2} = \alpha \langle r^2 \rangle_0^{1/2}$$

(3.16)

It will be shown later (Section 3.1.7) that in good solvents the coil is expanded and α is large whereas in the Θ condition α is equal to unity. It is worth pointing out at this stage that it has recently been shown from neutron scattering measurements that α is also close to 1 in an amorphous polymer in the solid state since in this condition the effect of long-range interactions on the chain dimensions is small.

TABLE 3.1 *Typical values of σ and characteristic ratio $\langle r^2 \rangle_0/nl^2$ at infinite chain length for some common polymers*

Polymer	σ	$\langle r^2 \rangle_0/nl^2$
Polyethylene	1.85	7
Polystyrene (atactic)	2.2	10
Poly(methyl methacrylate)		
(atactic)	1.9	8
(isotactic)	2.2	10
(syndiotactic)	1.9	7
Polypropylene		
(atactic)	1.7	5.5
(isotactic)	1.6	5
(syndiotactic)	1.8	6

3.1.4 *Thermodynamics of polymer solutions*

A solution can be defined as a mixture of two components when the mixing is on an atomic or molecular scale. For the majority of non-cross-linked polymers it is possible to dissolve the polymer in a suitable solvent to give a polymer solution. These systems are of considerable practical and theoretical interest since it is possible to gain a great deal of information about the dimensions of the polymer molecules when they are in the solution. Such information cannot normally be obtained when the molecules are in a solid polymer. The properties of the solution depend both upon the nature and size of the polymer chains and also upon the interactions between the polymer and solvent molecules. Solvents for polymers are usually termed either 'good' or 'poor'. In good solvents there is an affinity between the two types of molecule and the polymer coil is consequently expanded. For poor solvents there is less affinity between the polymer and solvent molecules and the polymer coil is more tightly coiled.

The thermodynamic properties of polymer solutions can be used to determine the molar mass of the polymer molecules. Measurements are normally done at constant temperature, T, and pressure, p, and so the fundamental thermodynamic property that is used to monitor the behaviour of the system is the Gibbs free energy, G, which is defined as

$$G = H - TS \qquad (3.17)$$

where H is the enthalpy and S the entropy. In thermodynamic terms a solvent of Gibbs free energy G_1 will dissolve a polymer of Gibbs free energy G_2 as long as the Gibbs free energy of the solution G_{12} is less than that of the sum of the Gibbs free energies of the components in isolation.

i.e. $\quad G_{12} < G_1 + G_2 \quad$ (for a solution)

We normally define a parameter ΔG_m called the Gibbs free energy of mixing as

$$\Delta G_m = G_{12} - (G_1 + G_2) \qquad (3.18)$$

and the condition for two components to form a solution is $\Delta G_m < 0$, i.e. the free energy of mixing is negative. The parameter ΔG_m is often thought of as being made up of two other parameters, ΔH_m the enthalpy of mixing and ΔS_m the entropy of mixing and ΔG_m is given by Equation (3.17) as

$$\Delta G_m = \Delta H_m - T\Delta S_m \qquad (3.19)$$

For any solution ΔS_m will be positive since mixing of different types of species on an atomic or molecular scale will always increase the number of distinguishable arrangements of the system and hence increase the entropy. If there are no interactions between the solvent and solute atoms or molecules then ΔH_m will be zero (ideal solution). It is possible to determine ΔG_m in this case by calculating only ΔS_m using a simple lattice model shown schematically in Fig. 3.8(a). It is assumed that there are N_1 atoms or molecules of solvent and N_2 atoms or molecules of solute and that they fit randomly into a three-dimensional lattice with $(N_1 + N_2)$ sites. It is also assumed that the solvent and solute particles are of the same size and have no preference as to where they are sited in the lattice. The number of distinguishable arrangements or configurations of the particles in the lattice, Ω, is number of combinations of $(N_1 + N_2)$ things taken N_1 at a time

i.e. $\quad \Omega = \dfrac{(N_1 + N_2)!}{N_1! N_2!}$ \hfill (3.20)

As there are no other interactions in this idealised system this must be the only source of increased entropy and so the entropy of mixing can be found using the Boltzmann relation from statistical thermodynamics,

$$\Delta S_m = k \ln \Omega \tag{3.21}$$

where k is Boltzmann's constant.

Combining Equations (3.20) and (3.21) gives

$$\Delta S_m = k \ln \left[\frac{(N_1 + N_2)!}{N_1! N_2!} \right]$$

and introducing Stirling's approximation $\ln N! = N \ln N - N$ (for large N) leads to

$$\Delta S_m = -k\{N_1 \ln[N_1/(N_1 + N_2)] + N_2 \ln[N_2/(N_1 + N_2)]\} \tag{3.22}$$

This equation is written in terms of the number of molecules. We normally deal with the number of moles, n, and mole fractions, X, which are defined as $n_1 = N_1/N_A$ and $X_1 = n_1/(n_1 + n_2)$ etc., where N_A is the Avagadro number and so Equation (3.22) becomes

$$\Delta S_m = -R[n_1 \ln X_1 + n_2 \ln X_2] \tag{3.23}$$

since the universal gas constant is given by $R = N_A k$. A solution which obeys Equation (3.23) and for which ΔH_m is zero is said to be *ideal*. Using Equation (3.19) it follows that

$$\Delta G_m = RT[n_1 \ln X_1 + n_2 \ln X_2] \tag{3.24}$$

It is found that polymer solutions only approach ideality at infinite dilution although under certain circumstances (Θ condition) they can exhibit pseudo-ideal behaviour.

Polymer solutions normally behave in a non-ideal manner because both ΔH_m and ΔS_m are usually different from their ideal values. There are normally interactions between the polymer and solvent molecules which

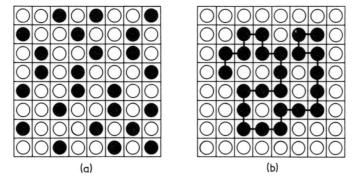

Fig. 3.8 Schematic representation of a liquid lattice. (a) Atomic or monomeric liquid. (b) Polymer molecule dissolved in a low-molar-mass solvent. (After Flory).

means that $\Delta H_m \neq 0$. The simple lattice model for the calculation of ΔS_m outlined earlier assumes that the polymer and solvent molecules are the same size which is clearly not the case. Also the model assumes that any site in the lattice has an equal probability of occupation by either species which again is not possible for polymer solutions and so ΔS_m also has a non-ideal value. Because of these factors one could hardly expect a polymer solution to behave in an ideal way and a great amount of effort has been put into theories to predict the thermodynamic properties of polymer solutions and these are discussed in the next sections.

3.1.5 *Flory–Huggins theory*

This theory which was proposed independently by Flory and Huggins can be used to predict ΔG_m for a polymer solution from the parameters ΔH_m and ΔS_m which are determined separately.

For the calculation of ΔS_m a modified lattice model is used as illustrated schematically in Fig. 3.8(b). It takes into account an obvious problem with the ideal lattice that polymer molecules are very much larger than solvent molecules. In the theory it is assumed that each space in the lattice is occupied by either one solvent molecule or a segment of a polymer molecule. Moreover, it takes into account the fact that each site occupied by a polymer segment must have two adjacent polymer sites so that there is a continuous path of polymer segments. If there are N_1 solvent molecules and N_2 polymer molecules made up of x segments then the total number of lattice sites is $(N_1 + xN_2)$. The calculation of the configurational entropy of mixing in this case is more complex than in the ideal case, but it has been shown by Flory that ΔS_m can be given by a remarkably simple expression which is similar to Equation (3.22),

$$\Delta S_m = -k\{N_1\ln[N_1/(N_1 + xN_2)] + N_2\ln[xN_2/(N_1 + xN_2)]\} \qquad (3.25)$$

This equation can be further simplified by multiplication by the Avagadro number and substituting for the volume fractions of solvent and polymer, ϕ_1 and ϕ_2 where

$$\phi_1 = N_1/(N_1 + xN_2) \quad \text{and} \quad \phi_2 = xN_2/(N_1 + xN_2)$$

The configurational entropy of mixing is then given by

$$\Delta S_m = -R[n_1\ln \phi_1 + n_2\ln \phi_2] \tag{3.26}$$

It must be stressed that this is only the configurational entropy of mixing and it is also possible that there may be entropy contributions from specific interactions between polymer and solvent molecules. These will be discussed later.

The Flory–Huggins theory also allows the enthalpy of mixing to be calculated in a simple way. It assumes that $\Delta H_m \neq 0$ because the polymer–solvent interaction energies in the solution are different from the polymer–polymer and solvent–solvent interaction energies in the pure unmixed states. As the forces between uncharged molecules decrease rapidly with the separation distance it is possible to restrict the calculation to nearest neighbour molecules. In the solution there will be in general three types of contacts, each with its own interaction energy.

Type of contact	Interaction energy
solvent–solvent	w_{11}
polymer–polymer	w_{22}
polymer–solvent	w_{12}

For every two polymer–solvent contacts made in solution one solvent–solvent and one polymer–polymer contact of the pure components are broken. The energy difference per polymer–solvent contact Δw between the mixed and unmixed states will therefore be

$$\Delta w = w_{12} - \tfrac{1}{2}(w_{11} + w_{22}) \tag{3.27}$$

Using the liquid lattice model in Fig. 3.8(b) it can be readily shown that the number of such contacts p is given by

$$p = xN_2z\phi_1 \tag{3.28}$$

where z is the co-ordination number of the lattice. The enthalpy of mixing ΔH_m is simply the energy per contact multiplied by the number of contacts and so

$$\Delta H_m = xN_2z\phi_1\Delta w \tag{3.29}$$

Since $xN_2\phi_1 = N_1\phi_2$ then x can be eliminated and Equation (3.29) written as

$$\Delta H_m = N_1\phi_2z\Delta w \tag{3.30}$$

This expression can be put into a different form by defining a new parameter χ_1, the *polymer–solvent interaction parameter* which is a dimensionless quantity given by

$$\chi_1 = z\Delta w / kT \tag{3.31}$$

and represents the interaction energy per solvent molecule divided by kT. Combining Equations (3.30) and (3.31) gives a final expression for the enthalpy of mixing

$$\Delta H_m = N_1 \phi_2 \chi_1 kT = n_1 \phi_2 \chi_1 RT \tag{3.32}$$

Having determined both ΔH_m (Equation 3.32) and ΔS_m (Equation 3.26) it is possible to calculate the free energy of mixing by using Equation (3.19) and so ΔG_m is given by

$$\Delta G_m = RT(n_1 \ln \phi_1 + n_2 \ln \phi_2 + \chi_1 n_1 \phi_2) \tag{3.33}$$

This equation becomes equivalent to that for an ideal solution (Equation 3.24) when $x = 1$ and $\chi_1 = 0$.

Although the Flory–Huggins approach is an important improvement upon ideal-solution theory it does not fit the behaviour of dilute polymer solutions very well. It is worth considering at this stage the reason why this is so. The theory assumes that the occupation of the lattice sites is purely statistical and ignores the tendency of polymer molecules to form isolated coils especially in dilute solutions. Also the location of the polymer segments or solvent molecules on particular sites will be influenced by specific interactions and will only be purely statistical when $\Delta w = 0$. There have been numerous attempts at improving the theory particularly by introducing more adjustable parameters. However, the physical justification for these changes is sometimes questionable and the complications introduced do not lead to significant improvements in the fit of experimental data.

A simple modification can be made by examining the significance of χ_1 (or Δw). In the calculation of ΔS_m specific interactions between polymer and solvent molecules were ignored and Equation (3.26) describes only the configurational entropy change. The specific interactions have only been considered as contributing to ΔH_m but there is also a possibility that they will contribute to the true entropy of mixing. Because of this, it is better to consider the term $n_1 \phi_2 \chi_1 RT$ (Equation 3.32) as a *standard state free energy change* rather than ΔH_m. This would be the change in free energy of the system involved in forming one particular configuration (microstate) of the solution and neglecting any other configurations. The entropy change due to these other configurations is given by Equation (3.26). The consequence of this is that Δw (and hence χ_1) is really made up of an enthalpy parameter

Δw_H and an entropy parameter Δw_S such that

$$\Delta w = \Delta w_H - T\Delta w_S \qquad (3.34\text{a})$$

$$\text{or} \quad \chi_1 = \chi_H + \chi_S \qquad (3.34\text{b})$$

It is possible to calculate ΔH_m and ΔS_m from Equation (3.33) using the standard thermodynamic relationships:

$$H = -T^2\left(\frac{\partial G/T}{\partial T}\right)_p \quad \text{and} \quad S = -\left(\frac{\partial G}{\partial T}\right)_p$$

It follows from Equation (3.33) that the most general expressions for ΔH_m and ΔS_m are

$$\Delta H_m = -RT^2\left(\frac{\partial \chi_1}{\partial T}\right)n_1\phi_2 \qquad (3.35)$$

and

$$\Delta S_m = -R\left[n_1\ln\phi_1 + n_2\ln\phi_2 + \frac{\partial(\chi_1 T)}{\partial T}n_1\phi_2\right] \qquad (3.36)$$

Since χ_1 is defined as $\chi_1 = z\Delta w/kT$ then if Δw is independent of temperature and hence contains no entropy contribution

$$\left(\frac{\partial \chi_1}{\partial T}\right) = \frac{-\chi_1}{T} \qquad (3.37)$$

and so Equations (3.32) and (3.35) become identical. In the same way $(\partial(\chi_1 T)/\partial T)$ is equal to zero when Δw is independent of T and Equations (3.26) and (3.36) also become identical.

Unfortunately these refinements do not lead to significant improvements in the agreement between the theory and experimental observations, but even so the Flory–Huggins theory can be applied with some success to studies of phase equilibria in polymer solutions and it goes some way towards explaining other phenomena.

3.1.6 *Partial molar quantities and chemical potential*

In thermodynamic systems it is often necessary to distinguish between *extensive* and *intensive* properties. The value of an extensive property increases with the number of moles of material in the system whereas the value of an intensive property is generally independent of the amount of material in the system. Examples of extensive properties are G, S and volume, V, whereas pressure, p, and temperature, T, are intensive properties. Generally speaking, ~~extensive properties can be used to describe the state of the whole system whereas it is possible to talk of the value of an intensive property at a point in the system.~~

In multicomponent systems such as polymer solutions we are often concerned with the way in which one thermodynamic parameter, which we will call Z, changes when a small amount of component, i, is added normally under conditions of constant temperature and pressure and keeping the amounts of all other components constant. The rate at which Z changes with the number of moles i, n_i is designated $\bar{Z}_i$ and defined by

$$\bar{Z}_i = \left(\frac{\partial Z}{\partial n_i}\right)_{p,T,n_j} \tag{3.38}$$

Such quantities $\bar{Z}_i$ are called ~~partial molar quantities~~ and with polymer solutions we are normally concerned with ~~partial molar Gibbs free energies~~ $\bar{G}_i$ and partial molar volumes $\bar{V}_i$. In single component systems, such as pure solvent, the partial molar quantities are identical to the corresponding ~~molar quantity which we will denote by a lower case symbol, z_i.~~

One partial molar quantity that is of particular use in discussion of the equilibrium of multicomponent systems at constant temperature and pressure is the partial molar Gibbs free energy $\bar{G}_i$ of the component i. This is also called the *chemical potential,* μ_i of the species i and so it follows that

$$\mu_i = \left(\frac{\partial G}{\partial n_i}\right)_{T,p,n_j} \tag{3.39}$$

It is also possible to relate differences in chemical potential to osmotic pressure which can be used to measure number-average molar masses of polymers.

3.1.7 *Dilute polymer solutions*

As mentioned earlier, the random distribution of chain segments over the liquid lattice, which is assumed by the Flory–Huggins theory, leads to serious discrepancies between experimental observations and predictions of the theory. For dilute polymer solutions it can be readily shown that there will not be a random distribution of polymer segments, but that it is more likely that the solution will be made up of isolated coiled chains occupying a volume from which all other chains are excluded. Flory and Krigbaum assumed that the solutions were discontinuous and made up of isolated chains in which there was a Gaussian distribution of polymer segments about the centre of mass of each coil and separated from each other by solvent molecules. They assumed that within the polymer coil the Flory–Huggins theory is still valid and went on to calculate the thermodynamic properties of the solution in terms of the *excess partial molar Gibbs free energy of mixing* which is given by

$$\Delta \bar{G}_1^E = RT(\kappa_1 - \psi_1)\phi_2^2 \tag{3.40}$$

where κ_1 and ψ_1 are enthalpy and entropy parameters respectively. It follows that since $\Delta \bar{G}_1^E = \Delta \bar{H}_1^E - T\Delta \bar{S}_1^E$ then

$$\Delta \bar{H}_1^E = RT\kappa_1 \phi_2^2 \tag{3.41}$$
$$\Delta \bar{S}_1^E = R\psi_1 \phi_2^2 \tag{3.42}$$

which are the excess partial molar enthalpies and entropies of mixing. The subscripts 1 and 2 refer to solvent and polymer respectively. The partial molar Gibbs free energy of mixing is given by the Flory–Huggins theory and can be obtained by differentiating Equation (3.33) with respect to n_1 and so

$$\Delta \bar{G}_1 = \left(\frac{\partial(\Delta G_m)}{\partial n_1} \right) = RT\left[\ln(1 - \phi_2) + \left(1 - \frac{1}{x} \right)\phi_2 + \chi_1\phi_2^2 \right] \tag{3.43}$$

For dilute solutions the volume fraction of polymer ϕ_2 is small and so

$$\ln(1 - \phi_2) \approx -\phi_2 - \frac{\phi_2^2}{2} - \frac{\phi_2^3}{3} - \dots$$

which leads to

$$\Delta \bar{G}_1 = -RT\phi_2/x + RT(\chi_1 - \tfrac{1}{2})\phi_2^2 - RT\phi_2^3/3 - \dots \tag{3.44}$$

The first term in Equation (3.44) can be considered the 'ideal' term as this is equivalent to the value of $\Delta \bar{G}_1$ for an ideal solution. The other terms, therefore, represent deviations from ideality and contribute to the excess partial molar free energy of mixing. For dilute solutions terms in ϕ^3 or higher powers will be negligible and so it follows by comparing Equation (3.40) with non-ideal terms in Equation (3.44) that

$$(\kappa_1 - \psi_1) = (\chi_1 - \tfrac{1}{2}) \tag{3.45}$$

It has been mentioned earlier that polymer solutions can sometimes exhibit pseudo-ideal behaviour and this will obviously be when the excess partial molar free energy of mixing is zero. For this to happen $\Delta \bar{H}_1^E$ must be equal to $T\Delta \bar{S}_1^E$ and therefore κ_1 equal to ψ_1. The temperature at which this is so is called the *Flory* or *Theta* temperature, Θ, which can be defined as

$$\Theta = \kappa_1 T/\psi_1 \tag{3.46}$$

Substitution in Equation (3.45) gives

$$(\kappa_1 - \psi_1) = \psi_1(\Theta/T - 1) = (\chi_1 - 1/2) \tag{3.47}$$

and it can be seen that at the Theta temperature $\chi_1 = 1/2$. The excess partial molar free energy can be written as

$$\Delta \bar{G}_1^E = RT\,\psi_1(\Theta/T - 1)\phi_2^2 \tag{3.48}$$

and so deviations from ideality disappear at $T = \Theta$. The Theta temperature will only have meaningful values when both κ_1 and ψ_1 have the same sign. In normal non-polar polymer solutions they are both usually positive. Some values of theta temperatures are given for a series of different polymer–solvent systems in Table 3.2.

It is now possible, at least qualitatively, to discuss the relationship between the theta temperature, Θ, and the expansion factor, α, for the polymer coil (Equation 3.16). As explained earlier, there is a tendency for the polymer coil to expand in a polymer solution to a size greater than that given by the unperturbed dimensions, $\langle r^2 \rangle_0$ which is controlled only by short-range interactions. The expansion is caused by long-range interactions due to the tendency for the segments of the polymer chain to interact with each other by trying to occupy the same part of space. These long-range interactions are often called *excluded-volume* effects and lead to the deviations from ideality found for polymer solutions. In good solvents there is also a tendency for there to be strong polymer–solvent interactions which also tend to expand the coil. In poor solvents, polymer–polymer contacts are favoured and these tend to cause the coil to contract. As the temperature of the solution is decreased the polymer–polymer contacts increase until at $T = \Theta$ the contraction due to polymer–solvent interactions exactly balances expansion due to excluded-volume effects. At this point α is equal to unity, the molecular coil has its unperturbed dimensions and the solution behaves in an ideal manner. If the temperature is reduced much below the Theta temperature precipitation takes place. In fact, theories predict that, if the molar mass of the polymer is infinite, precipitation will occur at the Theta temperature. It is possible to calculate a theoretical relationship between α and Θ and details of this are given in more advanced texts.

TABLE 3.2 *Values of Theta temperature for different polymer–solvent systems*

Polymer	Solvent	Θ/K
Polyethylene	Biphenyl	398
Polystyrene	Cyclohexane	308
Polystyrene	Decalin	300
Polypropylene	Diphenyl ether	406
Poly(methyl methacrylate)	Butanone/isopropanol (50/50 by vol.)	298
Poly(methyl methacrylate)	4-Heptanone	307

3.1.8 *Polymer solubility*

It is very important in practice to know what solvents will dissolve a particular polymer. It is often found that certain polymers will become swollen when they are in contact with particular liquids. Also many polymers craze (see Section 5.6.7) when they are subjected to certain liquids. It is possible to go some way towards explaining these phenomena by using a semi-empirical approach which was first suggested by Hildebrand. The main principle that is used is that polymers dissolve in chemically similar solvents and the two parameters that are used are the *cohesive energy density C* and *solubility parameter* δ and these have characteristic values for each polymer and solvent. The cohesive energy density is defined for a liquid by

$$C = (\Delta h_v - RT)/v \qquad (3.49)$$

where Δh_v is the molar latent heat of vapourization and v the molar volume. The solubility parameter δ is given by $\delta = C^{1/2}$. The molar enthalpy of mixing for a solution Δh_m can be calculated from the solubility parameters of the solvent and solute, δ_1 and δ_2 respectively and is given by

$$\Delta h_m = (\delta_1 - \delta_2)^2 v \phi_1 \phi_2 \qquad (3.50)$$

where v in this case is the molar volume of the solution. It is possible to calculate C and δ for simple liquids by measuring the enthalpy of vapourization and molar volume. For polymers it is not possible to measure Δh_v directly and indirect methods are used such as by finding a solvent which causes the maximum swelling of a polymer network. The polymer is then given a similar value of δ to the solvent.

It is possible to gain an idea of the critical conditions for solubility by comparing Equations (3.50) and (3.32) for one mole of solvent. It follows that

$$\chi_1 = (\delta_1 - \delta_2)^2 v \phi_1 / RT \qquad (3.51)$$

For a dilute solution $\phi_1 \approx 1$ and so since $\chi_1 < 1/2$ for solubility (Equation 3.47) then the critical condition for solubility is

$$(\delta_1 - \delta_2)^2 < 0.5 \, RT/v \qquad (3.52)$$

Values of δ_1 and δ_2 are tabulated in Table 3.3 and the inequality in Equation (3.52) shows that the solubility parameters for the polymer and solvent must be close for a solution to be formed. The theory is quite good for some amorphous polymers in non-polar solvents, but there are severe limitations of this approach. For example, it is found that some crystalline polymers are not soluble in solvents with similar values of δ at room temperature. The presence of hydrogen bonding in either the polymer or

TABLE 3.3 *Solubility parameters for various common solvents and polymers. (Taken from the 'Polymer Handbook' edited by Brandrup and Immergut, reproduced with permission)*

Solvent	$\delta/10^3 J^{1/2} m^{-3/2}$
Acetone	20.3
Carbon tetrachloride	17.6
Chloroform	19.0
Cyclohexane	16.8
Methanol	29.7
Toluene	18.2
Water	47.9
Xylene	18.0

Polymer	$\delta/10^3 J^{1/2} m^{-3/2}$
Polyethylene	16.4
Polystyrene	18.5
Poly(methyl methacrylate)	19.0
Polypropylene	19.0
Poly(vinyl chloride)	20.0

solvent also lead to further discrepancies. However, comparison of solubility parameters can be useful as a rough guide in choosing solvents for a particular polymer.

3.2 Number average molar mass

The knowledge of the distribution of molar mass of a polydisperse polymer is of extreme importance. Many of the physical properties of polymers depend upon this and a correct distribution is essential in any polymer which is used in a practical situation. One of the most important points in the distribution is the *number average molar mass* $\bar{M}_n$, and there are several techniques which can be used to measure it. All of these methods involve determining the number of molecules in a given mass of polymer. The polymer is usually dissolved in a solution of known concentration and hence the mass of polymer per unit volume is defined. The most important methods involve the use of the colligative properties of dilute solutions although for a limited number of polymers it is possible to measure $\bar{M}_n$ by end-group analysis.

The *colligative properties* of a solution are those which depend only upon the number of solute molecules in unit volume of the solution and not upon the nature of the molecules. They are insensitive to variables such as branching or copolymer composition and provide a useful method of measuring $\bar{M}_n$. Colligative properties include osmotic pressure, vapour pressure lowering, boiling-point elevation and freezing-point depression and the choice of the individual technique depends upon the type of polymer being studied and the expected value of $\bar{M}_n$.

3.2.1 *Chemical potential and osmotic pressure*

Osmosis is a phenomenon of great importance in many applications and particularly so in polymer science. It is a very accurate and sensitive method of determining $\bar{M}_n$ and provides a useful application of the theories of solution thermodynamics developed in Section 3.1. The apparatus used for measuring osmotic pressure is shown schematically in Fig. 3.9. It consists essentially of a chamber with two compartments separated by a *semi-permeable* membrane. In one compartment there is pure solvent and in the other there is a polymer solution and both are at the same temperature. The membrane is permeable only to solvent molecules and is *not* permeable to polymer molecules. If both of the compartments are at the same pressure initially it is found that solvent molecules tend to diffuse from the pure solvent through the membrane and the diffusion stops only when the pressure in the solution compartment is increased by either applying an external pressure or allowing a pressure head to develop. The pressure which is required to stop solvent diffusion across the membrane is called the *osmotic pressure.*

The reason for the diffusion taking place can be found by considering the chemical potential of the solvent in the two compartments in Fig. 3.9. If compartment A contains pure solvent and is initially at standard atmospheric pressure (101 kNm^{-2}) p_o, then the chemical potential of the solvent in A will be μ_1^A and given by

$$\mu_1^A = \mu_1^0 \tag{3.53}$$

where μ_1^0 is the standard chemical potential of the solvent. If compartment B contains polymer solution which is initially at standard atmospheric pressure then the chemical potential of the solvent in chamber B, μ_1^B will be less than the standard chemical potential of the solvent μ_1^0 (i.e. $\mu_1^B < \mu_1^0$). Because of this difference in chemical potential solvent diffuses through the membrane from a region of high chemical potential (A) to a region of low chemical potential (B). As mentioned earlier an increase in pressure in compartment B will stop the flow of solvent through the membrane and this is because the increase in pressure has the effect of increasing the chemical potential of solvent in the solution. When the flow of solvent ceases an equilibrium situation will be established and the chemical potential of the solvent in the two compartments will be equal and so

$$\mu_1^A = \mu_1^B + \int_{p_0}^{p} \left(\frac{\partial \mu_1}{\partial p} \right) \mathrm{d}p \tag{3.54}$$

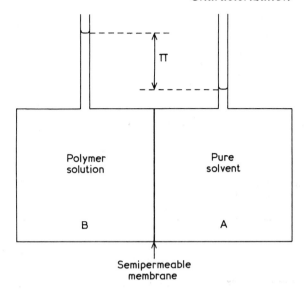

Fig. 3.9 *Schematic representation of an osmotic pressure cell consisting of a polymer solution in equilibrium with pure solvent across a semi-permeable membrane. The osmotic pressure* Π *can be determined from the difference of the height of the liquids in the capillaries.*

where μ_1^A and μ_1^B are the chemical potentials of the solvent in compartments A and B at atmospheric pressure and $(\partial \mu_1 / \partial p)$ is the rate of increase in the chemical potential of the solvent with increasing pressure. By definition (Section 3.1.6) μ_1 is equal to the partial molar Gibbs free energy of the solvent, $\overline{G}_1$ then since $(\partial G / \partial p) = V$ it follows that

$$\left(\frac{\partial \mu_1}{\partial p}\right) = \left(\frac{\partial \overline{G}_1}{\partial p}\right) = \overline{V}_1 \tag{3.55}$$

Combining Equations (3.54) and (3.55) and integrating gives

$$\mu_1^A - \mu_1^B = (p - p_0)\overline{V}_1 \tag{3.56}$$

The term $(p - p_0)$ is known as the osmotic pressure Π and using Equation (3.53) it follows that

$$\mu_1^B - \mu_1^0 = -\Pi \overline{V}_1 \tag{3.57}$$

In this case the term $(\mu_1^B - \mu_1^0)$ is the difference in chemical potential of the solvent between the pure solvent and polymer solution which is equal to the difference in the partial molar Gibbs free energy of the solvent in the two liquids, $\Delta \overline{G}_1$ and so it can be written with complete generality that

$$\Delta \overline{G}_1 = -\Pi \overline{V}_1 \tag{3.58}$$

This equation is applicable to any solute/solvent system and is independent of whether or not the solution is ideal.

The osmotic pressure of an ideal solution can be obtained by using Equation (3.58) and by differentiating Equation (3.24). For an ideal solution

$$\Delta \bar{G}_1 = \left(\frac{\partial \Delta G_m}{\partial n_1} \right)_{T,n_2} = RT \ln X_1 \qquad (3.59)$$

and so the osmotic pressure $\prod$ is given by

$$\prod = -\frac{RT \ln X_1}{\bar{V}_1} \qquad (3.60)$$

Since the solution is dilute it follows that $\ln X_1 \approx -X_2$ and the partial molar volume of the solvent in the solution $\bar{V}_1$ is approximately equal to its molar volume v_1 and so

$$\prod = \frac{RTX_2}{v_1} \qquad (3.61)$$

With solutions it is more convenient to use the concentration of solute, c rather than the mole fraction X_2 and since it can be readily shown that for a dilute solution $c = X_2 M/v_1$ where M is the molar mass of the solute, Equation (3.61) becomes

$$\prod = RTc/M \qquad (3.62)$$

This equation immediately shows the usefulness of osmotic pressure measurements as the osmotic pressure can be directly related to the molar mass of the solute, M, and this can be readily calculated by measuring the osmotic pressure of a solution as a function of the concentration.

It is possible, in a similar way, to derive an expression relating the osmotic pressure of a polymer solution to the molar mass of the polymer. In Section 3.1.7 it was shown that the partial molar free energy of mixing is given by the Flory–Huggins theory as

$$\Delta \bar{G}_1 = -RT\phi_2/x + RT(\chi_1 - \tfrac{1}{2})\phi_2^2 - RT\phi_2^3/3 - \ldots \qquad (3.44)$$

Combining Equations (3.44) and (3.58) gives an expression for the osmotic pressure of a polymer solution

$$\prod = RT\phi_2/\bar{V}_1 x + RT(\tfrac{1}{2} - \chi_1)\phi_2^2/\bar{V}_1 + RT\phi_2^3/3\bar{V}_1 + \ldots \qquad (3.63)$$

If the solution is dilute then $\bar{V}_1 \approx v_1$. Also $\phi_2 = xn_2/(n_1 + xn_2) \approx xn_2/n_1$ for a dilute solution and so the concentration of polymer is given by

$$c = n_2 M/n_1 v_1 \approx \phi_2 M/x v_1 \qquad (3.64)$$

where M is the molar mass of the polymer. Therefore, the osmotic pressure is related to c and M through the equation

$$\prod = RTc/M + RT(\tfrac{1}{2} - \chi_1)c^2x^2v_1/M^2 + RTc^3x^3v_1^2/3M^3 + \ldots \quad (3.65)$$

The first term in this equation is identical to that for an ideal solution (Equation 3.62) and the subsequent terms represent deviations from ideality.

Equation (3.65) gives the osmotic pressure for a solution of a homodisperse polymer of molar mass, M. Normally synthetic polymers have a distribution of molar mass and the equation can be readily modified to account for the distribution. The concentration of polymer c is simply the mass of polymer per unit volume of solution and so

$$c = NM/N_A V \qquad (3.66)$$

where N is the total number of polymer molecules in the solution of volume V and N_A is the Avagadro number. The ratio (c/M) then becomes equal to $(N/N_A V)$ and so Equation (3.65) becomes

$$\prod = RTN/N_A V + RT(\tfrac{1}{2} - \chi_1)N^2x^2v_1/N_A^2 V^2$$
$$+ RTN^3x^3v_1^2/3N_A^3V^3 + \ldots \qquad (3.67)$$

This equation clearly shows that the osmotic pressure depends upon the number of polymer molecules per unit volume (N/V). This behaviour is typical of a colligative property and will be discussed in detail later. If a polydisperse polymer is used in which there are N_i molecules of molar mass M_i then the concentration is given by

$$c = \frac{\sum (N_i/N_A)M_i}{V} = \frac{\sum (N_i/V)M_i}{N_A} \qquad (3.68)$$

and by definition

$$\overline{M}_n = \frac{\sum N_i M_i}{\sum N_i} = \frac{\sum (N_i/V)M_i}{N/V} \qquad (3.69)$$

Combining Equations (3.68) and (3.69) gives

$$N/N_A V = c/\overline{M}_n \qquad (3.70)$$

Substituting Equation (3.70) into Equation (3.67) and rearranging gives

$$\frac{\prod}{c} = \frac{RT}{\overline{M}_n} + RT(\tfrac{1}{2} - \chi_1)x^2v_1\frac{c}{\overline{M}_n^2} + RTx^3v_1^2\frac{c^2}{3\overline{M}_n^3} + \ldots \qquad (3.71)$$

This equation is of great importance since although the Flory–Huggins theory does not give a very accurate prediction of the thermodynamic properties of polymer solutions, Equation (3.71) which is derived from the theory gives a good representation of the form of the variation of the osmotic pressure of a polymer solution with c and $\bar{M}_n$. Equation (3.71) can be conveniently written in the form

$$\frac{\Pi}{c} = \frac{RT}{\bar{M}_n} + Bc + Cc^2 + \ldots \tag{3.72}$$

or alternatively

$$\frac{\Pi}{c} = RT\left[\frac{1}{\bar{M}_n} + A_2c + A_3c^2 + \ldots\right] \tag{3.73}$$

These types of expressions are known as virial equations and B (or A_2) and C (or A_3) are known as the second and third virial coefficients respectively. Virial equations are often used to describe the thermodynamic behaviour of real solutions and real gases and they always have the property that at infinite dilution $(c \to 0)$ the behaviour of the solutions or gases approaches ideality.

For dilute polymer solutions the terms in c^2 and higher become negligible and Equation (3.71) can be approximated to

$$\frac{\Pi}{c} = \frac{RT}{\bar{M}_n} + RT(\tfrac{1}{2} - \chi_1)x^2v_1\frac{c}{\bar{M}_n^2} \tag{3.74}$$

It was shown earlier (Section 3.1.7) that at the Theta temperature $\chi_1 = \tfrac{1}{2}$ and so it follows that at $T = \Theta$, $A_2 = 0$ and so

$$\Pi/c = RT/\bar{M}_n \tag{3.75}$$

which means that in the Theta condition Π/c is independent of concentration and the solution behaves in a pseudo-ideal manner.

3.2.2 *Osmotic pressure measurements*

It is possible in principle to measure the osmotic pressure of a polymer solution using simple apparatus similar to that shown schematically in Fig. 3.9. In this case a head of liquid is allowed to develop by the diffusion of solvent molecules through the membrane. In order to minimize the amount of material transported across the membrane, capillaries are used for the pressure head. If there is a large amount of diffusion the polymer solution will become diluted and the equilibrium will take a long time to be established. The rate at which this happens is controlled by the kinetics of diffusion through the membrane and it can take several hours for equilibrium to be established. It is possible, at least in theory, to speed up the rate at which equilibrium is established by using narrow capillaries and having a large membrane area, but these modifications can lead to serious problems. If the capillaries used have diameters that are very small there can be problems with the surface tension and viscosity of the solvent and solution affecting measurements of the osmotic pressure. Also thermal expansion due to minor fluctuations in temperature can lead to serious errors as the height of the liquids in a fine capillary will be very sensitive to minor changes in the total volume of solution. Since the membranes for osmotic pressure cells are usually made of soft flexible material there can be a tendency for 'ballooning' to occur under the action of the osmotic pressure if the area of the membrane is large. Although this has no effect upon the equilibrium pressure it can slow down the rate at which equilibrium is reached.

Most modern osmometers do not require a static equilibrium to be reached and they use the principle of stopping the flow of solvent through the membrane by applying an external pressure to the polymer solution. This method has the advantage of much greater speed and there are several commercial instruments now available that use this dynamic equilibrium principle.

When osmometry is being used to measure $\bar{M}_n$ for a particular polymer sample it is desirable to measure $\prod$ as a function of c over as wide a range of concentration as possible. It is often useful to choose a solution which is close to its Theta temperature as this can lead to simplifications in Equation (3.73) which becomes

$$\frac{\prod}{c} = RT\left[\frac{1}{\bar{M}_n} + A_2 c\right] \tag{3.76}$$

as the third virial coefficient, A_3 becomes insignificant. This means that a plot of $\prod/c$ versus c should give a straight line with an intercept at $c = 0$ of $RT/\bar{M}_n$ as shown in Fig. 3.10. If measurements are taken at $T = \Theta$ then $\prod/c$ will be independent of c and a line should be obtained which is parallel to the c axis. The value of the second virial coefficient A_2 can be obtained from the slope of the plot in Fig. 3.10. It is also possible to estimate the Theta temperature for a polymer–solvent system by measuring A_2 as a function of temperature and extrapolating to $A_2 = 0$.

Great care must be taken with solutions where the value of the third virial coefficient is significant. A plot of $\prod/c$ versus c in this case will not be

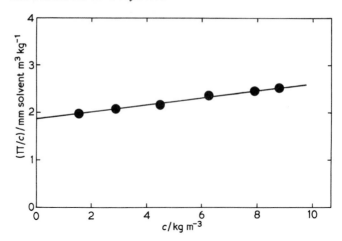

Fig. 3.10 *Plot of* $\prod/c$ *versus c for polystyrene dissolved in methyl ethyl ketone at 310 K (After Billingham).*

a straight line and will be curved as shown in Fig. 3.11. However, it is often possible even in this case to obtain a plot which approximates to a straight line especially if only a short range of concentrations is used. This can lead to the mistaken assumption that $A_3 \simeq 0$ and hence erroneous estimates of $\bar{M}_n$.

3.2.3 *Other colligative properties*

In theory any colligative property can be used to measure $\bar{M}_n$ for a polymer sample, but in practice it is found that with synthetic polymers the best results are obtained by using osmometry for reasons which will be explained later. The relationship between the particular colligative property and the concentration of the polymer solution or the number of molecules per unit volume can all be expressed in the form of virial equations. For example it was shown that the osmotic pressure could be given in the form of a virial equation as a direct consequence of the Flory–Huggins theory. Equation (3.67) can be written in the general form

$$\prod = \frac{RT}{N_A}\left(\frac{N}{V}\right) + K_2'\left(\frac{N}{V}\right)^2 + K_3'\left(\frac{N}{V}\right)^3 + \ldots \qquad (3.77)$$

where K_2' and K_3' are second and third virial coefficients and (N/V) is the number of polymer molecules per unit volume of solution. The presence of the solute in a polymer solution also has the effect of *lowering the vapour*

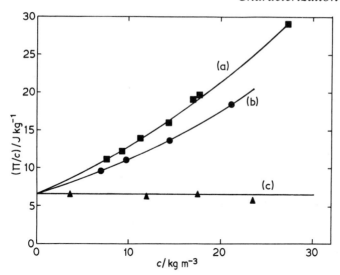

Fig. 3.11 *Plot of $\prod/c$ versus c for poly(methyl methacrylate) dissolved in three different solvents, (a) toluene, (b) acetone and (c) acrylonitrile. (Data taken from Fox, Kinsinger, Mason and Schuele, Polymer 3 (1962), 71.*

pressure of the solvent in the solution by an amount Δp_1 and this can be expressed in the form of a virial equation as

$$\Delta p_1 = \frac{p_1^0 v_1}{N_A}\left(\frac{N}{V}\right) + K_2''\left(\frac{N}{V}\right)^2 + K_3''\left(\frac{N}{V}\right)^3 + \ldots \tag{3.78}$$

where v_1 is the molar volume of the solvent and p_1^0 is the vapour pressure of pure solvent. For a polymer solution the boiling point of the solution is greater and the freezing point lower than that for pure solvent and these two effects form the basis of ~~ebullioscopy~~ and ~~cryoscopy~~ which are alternative methods of measuring $\bar{M}_n$. The increase in boiling point ΔT_v is given by

$$\Delta T_v = \frac{RT_v^2 v_1}{\Delta h_v N_A}\left(\frac{N}{V}\right) + K_2'''\left(\frac{N}{V}\right)^2 + K_3'''\left(\frac{N}{V}\right)^3 + \ldots \tag{3.79}$$

and the depression of the freezing point ΔT_m given by a similar expression

$$\Delta T_m = \frac{RT_m^2 v_1}{\Delta h_m N_A}\left(\frac{N}{V}\right) + K_2''''\left(\frac{N}{V}\right)^2 + K_3''''\left(\frac{N}{V}\right)^3 + \ldots \tag{3.80}$$

In these equations, T_v and T_m are the boiling and freezing points of the pure solvent and Δh_v and Δh_m are the molar enthalpy changes (latent heats) of vapourization and melting for the pure solvent. It can be seen that

Equations (3.77–3.80) all have the same form and can be written as a general virial equation

$$\Delta Q = K_1\left(\frac{N}{V}\right) + K_2\left(\frac{N}{V}\right)^2 + K_3\left(\frac{N}{V}\right)^3 + \ldots \tag{3.81}$$

where ΔQ is the difference in a particular quantity between the solution and pure solvent. The first virial coefficient K_1 corresponds to the ideal case since for an ideal solution (or a polymer solution in the Theta condition) when K_2, K_3 etc. are all zero. The number of polymer molecules per unit volume can be related to the concentration of the solution, c, and $\bar{M}_n$ through Equation (3.70).

$$N/N_A V = c/\bar{M}_n \tag{3.82}$$

The general procedure for relating any colligative property to $\bar{M}_n$ will be to measure ΔQ as a function of c and plot a graph of $\Delta Q/c$ versus c. The intercept at $c = 0$ will then enable $\bar{M}_n$ to be calculated since K_1 is normally known. As with osmotic pressure measurements (Section 3.2.2) it is usual with all colligative property measurements to use solutions which are close to the Theta condition so that the third virial coefficient becomes negligible and curvature in the plots is minimized.

A useful comparison of the various techniques can be made by calculating ΔQ for the different colligative properties measured under identical conditions. Table 3.4 gives approximate values of ΔQ for polymers of different $\bar{M}_n$ in benzene. The values given are for the same concentration assuming ideal behaviour. Deviations from ideality will, of course, lead to higher values of ΔQ. This table clearly shows the advantages of using osmotic pressure for determining $\bar{M}_n$. All of the methods have measurable values of ΔQ when $\bar{M}_n < 10\ 000$ g mol^{-1} but only the osmotic pressure is still measurable when $\bar{M}_n$ is 10^6 g mol^{-1} since the changes in the other colligative properties are much too small to be determined for this value of $\bar{M}_n$. Osmotic pressures can be measured to an accuracy of $\pm$ 0.1 mm of solvent. In fact problems are encountered with membrane osmometry if $\bar{M}_n$ is too small and 10^4 g mol^{-1} is usually the lower limit for commercially available membranes supplied for use with polymer solutions. This is because there can be significant permeation of low-molar-mass polymer through the membrane. It is clear therefore that osmotic-pressure measurements are by far the most useful methods of determining $\bar{M}_n$ for synthetic polymers since most polymers with useful physical properties have $\bar{M}_n > 10^4$ g mol^{-1}.

3.2.4 End-group analysis

This technique involves methods of counting the number of end-groups on a known mass of polymer molecules normally in solution. The number

TABLE 3.4 *Values of ΔQ for different colligative properties for a polymer dissolved in benzene at a concentration of* 20 *kg* m^{-3}. *(Ideal behaviour is assumed.)*

Property	ΔQ $\bar{M}_n = 10\ 000$	ΔQ $\bar{M}_n = 1\ 000\ 000$
Vapour-pressure lowering (25°C)	1.8×10^{-2}mm Hg	1.8×10^{-4}mm Hg
Freezing-point depression	1.2×10^{-2}K	1.2×10^{-4}K
Boiling-point elevation	6×10^{-3}K	6×10^{-5}K
Osmotic pressure (25°C)	600 mm solvent	6 mm solvent

average molar mass $\bar{M}_n$ can then be determined if the chemical structure of the molecules is known. This method is particularly useful for measuring $\bar{M}_n$ in the range of 10 000–40 000 g mol^{-1} which is somewhat inaccessible for colligative-property measurements. There is often significant permeation of low-molar-mass polymer through the membrane in this range during osmotic pressure measurements and the changes in the other properties are usually too small to be measured for solutions of reasonable concentration. It is found that many polymers produced during step-growth (or condensation) polymerization (Section 2.1) often have $\bar{M}_n$ in the range 10 000–50 000 g mol^{-1}. It is fortunate that they also have end-groups which are normally amenable to analysis and so this method is sometimes applied to such polymers.

The methods of analysis can be conveniently grouped into three main categories; *chemical, radiochemical* and *spectroscopic.* The simple chemical method usually involves titration of a solution of a strong acid or base into a solution of the polymer to determine the concentration of carboxyl or amino end-groups in the polymer. The technique is relatively simple and modern apparatus is often automated. There can be problems in using standard acid–base indicators as the behaviour in organic solvents can be different from that in water. However, such problems can be readily overcome by using electrochemical methods to determine the end-point of the titration. Radiochemical analysis involves incorporating radioactive groups at the ends of the polymer chains either during or after polymerization and measuring the radioactivity of the polymer produced. The sensitivity of this technique can be very high. Most standard spectroscopic techniques can be adapted for use in end-group analysis. They usually require prior calibration and will measure the concentration of groups or atoms situated at the ends of the molecules, if they are sufficiently different from those on the main chain.

End-groups analysis is an extremely useful method of measuring $\bar{M}_n$ for condensation polymers such as polyesters and polyamides. For these materials the titration technique is normally used. There can be problems in dissolving many of these polymers in suitable solvents especially if they contain aromatic groups in the main chain or are crystalline. Problems can also be encountered with polymer degradation and the presence of impurities, but, nevertheless, end-group analysis is a very powerful tool for determining $\bar{M}_n$ for condensation polymers. On the other hand, it is rarely used to measure $\bar{M}_n$ for addition polymers, because the molar masses of such polymers are usually relatively high and the chains often have more than two ends because of branching reactions. In fact if end-group analysis is combined with another technique such as membrane osmometry it can be used to monitor the number and type of chain ends and hence give information on the mechanism of polymerization and branching reactions. Normally end-group analysis is of very little use for measuring $\bar{M}_n$ for addition polymers, but 'living' polymers (Section 2.3.6) are a notable exception. In the absence of impurities there is no termination during the reaction in which they are formed and particular groups can be placed in a controlled way at the chain ends. Branching is also absent in such reactions and so such polymers are very amenable to end-group analysis.

3.3 Light Scattering

As mentioned earlier it is desirable that some idea of the distribution of molar mass of a polymer is known when the polymer is used in a practical situation. In the last section methods of determining $\bar{M}_n$ have been discussed and in this section we will be concerned with a technique for determining another point on the molar mass distribution curve, the *weight average molar mass*, $\bar{M}_w$. By far the most important and widely used method of determining $\bar{M}_w$ is by measuring the light scattering from a solution of the polymer. The theory of light scattering, originally developed to explain scattering from gases, has been successfully adapted for polymer solutions. Light scattering is potentially a powerful technique for polymer characterization as, in principle, it allows the size and shape of the polymer molecules to be determined in solution. Its sensitivity increases as the size of the molecules increases in contrast to colligative properties which generally become less sensitive as $\bar{M}_n$ increases.

3.3.1 *Light scattering by polymer solutions*

The phenomenon of light scattering is encountered widely in everyday life. For example, the dust particles in the atmosphere of a room will cause a beam of light coming through a window to be seen as a shaft of light. It also enables us to see a laser as a pencil of coloured light through scattering with atmospheric particles. Light scattering even gives rise to the blue colouration of the sky during the day and the spectacular colours that can sometimes be seen at sunset.

When a beam of radiation interacts with matter two things, in general, can happen. The radiation can either be absorbed or scattered. The absorption of radiation is used in the spectroscopic techniques that will be discussed later. The scattered radiation can either have the same or a different wavelength (or frequency) as the incident radiation. If the wavelength is the same the scattering is said to be *elastic* whereas if there is a difference in wavelength the scattering is termed *inelastic*. Raman spectroscopy is based on the inelastic scattering of light and is becoming a widely used technique for studying the structure of polymers. However, as far as determinations of $\bar{M}_w$ are concerned we are interested in only the elastic scattering of light by dilute polymer solutions.

The theory of light scattering was first developed by Lord Rayleigh in the last century during his studies on the properties of gases. In this case the particles are very much smaller than the wavelength of the light λ and the theory can be readily adapted to interpret the scattering by polymer solutions when the dimensions of the polymer coils are $\ll\lambda$. A schematic diagram of the essential features of the apparatus for a light scattering experiment is given in Fig. 3.12. The incident beam of light has a wavelength λ and an intensity I_0. The light is scattered at all angles θ and the intensity, per unit volume of the scattering medium, of the light scattered at an angle θ is i_θ. It is convenient to use the *reduced scattering intensity* R_θ which is given by

$$R_\theta = i_\theta r^2/I_0 \tag{3.83}$$

where r is the distance between the centre of the scattering medium and the detector. Light will be scattered from both the polymer solution and pure solvent. In order to obtain information from the polymer molecules only it is necessary to measure the scattering from the polymer solution and then subtract the scattering due to the solvent molecules. This leads to a reduced scattering intensity for polymer molecules only which is given by

$$R_\theta = R_\theta(\text{solution}) - R_\theta(\text{solvent}) \tag{3.84}$$

Developments and modifications of Rayleigh's original theory allow R_θ to be related to the molar mass of a homodisperse polymer M through the equation

$$R_\theta = \frac{2\pi^2\bar{n}_0^2}{\lambda^4}\left(\frac{d\bar{n}}{dc}\right)^2(1+\cos^2\theta)\left(\frac{N}{V}\right)\frac{M}{N_A^2}\frac{RT}{\left(d\prod/dc\right)_T} \tag{3.85}$$

where $\bar{n}_0$ is the refractive index of the solvent and $(d\bar{n}/dc)$ is the rate of change of refractive index of the solution with concentration. The number of polymer molecules per unit volume is (N/V) and $\prod$ is the osmotic pressure of the solution. The other parameters have their usual meanings.

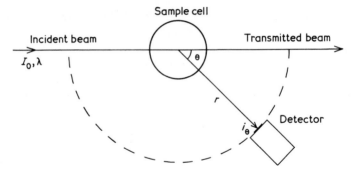

Fig. 3.12 *Main features of the apparatus required to measure the light scattered from a polymer solution (After Billingham).*

The equation can be simplified since differentiating Equation (3.72) with respect to c for a monodisperse polymer sample gives

$$\left(\frac{d\Pi}{dc}\right)_T = \frac{RT}{M}(1 + 2Bc + \ldots)\tag{3.86}$$

If the solvent is not too 'good' the higher terms in the expansion can be ignored and so substituting Equation (3.86) into Equation (3.85) gives

$$R_\theta = \frac{2\pi^2\bar{n}_0^2}{\lambda^4}\left(\frac{d\bar{n}}{dc}\right)^2\frac{(1 + \cos^2\theta)}{(1 + 2Bc)}\frac{N\,M^2}{V\,N_A^2}\tag{3.87}$$

It is customary to group the parameters which are constant for a given scattering experiment in one term K which is given by

$$K = \frac{2\pi^2\bar{n}_0^2}{\lambda^4 N_A}\left(\frac{d\bar{n}}{dc}\right)^2\tag{3.88}$$

Equation (3.87) can also be generalized for a polydisperse polymer. In this case NM^2/V can be replaced by $\sum N_iM_i^2/V$ but since, by definition,

$$\bar{M}_w = \sum N_iM_i^2 \Big/ \sum N_iM_i\tag{3.89}$$

then using Equation (3.68) it follows that NM^2/VN_A can be replaced by $\bar{M}_wc$ for a polydisperse sample. Substituting Equations (3.88) and (3.89) into (3.87) and rearranging gives

$$\frac{K(1 + \cos^2\theta)c}{R_\theta} = \frac{1}{\bar{M}_w}(1 + 2Bc)\tag{3.90}$$

as the general scattering equation for a polymer solution. Measurements are often made at $\theta = 90°$ and R_{90} is often called the *Rayleigh ratio.* A plot of Kc/R_{90} versus c will be a straight line if the solvent is not too 'good' and the intercept as $c \to 0$ will yield the value of $\bar{M}_w$. The slope of such a plot will be $2B/\bar{M}_w$ but B will only be related to the second virial coefficient obtained from osmotic-pressure measurements for a monodisperse polymer sample. If there is a distribution of molar mass great care must be taken relating light-scattering data to thermodynamic parameters and this can only be done if the exact form of the distribution is known.

At this stage it is important to consider some of the practicalities of light-scattering experiments upon polymer solutions. For example, dust particles can also give rise to scattering and great care must be taken to ensure that the polymer solutions are dust-free. This is often done by centrifugation or carefully filtering the polymer solutions. The accuracy to which $\bar{M}_w$ can be measured depends critically upon the accuracy to which K is known. Two variables which make up K are the refractive index of the solvent $\bar{n}_0$ and the rate of change of the refractive index of the solution with concentration $(d\bar{n}/dc)$ and it is essential that both are measured accurately by separate experiments. It is also important that there is a change in refractive index with concentration so that $(d\bar{n}/dc)$ is finite otherwise there will be no excess scattering from the polymer molecules in the solution. It may also be noted that the intensity of scattering is inversely proportional to λ^4. A mercury-vapour lamp is often used as a source for a light-scattering experiment. This has characteristic wavelengths of 365, 436 and 546 nm and the greatest scattering will be obtained for a given intensity level using the shortest wavelength which can be selected using a filter.

3.3.2 *Light scattering from large molecules*

The equations derived in the previous section for scattering from polymer solutions are only valid if the polymer coils have dimensions of less than $\lambda'/20$ where λ' is the wavelength of the light in the solution $\approx \lambda/\bar{n}_0)$. This is because if the polymer coils are too large there can be a difference in phase between light scattered from different parts of the same coil arriving at the same point in the detector. This effect is illustrated schematically in Fig. 3.13 and it is termed *internal interference*. The magnitude of the interference depends upon both the size and shape of the polymer coils and the scattering angle θ. There is no internal interference at any angle for small particles and at $\theta = 0$ for large particles. This effect is shown in the form of intensity distributions for light scattered as a function of θ for different size particles in Fig. 3.14. Interference effects increase as θ increases and are a maximum at $\theta = 180°$. It is convenient to define a parameter $P(\theta)$ which is termed the *particle scattering factor* and is given by the ratio

$$P(\theta) = \frac{R_\theta(\text{observed})}{R_\theta(\text{no interference})}$$

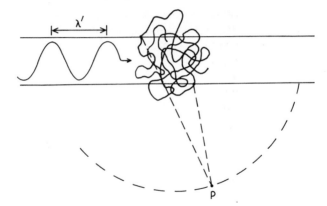

Fig. 3.13 *Schematic illustration of the origin of internal interference during light scattering from a polymer molecule in solution. The light scattered from different parts of the molecule arrives out of phase at point P.*

The scattering Equation (3.90) which was calculated previously can be modified to include the effect of internal interference and becomes

$$\frac{K(1 + \cos^2\theta)c}{R_\theta} = \frac{1}{\overline{M}_w P(\theta)}(1 + 2Bc) \qquad (3.91)$$

where R_θ in this case is the value measured when there is interference. At $\theta = 0$, $P(\theta) = 1$ and Equation (3.91) becomes

$$\frac{2Kc}{R_\theta} = \frac{1}{\overline{M}_w}(1 + 2Bc) \qquad (3.92)$$

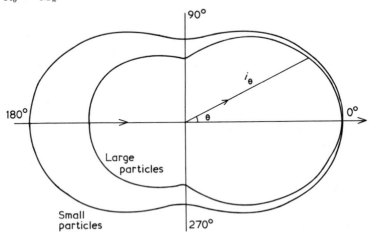

Fig. 3.14 *Effect of particle size upon the intensity distribution of light scattered at different angles. The distribution is symmetrical for small particles ($<\lambda'/20$) but for larger particles the intensity is reduced at all angles except zero.*

However, measurements cannot be made directly at $\theta = 0$ since the intensity of the scattered beam is very much less than the transmitted beam. Measurements can be made at different values of θ and then extrapolated to $\theta = 0$. The value of $\bar{M}_w$ can be determined by doing this for various concentrations and then extrapolating to zero concentration.

It is not always desirable to eliminate internal interference as it can be used to advantage to obtain information concerning the conformations of the polymer coils. It is possible theoretically to calculate $P(\theta)$ as a function of the size and shape of the coil. Such calculations are beyond the scope of this book and it will suffice to say that it can be shown that for a random Gaussian coil when θ is small

$$P(\theta)^{-1} = 1 + \frac{8\pi^2 \langle \bar{r^2} \rangle}{9\lambda'^2} \sin^2\theta/2 \tag{3.93}$$

where $\langle \bar{r^2} \rangle$ is the mean square displacement length of the polymer molecules. Substitution of Equation (3.93) into (3.92) gives a general scattering equation

$$\frac{K(1 + \cos^2\theta)c}{R_\theta} = \frac{1}{\bar{M}_w}(1 + 2Bc)\left(\frac{1 + 8\pi^2\langle\bar{r^2}\rangle\sin^2\theta/2}{9\lambda'^2}\right) \tag{3.94}$$

This equation enables $\bar{M}_w$, B and $\langle \bar{r^2} \rangle$ to be determined if suitable measurements are made. A plot of $K(1 + \cos^2\theta)c/R_\theta$ versus $\sin^2\theta/2$ at constant c should give a moderately good straight line which may be extrapolated to $\theta = 0$. The slopes of such lines at constant c should give $\langle\bar{r^2}\rangle$ whereas the intercepts at $\sin^2\theta/2 = 0$ will give $2Kc/R_\theta$. The intercepts can then be extrapolated to zero concentration to yield $\bar{M}_w$. The slope of this line give B.

Normally these two extrapolations are made on the same graph using an elegant technique devised by Zimm. In the Zimm plot, $K(1 + \cos^2\theta)c/R_\theta$ is plotted against $\sin^2\theta/2 + kc$ where k is an arbitrary constant chosen to spread the data to produce a grid-like plot similar to that shown schematically in Fig. 3.15. A graph is obtained consisting of two sets of parallel lines, one joining points of constant c and the other joining points with the same value of θ. It is then possible to do a double extrapolation by taking points for the same concentration and extrapolating to $\theta = 0$ and then taking points of equal θ and extrapolating to zero concentration. The two sets of extrapolated points give the $c = 0$ and $\theta = 0$ lines which both intercept the axis at the same point. This intercept is $1/\bar{M}_w$ and the parameters $\langle\bar{r^2}\rangle$ and B can be obtained from the slopes of the $c = 0$ line and the $\theta = 0$ line respectively.

It is possible to gain information upon the shape of the polymer coils in a solution from the form of $P(\theta)$. Polymer molecules are not always in the shape of a Gaussian coil in solution and can adopt a variety of alternative

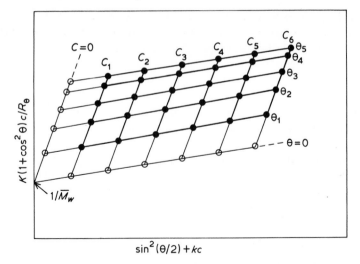

The plot shows axis label $K(1+\cos^2\theta)c/R_\theta$ on the vertical axis and $\sin^2(\theta/2)+kc$ on the horizontal axis, with $1/\overline{M}_w$ marked at the intercept. Data columns are labelled $C=0$, C_1, C_2, C_3, C_4, C_5, C_6 and rows labelled $\theta=0$, θ_1, θ_2, θ_3, θ_4, θ_5.

Fig. 3.15 *Schematic illustration of a Zimm plot for analysing light-scattering data. The solid points (●) represent experimental measurements, the open circles (○) are extrapolated points.*

shapes such as rigid rods, helices, discs, spheres etc., depending upon such factors as the type of bonding in the polymer backbone, branching, side-groups etc. It is fairly easy to determine the size and shape of molecules of a monodisperse polymer, however complications can arise with polydisperse samples as polydispersivity can also affect $P(\theta)$.

3.4 Transport measurements

The techniques that have been described so far to measure the molar masses of polymers in solution depend upon the equilibrium properties of the polymer solution. It is possible to relate the molar mass of the polymer to the solution properties through theoretical (e.g. thermodynamic) equations and the measurements are normally extrapolated to zero concentration where the solutions exhibit ideal behaviour. It is also possible to determine molar masses by studying the *transport properties* of polymer solutions which are usually analysed in terms of hydrodynamic models. These properties can be divided into two categories; one of which involves the motion of the molecules through a solvent which is itself stationary (e.g. ultracentrifuge) and the other deals with the effect of polymer molecules upon the motion of the whole solution (e.g. solution viscosity). The theoretical models which have been devised to explain the transport properties are by no means as well developed as those used to explain, for example, the thermodynamic properties of polymer solutions and so transport properties are normally analysed using semi-empirical approaches. Nevertheless, the ultracentifuge is widely used in the study of biological macromolecules and solution viscosity provides a quick and simple method of determining the molar mass of a synthetic polymer.

3.4.1 *Viscosity of polymer solutions*

The presence of polymer molecules in particular solvents can give rise to a dramatic increase in viscosity which is very much greater than that found for equivalent concentrations of low-molar-mass solutes. This is because of the enormous difference in dimensions between the polymer and solvent molecules and in good solvents the polymer coils are expanded even further. In general the increase in viscosity depends upon a number of factors which are as follows.

(i) The nature of the solvent.
(ii) The type of polymer.
(iii) The molar mass of the polymer.
(iv) The concentration of the polymer.
(v) The temperature.

Polymer scientists are normally interested in the variation of the solution viscosity with molar mass. Unfortunately no theories have been developed to relate them to each other and empirical relationships are used to determine the molar mass of a polymer sample from measurements of solution viscosity. However, this method does have the distinct advantage of being a rapid technique that can be carried out quickly using relatively simple apparatus.

The two main pieces of apparatus that are generally used for solution viscosity measurements are the Ostwald viscometer and a modified version called the Ubbelohde viscometer and they are both shown schematically in Fig. 3.16. The Ostwald viscometer consists of a measuring bulb, a capillary through which the solution flows and a reservoir. The pressure head driving the liquid through the capillary is determined by the difference in height between the liquid in the measuring bulb and reservoir and therefore depends upon the total volume of liquid in the viscometer. Because of this it is essential to use the same volume of liquid for each measurement during an experiment and a mark is usually put on the arm of the viscometer above the reservoir so that the viscometer is filled with the same volume of liquid each time it is used. The two marks above and below the measuring bulb give two reference points for measuring the time of flow (and hence viscosity) of the liquid through the capillary.

The necessity of using identical volumes of liquid in the Ostwald viscometer clearly leads to a potential source of error and this problem is overcome in the Ubbelohde or suspended-level viscometer also shown in

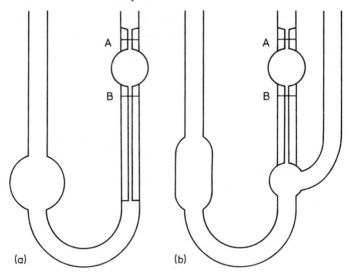

Fig. 3.16 *Schematic illustrations of different viscometers. The flow time is determined from the time taken for the level of the liquid to fall from A to B. (a) Ostwald viscometer. (b) Ubbelohde or suspended-level viscometer.*

Fig. 3.16. In this viscometer the flow of liquid leaving the capillary is broken by having a third arm which is open to the atmosphere. Because of this the pressure head depends only upon the volume of liquid in and above the capillary and is independent of the total volume of liquid in the viscometer. Since the same marks are always used to monitor the flow of liquid through the measuring bulb, the pressure head and volume of liquid flowing through the capillary will always be the same. Viscosity measurements are normally made on a series of solutions of the polymer sample of different concentrations in a particular solvent. With the Ubbelohde viscometer it is possible to dilute the solution undergoing measurement *in situ* by adding more solvent whereas when the Ostwald viscometer is used it is necessary to empty, dry and refill the viscometer with new solution each time the concentration is changed.

From a practical viewpoint it is necessary with all viscometers to make sure that the solutions and capillaries are clean and free from dust or dirt. It may be necessary to filter the solutions and flush out the capillary with solvent from time to time otherwise serious errors may occur in the results. The viscometer is usually kept in a water bath during measurements and sufficient time must be allowed for the temperature of the solution and apparatus to equilibriate as the viscosity of the solution varies strongly with temperature.

When the liquid flows through the capillary its velocity is not constant across a section of the capillary. The liquid near to the sides is virtually stationary and the velocity increases towards the middle where it is maximum. The rate of flow of liquid through a capillary of radius r and length l is given by Poiseuille's equation as

$$\frac{dV}{dt} = \frac{\pi p r^4}{8\eta l} \qquad (3.95)$$

where η is the viscosity of the liquid, p the pressure head causing the flow and dV/dt is the volume of liquid flowing through the capillary in unit time. The pressure head, p, continually decreases during an experiment and so it is convenient to define an average pressure $\bar{p}$. A constant volume of liquid V is normally used and so Equation (3.95) can be approximated to

$$V/t = \pi\bar{p}r^4/8\eta l \qquad (3.96)$$

where t is the time of flow of the liquid through the capillary. We are normally interested in the difference between the flow times of the solution and pure solvent and so if the subscript $_0$ is used to designate the solvent then Equation (3.96) can be written as

$$V/t = \pi\bar{p}r^4/8\eta l \qquad \text{(solution)}$$
$$\text{or} \quad V/t_0 = \pi\bar{p}_0r^4/8\eta_0 l \qquad \text{(solvent)}$$

where η is the viscosity of the solution and η_0 is that of the pure solvent. The average pressure producing flow is given by the standard relation

$$\bar{p} = \rho g\bar{h} \qquad (3.97)$$

where ρ is the density of the liquid, $\bar{h}$ the average head of liquid and g the acceleration due to gravity. Equation (3.96) can clearly be used to determine the viscosity η of a polymer solution if the viscometer is calibrated with liquids of known viscosity. However, we are not normally interested in the absolute viscosity of the solution, but are more concerned with the increase in viscosity of the solvent caused by the presence of the polymer molecules. A parameter of considerable importance is the *viscosity ratio* or *relative viscosity*, η_r, which is defined as (η/η_0) and can be related to the flow times t and t_0 through Equations (3.96) and (3.97) which lead to

$$\eta_r = \eta/\eta_0 = t\rho/t_0\rho_0 \qquad (3.98)$$

This equation can be further simplified since for a dilute solution $\rho \simeq \rho_0$ and so η_r is normally taken as t/t_0. Since η_r becomes unity for an infinitely dilute solution it is more useful to define the *specific viscosity*, η_{sp}, which is given by

$$\eta_{sp} = (\eta_r - 1) = (t - t_0)/t_0 \qquad (3.99)$$

The specific viscosity then corresponds to the fractional increase in viscosity of the solvent due to the presence of the polymer molecules. It is obvious that this increase in viscosity will depend upon the concentraton of polymer molecules in the solution and it is found that it is possible to represent the variation of specific viscosity with concentration as a power series in concentration such as

$$\eta_{sp} = [\eta]c + k[\eta]^2c^2 + \dots$$

$$\text{or} \quad (\eta_{sp}/c) = [\eta] + k[\eta]^2c + \dots \tag{3.100}$$

where k is a constant.

The term (η_{sp}/c) is often called the *viscosity number*. Equation (3.100) is of a similar form to those used to describe the colligative properties of polymer solutions (virial equations) and can be considered in a similar way. The *limiting viscosity number* (or *intrinsic viscosity*) $[\eta]$ describes the ability of the polymer molecules to increase the viscosity of the solvent in the absence of any intermolecular interactions. The second term represents the interactions between different molecules in the solution. Measurements of (η_{sp}/c) are normally made for a series of solutions of different concentration by determining flow times. Fig. 3.17 shows a typical plot of $(t - t_0)/t_0c$ or (η_{sp}/c) against c for a dilute polymer solution. Although a straight-line relationship is obtained in this case, curvature can be found using more concentrated solutions because of contributions from terms in higher powers of c.

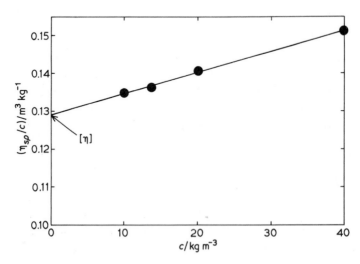

Fig. 3.17 *Plot of (η_{sp}/c) versus c measured at 303 K for a sample of polystyrene dissolved in toluene.*

3.4.2 *Limiting viscosity number and molar mass*

The limiting viscosity number $[\eta]$ can be readily related to the molar mass of a monodisperse polymer, M, through a semi-empirical equation of the form

$$[\eta] = KM^a \tag{3.101}$$

This is often called the Mark–Houwink equation and is usually treated as a semi-empirical equation although it will be shown later that it can be justified theoretically. The Mark–Houwink constants are characteristic for a given polymer–solvent system. Each system is normally calibrated by measuring $[\eta]$ as a function of M for a series of monodisperse samples and plotting log $[\eta]$ against log M which should give a straight line. The constants K and a can be determined from the intercept and slope of the line. Table 3.5 gives a series of values of the constants for different polymer–solvent systems determined at different temperatures. It can be seen that the value of a is typically between 0.5 and 1. In fact it is found to be about 0.5 for a solution in the Theta condition and is higher for better solvents.

Since Equation (3.101) is normally calibrated using monodisperse polymer, care must be taken when $[\eta]$ is measured for a polydisperse sample. The value of M obtained in this case is an average one and it is often called the viscosity average molar mass $\bar{M}_v$. It can be shown that $\bar{M}_v$ is given by

$$\bar{M}_v = \left[\frac{\sum N_i M_i^{1+a}}{\sum N_i M_i} \right]^{1/a} \tag{3.102}$$

for a polydisperse polymer sample. Because a normally lies between 0.5 and 1 the value of $\bar{M}_v$ is between $\bar{M}_n$ and $\bar{M}_w$. It is usually close to $\bar{M}_w$ and in the special case when $a = 1$, $\bar{M}_v \simeq \bar{M}_w$. Also since $\bar{M}_v$ is closer to $\bar{M}_w$ than $\bar{M}_n$ it is usually better to calibrate the monodisperse fractions for a particular polymer–solvent system by light scattering rather than osmometry.

The increase in viscosity of the polymer solution over that of pure solvent is thought to be due principally to frictional interactions between the polymer coil and the flowing solvent. Because of this the behaviour is usually explained in terms of hydrodynamic models. Using this approach Flory and Fox suggested that the viscosity of a polymer solution should depend upon the volume occupied by the polymer molecule. They produced a theory relating to limiting viscosity number to the dimensions of the polymer coil and the molar mass of the polymer. The details of the theory are beyond the scope of this book, but the important result is that it

TABLE 3.5 *Typical values of K and a for different polymer–solvent systems*

Polymer	Solvent	T/K	$K/10^{-5} m^3 kg^{-1}$	a
Polyethylene	Decalin	408	6	0.7
Polystyrene	Cyclohexane	308*	8	0.5
Polystyrene	Toluene	298	1	0.73
Poly(methyl methacrylate)	Butanone/isopropanol (50/50)	298*	6	0.5
Poly(methyl methacrylate)	Benzene	298	0.6	0.76

*$T = \Theta$

N.B. These values of K and a are for relatively high molar mass polymer. The exact value depends upon the molar mass of the polymer being characterized.

can be shown that the limiting viscosity number in a Theta solvent, $[\eta]_\theta$ is given by

$$[\eta]_\theta = K_\theta M^{1/2} \tag{3.103}$$

where

$$K_\theta = \Phi(\langle \overline{r^2} \rangle_0 / M)^{3/2} \tag{3.104}$$

The parameter Φ comes out of the theory as a universal constant with a theoretical value in c.g.s. units of 2.86×10^{23}. Experimentally it is found that Φ is not quite constant and can vary slightly with molar mass and other factors and it is normally taken to be about 2.5×10^{23} (c.g.s. units). Equation (3.103) is of the same form as the Mark–Houwink equation and since $(\langle r^2 \rangle_0 / M)$ is constant for a given polymer, the theory predicts that a should be 0.5 for a solution in the Theta condition, which is exactly what is observed experimentally.

For solutions not in the Theta condition Equation (3.103) can be modified and it becomes

$$[\eta] = K_\theta M^{1/2} \alpha^3 \tag{3.105}$$

where α is the expansion factor for the polymer coil. It was shown in Section 3.1.7 that the expansion of the coil is due principally to long-range interactions and that α will be unity in the Theta condition and larger in better solvents. For good solvents the Mark–Houwink parameter a is found to be greater than 0.5. It is possible to rationalize this with Equation (3.105) because α is a function of the molar mass of the polymer. Extensions of the Flory–Krigbaum theory (Section 3.1.7) have enabled a relationship between α and M to be established and it can be shown theoretically that a will increase as α increases, which is again consistent with experimental observations.

3.4.3 *Ultracentifugation*

The underlying principle behind the use of the ultracentrifuge in polymer science is that it is possible to determine the size and mass of particles from the way in which they settle out in a liquid under the action of a gravitational field. The rate at which macromolecules settle out in a polymer solution under the action of the specific force of gravity, g, is very slow because of Brownian motion. This effect can be overcome by subjecting the solution to the high gravitational field experienced in the cell of an ultracentrifuge. In modern instruments rotation speeds of up to 70 000 rev/min can be employed producing up to 400 000 g. The ultracentrifuge cell is in the form of a truncated cone which would have a peak at the centre of the axis of rotation. It also has windows which allows the point-to-point variation in the concentration c of the polymer in the solution to be monitored optically while rotation is taking place. This is normally done by measuring the refractive index, $\bar{n}$, of the solution and relating this to c, having previously determined the relationship between $\bar{n}$ and c (cf. light scattering, Section 3.3.1).

Ultracentifugation techniques, although widely used for biological macromolecules, have found only limited use with synthetic polymers. This is due to the compact nature of the conformations of biological polymers which often exist in solution as spheres or rigid rods. On the other hand synthetic polymers tend to be in the form of random coils which have a more open structure which may lead to serious deviations from ideality and entanglement of the polymer chains in solution, both of which make theoretical analysis rather difficult. In general, two basic methods of analysing the behaviour of macromolecules in the ultracentrifuge are used and only a brief non-mathematical outline will be given here. The analysis of data obtained in ultracentrifugation experiments is highly sophisticated and the reader is directed to more advanced texts for details.

In a *sedimentation equilibrium* experiment the cell is rotated at a relatively low speed and the molecules become distributed in the cell according to their size and the molar mass distribution, under equilibrium conditions. The centrifugal force just balances the tendency of the molecules to diffuse back against the concentration gradient that develops. Such systems can take days to reach equilibrium, but they do have the advantage of potentially being able to yield the complete molar mass distribution curve for the polymer being studied. The theories used are based on equilibrium thermodynamics and the method is an absolute one in that it does not require calibration although in practice the analysis is only relatively easy for solutions at the Theta temperature.

The *sedimentation velocity* method has the advantage of being a much more rapid technique, but is a semi-empirical method which normally requires calibration with fractions of the polymer of known molar mass. In this case the ultracentifuge is operated at very high speed. Initially the solution is of uniform concentration. Very quickly a region of pure solvent develops at the top of the cell as the molecules begin to sediment and the boundary between the pure solvent and the solution can be located from an abrupt change in the refractive index, $\bar{n}$. The rate of movement of the boundary can be used to determine the molar mass of the polymer.

3.5 Fractionation and gel permeation chromatography

The techniques of polymer synthesis outlined in Chapter 2 nearly always produce polymer samples which have a distribution of molar mass. In order to evaluate the nature of the distribution it is often desirable to split up the polymer into fractions which are, as far as is possible, monodisperse. This process is termed *fractionation* and the molar-mass distribution curve can be generated by determining the weight and molar mass of each fraction. The fractions can also be very useful for determining the effect of molar mass upon the physical properties of the polymer or testing solution theories. However, such fractions have been largely superseded by the use of 'living' polymers for this purpose. Also these days the distribution of molar mass is invariably determined by using *gel permeation chromatography*. This is a rapid and reliable technique in which the polymer is split up into fractions by passing the polymer solution through a column containing a rigid porous gel, normally highly cross-linked porous polystyrene beads.

In this section the theory of phase equilibria of polymer solutions is discussed as it is a simple practical illustration of the Flory–Huggins theory and can be extended to explain the principles behind the fractionation techniques which are used with polymer solutions. Finally the technique of gel-permeation chromatography, which is now widely used in polymer laboratories, is described in detail.

3.5.1 *Phase equilibria of polymer solutions*

A simple method of polymer fractionation is to dissolve the polymer in a poor solvent and cause precipitation to occur by lowering the temperature or adding a non-solvent. This method relies upon the fact that the solubility of the polymer depends upon its molar mass with small molecules generally being more soluble than large ones. This is a natural consequence of the Flory–Huggins theory and can be demonstrated by considering the equilibrium of a binary system which consists of a polymer and a poor solvent for that polymer. A polymer is capable of dissolving in a solvent if the Gibbs free energy of the solution is less than the combined free energies of the two components taken in isolation, i.e. if the Gibbs free energy of mixing ΔG_m, is negative. If the temperature of the solution is changed it will remain homogeneous as long as the free energy of the single-phase solution is less than that of any two other phases that may form. The variation of the Gibbs free energy of mixing of a polymer solution with volume fraction of polymer, ϕ_2, is shown schematically in Fig. 3.18. Two curves are given corresponding to different solution temperatures. The upper curve is for a relatively high temperature at which there is complete miscibility at all compositions. If the temperature is reduced the curve of the variation of ΔG_m with ϕ_2 may change in appearance and develop two minima as shown in the lower curve of Fig. 3.18. Although in this case ΔG_m is negative this is not a sufficient condition for the solution to remain homogeneous. If the solution is of a composition between ϕ_2' and ϕ_2'' and cooled to the lower temperature it will separate out into two liquid phases of compositions ϕ_2' and ϕ_2''. The exact values of ϕ_2' and ϕ_2'' are defined by the points of contact of the common tangent of the two minima. If the temperature of the solution is increased the two minima move closer together until at a critical temperature, T_c, they eventually merge. The variation of ϕ_2' and ϕ_2'' with temperature plots out the phase diagram shown schematically in Fig. 3.19 and the temperature at the peak of the curve when the polymer and solvent are just miscible in all proportions is known as the *critical solution temperature*, T_c.

A general principle of free energy-composition curves is that phase separation will occur if a tangent can be drawn which touches the curve at two points. This enables T_c to be calculated since it is the point at which the inflexion points merge. This can be expressed mathematically as the point

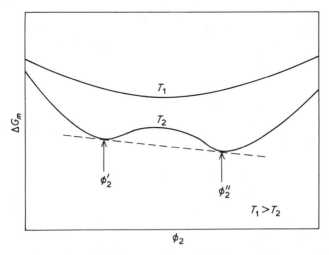

Fig. 3.18 *A schematic plot of the variation of the Gibbs free energy of mixing of a polymer solution ΔG_m with volume fraction of polymer ϕ_2 at two different temperatures.*

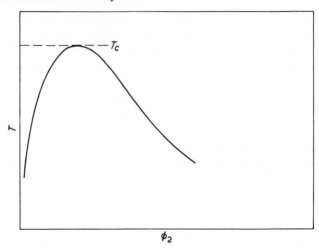

Fig. 3.19 *A typical phase diagram for a polymer–solvent mixture. The volume fraction of polymer is ϕ_2 and the critical solution temperature is T_c.*

where the first, second and third derivatives of ΔG_m with respect to ϕ_2 are all equal to zero.

i.e.
$$\left(\frac{\partial \Delta G_m}{\partial \phi_2}\right) = \left(\frac{\partial^2 \Delta G_m}{\partial \phi_2^2}\right) = \left(\frac{\partial^3 \Delta G_m}{\partial \phi_2^3}\right) = 0 \tag{3.106}$$

For the polymer–solvent system $\phi_1 + \phi_2 = 1$ and so $d\phi_1 = -d\phi_2$. Also the chemical potential of the solvent in the solution μ_1 can be related to ΔG_m through Equation (3.39) and so the critical conditions for miscibility can be written in terms of chemical potentials as

$$\left(\frac{\partial \mu_1}{\partial \phi_2}\right) = \left(\frac{\partial^2 \mu_1}{\partial \phi_2^2}\right) = 0 \tag{3.107}$$

The chemical potential of the solvent in the solution can be determined using the Flory–Huggins theory and is given by combining Equations (3.39) and (3.43) as

$$\Delta \bar{G}_1 = \mu_1 - \mu_1^0 = RT\left[\ln(1 - \phi_2) + \left(1 - \frac{1}{x}\right)\phi_2 + \chi_1\phi_2^2\right] \tag{3.108}$$

where μ_1^0 is the standard chemical potential of pure solvent. Application of the criteria in Equation (3.107) to Equation (3.108) leads to the critical conditions for phase separation to occur as

$$\phi_2^c = 1/(1 + x^{1/2}) \simeq 1/x^{1/2} \tag{3.109}$$

$$\text{and} \quad \chi_1^c = \tfrac{1}{2} + 1/x^{1/2} + 1/2x \tag{3.110}$$

where the superscript c refers to critical values of ϕ_2 and χ_1. The parameter x was originally introduced in Section 3.1.5 as the ratio of the molar volume of the polymer and solvent and this is clearly closely related to the number-average degree of polymerization $\bar{x}_n$ for a polydisperse sample and so x can be replaced in Equations (3.109) and (3.110) by $\bar{x}_n$.

There are some aspects of the predictions of Equations (3.109) and (3.110) that are worth considering. Firstly, they can be used in conjunction with the Flory–Huggins theory to measure the Theta temperature of a polymer solution from studies of phase separation. It is possible to relate T_c and $\bar{x}_n$ to Θ by combining Equations (3.47) and (3.110) which give

$$\psi_1(\Theta/T_c - 1) = 1/\bar{x}_n^{1/2} + 1/2\bar{x}_n \qquad (3.111)$$

If measurements are made of the critical solution temperature, T_c, as a function of $\bar{x}_n$, the Theta temperature can be found from a plot of $1/T_c$ against $(1/\bar{x}_n^{1/2} + 1/2\bar{x}_n)$. The Theta temperatures obtained from such measurements are found to be in good agreement with those obtained by determining the temperature at which the second virial coefficient becomes zero in osmotic pressure measurements. Inspection of Equation (3.111) shows that as $x_n \rightarrow \infty$, $T_c \rightarrow \Theta$ and it follows that for most polymers T_c will be just below the Theta temperature. Also Equation (3.109) shows that for polymer molecules ϕ_2^c will be small because $\bar{x}_n$ is relatively large. This means that the phase diagram (Fig. 3.19) will not be symmetrical, but will be distorted with the maximum at a low volume fraction and hence concentration of polymer.

3.5.2 *Fractionation*

The equations derived in the previous section are extremely useful for consideration of the conditions under which polymer fractionation can take place. Since Equation (3.110) implies that there is a unique relationship between $\bar{x}_n$ and χ_1^c it is possible to induce a particular molecular species to separate out by controlling the value of χ_1 for a polymer solution. It is possible to relate χ_1 to the solvent power and it will vary with the type of solvent and temperature. It is found to decrease approximately linearly as the temperature is increased. The most common methods of fractionation depend upon the variation of χ_1 with solvent power or temperature.

A simple method of fractionation is by the addition of a non-solvent to a dilute solution. This causes χ_1 to increase and eventually the solution becomes turbid when χ_1^c is reached. The turbidity is due to the solution separating out into two phases, one of which contains a higher proportion of polymer than the other. If the solution is polydisperse the analysis of the situation can be rather complex. However, it can be seen from Equation (3.110) that as χ_1 is increased the longest molecules will tend to separate out first of all with the shorter molecules remaining in solution. This phase separation is normally known as precipitation. The addition of non-solvent is normally done at constant temperature and when precipitation is first observed the addition is stopped and the solution warmed slightly. This has the effect of decreasing χ_1 and so the precipitate redissolves. It re-appears again on cooling to the original temperature, but this procedure has the effect of ensuring that low-molar-mass polymer, which may be precipitated where there are high local concentrations of non-solvent, is eliminated from the fraction. The precipitated phase is then allowed to settle and is separated from the supernatant liquid. The precipitate may be fairly dilute and can sometimes only contain about 10 percent of polymer. The polymer fraction that can be extracted from it may have a distribution of molar mass and will not be exactly homodisperse. The procedure can be repeated on the remaining solution and a series of fractions with progressively lower molar masses may be prepared by further addition of non-solvent.

A second method of fractionation involves a similar manipulation of χ_1, in this case by lowering the temperature of the solution. This is best done by dissolving the polymer in a relatively poor solvent so that precipitation can be induced by only small changes in temperature.

A technique of fractionation known as column chromatography involves the progressive dissolution of a polymer rather than precipitation. The polymer is coated very thinly on the surface of glass beads which are packed into a column along which there is normally a gradient of temperature. A mixture of solvent and non-solvent is streamed through the column. The liquid used initially is a poor solvent and the proportion of good solvent in the mixture is gradually increased causing χ_1 to decrease. With this system the low-molar-mass fractions are leached out of the polymer first of all and the liquid leaving the column contains progressively longer molecules as the solvent power increases.

With all of these techniques there will be distribution of molar mass in the fractions obtained. Because of this, care must be taken in determining the distribution curve of the polydisperse polymer sample from the weights and average molar masses of the fractions. This situation is shown schematically in Fig. 3.20 where there is considerable spread of molar mass in the fractions and overlap of the individual distribution curves. It is necessary to allow for the overlap when the distribution curve for the polymer is being constructed.

3.5.3 *Gel permeation chromatography*

Gel permeation chromatography, GPC, is an extremely powerful technique for fractionating a polymer and determining its molar mass distribution. It has only been used for synthetic polymers since the early

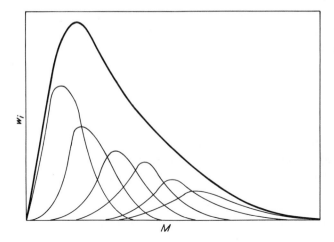

Fig. 3.20 *Schematic representation of typical molar mass distribution curves for the fractions obtained from a polydisperse sample (heavy curve).*

1960s, but it is now a routine technique of polymer characterization. The apparatus that is used consists essentially of a column packed with rigid porous beads as shown schematically in Fig. 3.21 with a continuous stream of solvent flowing through the column. The beads are often made of a highly cross-linked polystyrene and it is essential that they have the correct distribution of pore sizes (typically $10-10^5$nm). The principle behind GPC is very simple. As a polymer solution flows through the column the solvent will go both through and around the beads. The small molecules will be small enough to pass through the pores in the beads and so their flow through the column will be retarded. On the other hand, large molecules will be unable to enter the pores and will remain in the solvent going around the beads. In this way the large molecules will pass through the column relatively quickly whereas the small molecules will take longer.

In practice sample and reference columns are normally run side-by-side with pure solvent streaming through both of them. A dilute solution of the polymer sample is injected into the solvent stream of the sample column. Both streams are then passed through a differential refractive index detector, the output of which is fed to a chart recorder. At first the signal is zero as the refractive index of the two streams is identical. When the polymer molecules appear in the stream from the sample column there is a difference in refractive index and a signal is registered. In fact, it is possible to determine the concentration of polymer in the solvent from knowledge of the relationship between the refractive index and concentration but this information is often not required.

A typical gel chromatogram is given in Fig. 3.22. The recorder output is

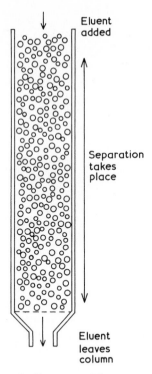

Eluent added

Separation takes place

Eluent leaves column

Fig. 3.21 *Schematic illustration of gel-permeation chromatography column packed with beads.*

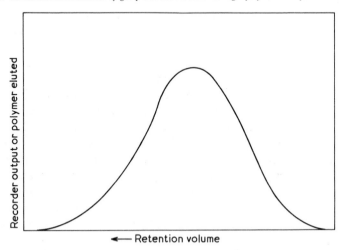

Fig. 3.22 *A typical gel permeation chromatograph. This can be related directly to the molar mass distribution for the polymer.*

normally plotted as a function of time or retention volume, V_r. This is the total volume of solvent leaving the column after injection of the polymer solution into the solvent stream. In modern instruments the procedure is highly automated and the output is often interfaced with a computer.

GPC is not an absolute method of polymer characterization and requires calibration with sharp fractions of known molar mass. It is often found for a given polymer–solvent system that the molar mass of the polymer can be related to the retention volume through an equation of the form

$$\log M = a + bV_r \qquad (3.112)$$

where a and b are constants. It is clear that the behaviour of a particular molecule in the GPC column will depend upon the 'size' of that molecule. This can be somewhat difficult to define, but for a flexible random-coil polymer it follows from the Flory–Fox theory of solution viscosity that the dimensions of the polymer coil will be related to $[\eta] M$ (Section 3.4.2). A universal calibration of a GPC instrument can be made by plotting $\log([\eta]M)$ against V_r as shown in Fig. 3.23. This calibration is found to be valid for many different linear and branched polymers.

GPC is now widely used both for analysis of molar-mass distributions and the preparation of sharp fractions of many polymers. There can be problems with instrumental broadening whereby sharp fractions will yield a broad curve on the chromatogram. However, with care, accurate and reliable results can be obtained for certain polymers in a short period of time (2–4 hours) using only a few milligrams of sample.

3.6 Spectroscopic analysis of polymers

Spectroscopic methods are finding increasing use in the characterization and analysis of polymers. All of the methods that are employed were developed initially for use with low-molar-mass materials and they have been extended for the analysis of polymers. The spectrum obtained for a particular polymer is often characteristic of that polymer and can therefore be used for identification purposes. Polymer spectra can be surprisingly simple given the complex nature of polymer molecules and they are often similar to the spectra obtained from their low-molar-mass counterparts. This can make the analysis of the spectra a relatively simple task allowing important spectral details to be revealed. In fact, certain information can only be obtained using spectroscopic methods. For example, nuclear magnetic resonance (n.m.r.) is the only technique that can be used to measure directly the tacticity of a polymer molecule.

Since the spectroscopic techniques applied to polymers are normally adaptations of conventional methods, the details of each method will not be discussed as they can readily be found in standard texts.

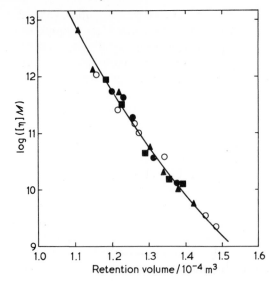

Fig. 3.23 *Universal GPC calibration curve for different types of polyethylene dissolved in o-dichlorobenzene at 403 K. The solid points are for linear polymer fractions and the open circles are for fractions of branched polymer. (Data taken from Wild and Guliana, J. Polym. Sci. A2,* **5** *(1967) 1087). The product [η] M has units of m³ mol⁻¹.*

3.6.1 *Infrared spectroscopy*

Most polymers absorb electromagnetic radiation in the wavelength range 1–50 μm, i.e. in the infrared region. This is because the molecules undergo transitions between vibrational states of different energies causing both the absorption and emission of radiation. The frequency, ν, and wavelength, λ, of the radiation are related to the difference in energy between the states, ΔE, by the equation

$$\nu = \Delta E/h = c/\lambda \qquad (3.113)$$

where h is Planck's constant and c the velocity of light. Organic molecules tend to have values of ΔE that correspond to the frequencies and wavelengths of infrared (i.r.) radiation. It is possible to relate the i.r. spectrum exactly to particular vibrational modes only for simple molecules containing a few atoms. But as soon as the molecules become complex and contain a large number of atoms such correlations become more difficult. In theory, therefore, the i.r. spectra of polymers could be extremely complicated, but since polymer molecules are made up of sequences of many identical units the spectra are often unexpectedly simple. This is because the vibrations of groups of atoms take place at a frequency which is often independent of the length of the polymer chain and also similar to the frequencies at which the same groups vibrate in short molecules.

The wavelengths and frequencies of the characteristic absorption bands of different groups of molecules commonly found in synthetic polymers are given in Fig. 3.24. The wavelengths are given in micrometres and the frequencies in wave numbers (the reciprocal of the wavelength in centimetres). This has been done because spectroscopists still tend to talk in terms of wave numbers even though they are not SI units. The correlation between the i.r. absorption frequencies and particular vibrational modes in Fig. 3.24 can sometimes be used to analyse the detailed microstructure of polymer molecules. The i.r. spectra can also be used to help to identify an unknown polymer sample. In fact i.r. spectra are often used as 'fingerprints' for the identification of polymers.

It is possible to use i.r. spectroscopy for structural characterization as well as identification. For example, in certain copolymers it is possible to determine quantitatively the composition of the copolymer from the intensities of specific absorption bands. The characteristic nitrile band can be used in the analysis of samples of acrylonitrile–butadiene–styrene copolymer. It is also possible to determine the relative amounts of *cis*-1,4, *trans*-1,4 and 1,2 addition in polybutadiene (Section 2.5.4) from the relative strengths of specific absorption bands.

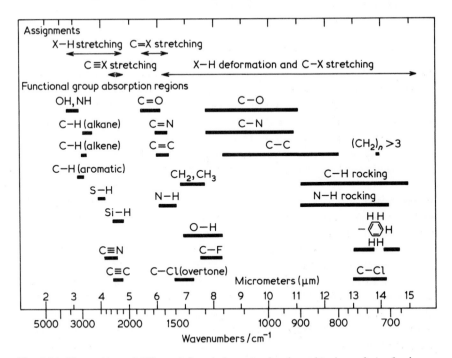

Fig. 3.24 *The position of different infrared absorption bands used in the analysis of polymers (Billmeyer, reproduced with permission).*

The i.r. spectra of crystalline polymers tend to be sharper and more well-defined than those of their amorphous counterparts. In the case of polyethylene terephthalate certain bands have been found to be character-istic of the crystalline form of the polymer and others are found to be due to vibrations in amorphous polymer. This is because the molecules are all *trans* in the crystals whereas *gauche* conformations can exist in amorphous regions. The different conformations give rise to bands with different frequencies of absorption in the i.r. spectrum.

The better definition of spectra from crystalline polymers can give an indication of polymer tacticity since atactic polymers are generally non-crystalline and there are certain polymers such as poly(methyl methacrylate) and polypropylene in which specific absorption bands can be assigned to the presence of molecules with particular types of tacticity. Some idea of the tacticity for samples of these polymers can, therefore, be obtained from measurements of the strength of the absorption of the relevant bands, but in general this method is not as accurate as n.m.r. for tacticity determination.

3.6.2 *Nuclear magnetic resonance spectroscopy*

N.m.r. is probably the most powerful technique available for the character-ization of the detailed microstructure of macromolecules. It utilizes the property of spin possessed by certain nuclei such as H^1, C^{13} and F^{19}. If molecules containing these nuclei are subjected to a powerful magnetic field the nuclei of these atoms will align themselves in the field. If a relatively weak alternating magnetic field of radio or microwave frequency is applied normal to the original field the nuclei can be induced to undergo resonance which leads to the absorption and emission of energy. This provides the basis of n.m.r. spectroscopy. The nuclei of different types of atoms resonate at widely different frequencies, but the most important aspect of n.m.r. is that nuclei of the same type of atom in a molecule will resonate at slightly different frequencies depending upon the chemical environments of the atoms. This gives rise to the phenomenon of *chemical shift*. Because there is no fundamental zero of chemical shift, the shifts are given relative to those in a standard reference chemical. Tetramethyl silane is taken as the standard reference substance for protons and the magnitudes of the shifts are given in terms of the parts per million relative change in field strength (either τ or δ). The chemical shifts of protons in groups commonly found in polymers are given in Fig. 3.25. It can be seen that there is a range of chemical shift for most groups. This is because the exact magnitude of the shift depends upon the detailed structure of the polymer molecule. Thus, the n.m.r. spectrum will be characteristic for a given polymer and so the technique can be used for identification purposes.

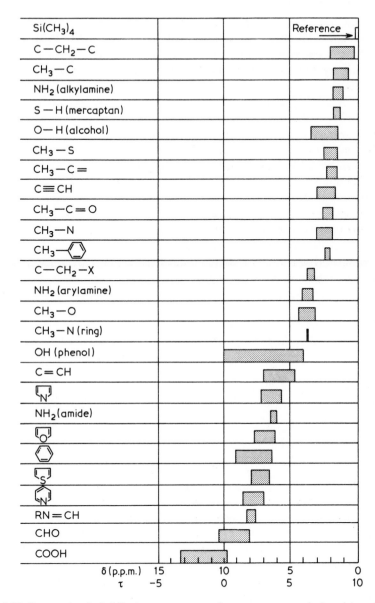

Fig. 3.25 *Proton chemical shifts in n.m.r. spectra for groups commonly found in polymer molecules (Billmeyer, reproduced with permission).*

Perhaps the most important application of n.m.r. is in the estimation of tacticity. Three n.m.r. spectra of samples of poly(methyl methacrylate) with different tacticities are given in Fig. 3.26. The tacticity of each sample can be estimated from the relative strengths of the resonances of the α-methyl protons in the vicinity of $\tau \approx 9$. It is known that these protons in isotactic, heterotactic and syndiotactic sequences will resonate at τ values of 8.78, 8.98, 9.13 respectively. In fact, the mole fraction of protons in each of these environments is proportional to the area of the relevant peak. It can be seen that the isotactic sample in Fig. 3.26 is made up almost exclusively of isotactic sequences whereas the other two have mixtures of heterotactic and syndiotactic sequences which are in different proportions in each sample.

The width of the lines in an n.m.r. spectrum is sensitive to molecular motion within the sample. The highest resolution is normally obtained with narrow-line spectra obtained from polymer molecules in solution where molecular motion is relatively easy. The lines are normally broader in solid samples and the line width can be used to study internal molecular motion or determine the degree of crystallinity in a semi-crystalline sample.

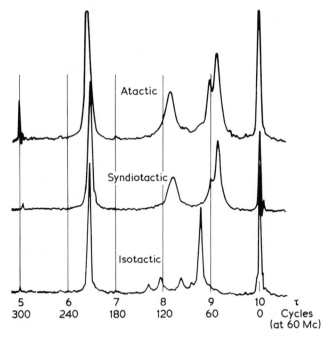

Fig. 3.26 *N.m.r. spectra for proton resonance in atactic, syndiotactic and isotactic poly(methyl methacrylate) in CDCl₃ at 331 K. (McCall and Slichter, in 'Newer Methods of Polymer Characterization' ed. B. Ke, Wiley-Interscience, New York, 1964, reproduced with permission).*

3.6.3 *Other spectroscopic methods*

Almost all the spectroscopic methods used by organic chemists for the study of low-molar-mass substances have been applied with varying degrees of success to polymers. Most conventional synthetic carbon-chain polymers tend to absorb most strongly in the infrared region, but certain polymers, in particular those containing conjugated multiple bonds, also absorb ultraviolet and visible radiation. They can, therefore, be studied by using ultraviolet spectroscopy. This technique also finds use in analysing additives such as anti-oxidants or residual monomer which are present in most commercial polymers.

Raman spectroscopy is frequently used to complement i.r. spectroscopy because certain vibrational modes which are inactive in the infrared are found in Raman spectra. The technique of Raman spectroscopy is particularly useful for studying carbon–carbon bond stretching modes in polymers and the use of lasers as high-intensity light sources has given the technique a considerable boost. Fluorescence is often encountered when attempting to obtain Raman spectra from commercial polymers and this can lead to problems. However, if impurities are stringently removed from the polymer the fluorescence can be kept to tolerable levels.

Electron spin resonance (e.s.r.) spectroscopy has also been applied with limited success to polymers. Its principal use is in the detection of free radicals generated during degradation of the polymer. E.s.r. has been used in the study of pyrolysis, the effects of high-energy irradiation and more recently in the generation of free radicals due to chain scission during deformation and fracture.

Further reading

Billingham, N.C. (1977), *Molar Mass Measurements in Polymer Science,* Kogan Page, London.

Billmeyer, F.W. (1971), *Textbook of Polymer Science,* Wiley-Interscience Inc., New York.

Cowie, J.M.G. (1973), *Polymers: Chemistry and Physics of Modern Materials,* International Textbook Company, Aylesbury, U.K.

Flory, P.J. (1953), *Principles of Polymer Chemistry,* Cornell University Press, Ithaca, N.Y.

Margerison, D. and East, G.C. (1967), *An Introduction to Polymer Chemistry,* Pergamon Press, London.

Morawetz, H. (1975), *Macromolecules in Solution,* Wiley-Interscience, New York.

Tanford, C. (1961), *Physical Chemistry of Macromolecules,* John Wiley & Sons Inc., New York.

Zbinden, R. (1964), *Infrared Spectroscopy of High Polymers,* Academic Press Inc., New York.

Problems

3.1 A sample of polymer is made up of a mixture of three sharp fractions of molar mass 10 000, 30 000 and 100 000. Calculate $\bar{M}_w$ and $\bar{M}_n$ for each of the following mixtures.

   (a) Equal numbers of molecules of each fraction.
   (b) Equal masses of each fraction.
   (c) The two fractions 10 000 and 100 000 are mixed in the ratio 0.145:0.855 by mass.

Comment on the value of $\bar{M}_w/\bar{M}_n$ for (c).

3.2 A solution of 10^{-3} kg of a polyester was neutralized with 0.012 dm^3 of an alkaline solution. The concentration of the alkali was 10^{-3} mol dm^{-3} and the alkali was monofunctional. If the polyester was made from an ω-hydroxy carboxylic acid, calculate the number average molar mass of the polymer, $\bar{M}_n$.

3.3 A sample of styrene was polymerized using radioactive AZBN as an initiator. The AZBN breaks down into free radicals during the reaction to form active centres which grow until they terminate by combination. The original AZBN had a radioactivity of 2.5×10^8 counts s^{-1} mol^{-1} in a liquid-scintillation counter. If 0.001 kg of the polystyrene had a radioactivity of 3.2×10^3 counts s^{-1} determine the number average molar mass of the polymer assuming that there was no branching during the polymerization reaction.

3.4 Show that the free energy of mixing n_1 moles of component 1 and n_2 moles of component 2 to form an ideal solution, ΔG_m, is given by

$$\Delta G_m = RT[n_1\ln X_1 + n_2\ln X_2]$$

where X_1 and X_2 are the mole fractions of components 1 and 2 respectively. You may assume that the increase in number of accessible microstates due to configurational changes, Ω_c, is given by

$$\Omega_c = \frac{(N_1 + N_2)!}{N_1!N_2!}$$

when N_1 molecules of component 1 are mixed with N_2 molecules of component 2. (When N is large $\ln N! = N \ln N - N$).

3.5 Using the expression for the free energy mixing of an ideal solution given in the previous question show that the osmotic pressure ($\prod$) for such a solution in equilibrium with pure solvent is

$$\prod = -\frac{RT}{\bar{V}_1}\ln X_1$$

where $\bar{V}_1$ is the partial molar volume of the solvent in the solution.

   N.B. $-\prod \bar{V}_1 = \mu_1 - \mu_1^0$

where μ_1^0 is the chemical potential of the pure solvent and μ_1 is the chemical potential of the solvent in the solution.

3.6 Show that the partial molar free energy of mixing $\Delta \bar{G}_1$ of a polymer solution is given by

$$\Delta \bar{G}_1 = RT\left[\ln(1 - \phi_2) + \left(1 - \frac{1}{x}\right)\phi_2 + \chi\phi_2^2 \right]$$

given that the free energy of mixing, ΔG_m, is given by

$$\Delta G_m = RT(n_1\ln \phi_1 + n_2\ln \phi_2 + \chi n_1\phi_2)$$

ϕ_1 is the volume fraction of solvent,
ϕ_2 is the volume fraction of polymer,
χ is the polymer–solvent interaction parameter,
x is the ratio of the molar volume of polymer to the molar volume of solvent.

3.7 Find the contour lengths of the following polymer molecules given that the length of the C—C bond is 1.54 Å.

(a) A polyethylene molecule of molar mass 14 000 g mol^{-1}
(b) A polystyrene molecule of molar mass 140 000 g mol^{-1}

Calculate the RMS displacement length of a polyethylene molecule of relative molecular mass 119 980 taking into account fixed bond angles, but not the effects of restricted rotation. Compare this value with its contour length.

3.8 Given that the probability that the distance between the ends of a *freely jointed* polymer chain is r is given by

$$W(r) = (\beta/\pi^{1/2})^3 \exp(-\beta^2 r^2)4\pi r^2$$

$$\text{where} \quad \beta = \frac{1}{l}\left(\frac{3}{2n}\right)^{1/2}$$

and l = link length
       n = number of links

Find:

(a) the most probable value of r,
(b) the root mean square value of r,
(c) the mean value of r,

N.B. For $I(n) = \int_0^\infty x^n \exp(-\alpha x^2)dx$

$I(0) = \frac{1}{2}\pi^{1/2}\alpha^{-1/2}$
$I(1) = \frac{1}{2}\alpha^{-1}$
$I(2) = \frac{1}{4}\pi^{1/2}\alpha^{-3/2}$
$I(3) = \frac{1}{2}\alpha^{-2}$
$I(4) = \frac{3}{8}\pi^{1/2}\alpha^{-5/2}$

3.9 The following experimental osmotic pressure data were obtained for polystyrene in toluene at 25°C.

c/kgm^{-3}	2.56	3.80	5.38	7.80	8.68
$\Pi/\text{mm toluene}$	3.25	5.45	8.93	15.78	18.56

Calculate the number average molar mass of the polystyrene sample and the value of the second virial coefficient for this system at 25°C.

Π osmotic pressure,
c concentration of polymer in toluene,
R molar gas constant,
ρ density of toluene at 25°C, 861.8 kg m^{-3},
g acceleration due to gravity 9.81 m s^{-2}.

3.10 The results obtained from osmosis and light-scattering experiments performed on a series of solutions of a polymer at 298.2 K are given by

c	Π	R_θ
1.30	7.08	0.383×10^{-2}
2.01	11.15	0.558×10^{-2}
3.01	17.10	0.767×10^{-2}
5.49	33.2	1.180×10^{-2}
6.62	41.1	1.325×10^{-2}

c = concentration of polymer in kg m^{-3}
Π = osmotic pressure in mm of solvent
R_θ = Rayleigh ratio in m^{-1}

The values of R_θ were obtained at $\theta = \pi/2$ and there were no interference effects observed.

Calculate $\bar{M}_w$ and $\bar{M}_n$ from this data and hence determine $\bar{M}_w/\bar{M}_n$.

What extra information could be obtained if interference effects had been found?

[Refractive index of solvent $(\bar{n}_0)$ = 1.513. Change of $\bar{n}_0$ with polymer concentration $(d\bar{n}/dc)$ may be taken as 1.11×10^{-4} m^{3}kg^{-1}
Wavelength of radiation (λ) = 4.358×10^{-7} m
Avagadro number (N_A) = 6.023×10^{23} mol^{-1}
π = 3.142 (*not* Π)
Density of solvent (ρ) = 903 kg m^{-3}
Acceleration due to gravity (g) = 9.81 m s^{-2}
Molar gas constant (R) = 8.314 JK^{-1} mol^{-1}]

3.11 During a light scattering experiment the following values of $R_\theta \times 10^4$ m^{-1} were obtained for a series of polystyrene solutions in benzene at 25°C with the detector situated at various angles, θ, to an incident beam of monochromatic light of wavelength 546.1 nm.

$c/\text{kg m}^{-3}$		θ		
	30°	60°	90°	120°
2.00	108.9	71.6	53.2	61.8
1.50	95.1	63.2	45.5	52.0
1.00	75.7	49.0	34.8	39.6
0.50	45.1	28.6	19.7	22.2

Determine the weight average molar mass of the polystyrene, $\bar{M}_w$, and the root mean square displacement length of the polymer chain, $\langle r^2 \rangle^{1/2}$.

In the scattering equation

$$\frac{K(1 + \cos^2\theta)c}{R_\theta} = \frac{1}{M_w}(1 + 2Bc + \ldots)(1 + K'\overline{r^2}\sin^2\theta/2)$$

K is given by $\quad K = 2\pi^2\bar{n}_0^2\left(\frac{d\bar{n}}{dc}\right)^2 \frac{1}{N_A\lambda^4}$

and K' by $\quad K' = \dfrac{8\pi^2}{9\lambda^2}\bar{n}_0^2$

$\pi = 3.1416$
$\bar{n}_o$ = refractive index of benzene, 1.502
$d\bar{n}/dc$ = change of $\bar{n}$ with c, 1.08×10^{-4} m^3 kg^{-1}
λ = wavelength of light, 546.1 nm
N_A = Avagadro number, 6.023×10^{23} mol^{-1}
c = concentration of solution
R_θ = Rayleigh ratio

3.12 The molar masses of a series of homodisperse polystyrene samples were measured by osmosis. The limiting viscosity number $[\eta]$ was found for each sample in toluene at 30°C using an Ubbelohde viscometer and the results are given below.

$M/\text{g mol}^{-1}$	$[\eta]/\text{m}^3 \text{ kg}^{-1}$
76 000	0.0382
135 000	0.0592
163 000	0.0696
336 000	0.1054
440 000	0.1292
556 000	0.165
850 000	0.221

Find K and a for polystyrene in toluene at 30°C.

Various amounts of each of the samples were mixed together and the flow times (t) in the viscometer for different concentrations of this mixture in toluene were determined as shown below.

$c/\text{kg m}^{-3}$	t/s
0	100
1.0	113.5
1.3	118
2.0	128
4.0	160

Calculate the average molar mass of the mixture. What type of average is it?

4 Structure

4.1 Polymer Crystals

It has been recognized for many years that polymer molecules possess the ability to crystallize. The extent to which this occurs varies with the type of polymer and its molecular microstructure. The main characteristic of crystalline polymers that distinguishes them from most other crystalline solids is that they are normally only *semi-crystalline*. This is self-evident from the fact that the density of a crystalline polymer is normally between that expected for fully crystalline polymer and that of amorphous polymer. Also X-ray-diffraction patterns from melt-crystallized polymers are usually in the form of rings superimposed on a diffuse background as shown in Fig. 4.1. This background indicates the presence of a non-crystalline (amorphous) phase and the rings indicate that there is a second phase consisting of small randomly oriented crystallites also present. The degree of crystallinity and the size and arrangement of the crystallites in a semi-crystalline polymer have a profound effect upon the physical and mechanical properties of the polymer and in order to explain these properties fully it is essential to have a clear understanding of the nature of these crystalline regions.

4.1.1 *Crystallinity in polymers*

The crystallization of polymers is of enormous technological importance. Many thermoplastic polymers will crystallize to some extent when the molten polymer is cooled below the melting point of the crystalline phase. This is a procedure which is done repeatedly during polymer processing and the presence of the crystals has an important effect upon polymer properties. There are many factors which can affect the rate and extent to which crystallization occurs for a particular polymer. They can be processing variables such as the rate of cooling, the presence of orientation in the melt and the melt temperature. Other factors include the tacticity and molar mass of the polymer, the amount of chain branching and the presence of any additives such as nucleating agents.

The obvious questions to ask concerning crystallinity in polymers are why and how do polymer molecules crystallize? The answer to the question why is given by consideration of the thermodynamics of the crystallization

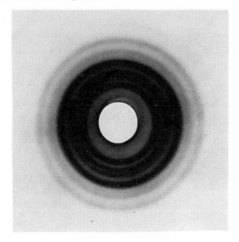

Fig. 4.1 *Flat-plate X-ray diffraction pattern obtained from an isotropic sample of melt-crystallized polypropylene.*

process. The free energy G of any system is related to the enthalpy H and the entropy S by the equation

$$G = H - TS \qquad (4.1)$$

where T is the thermodynamic temperature. The system is in equilibrium when G is a minimum. A polymer melt consists of randomly coiled and entangled chains. This gives a much higher entropy than if the molecules are in the form of extended chains since there are many more conformations available to a coil than for a fully extended chain. The higher value of S leads to a lower value of G. Now if the melt is cooled to a temperature below the melting point of the polymer, T_m, crystallization may occur. There is clearly a high degree of order in polymer crystals as in any other crystalline materials and this ordering leads to a considerable reduction in entropy, S. This entropy penalty is, however, more than offset by the large reduction in enthalpy that occurs during crystallization. If the magnitude of the enthalpy change ΔH_m (latent heat) is greater than that of the product of the melting temperature and the entropy change ($T_m \Delta S_m$) crystallization will be favoured thermodynamically since a lower value of G will result.

As with all applications of thermodynamics, it can only be strictly applied to processes which occur quasi-statically, that is very slowly. Polymers are often cooled rapidly from the melt, particularly when they are being processed industrially. In this situation crystallization is controlled by kinetics and the rate at which the crystals nucleate and grow becomes important. With many crystallizable polymers it is possible to cool the melt so rapidly that crystallization is completely absent and an amorphous glassy polymer results. With these systems crystallization can normally be induced by annealing the amorphous polymer at a temperature between the glass transition temperature and the melting-point, T_m, of the crystals.

The second question that was raised is how do polymer molecules crystallize? A useful way of imagining a molten polymer is as an enormous bowl of tangled wriggling spaghetti. The actual strands of polymer spaghetti would be very much longer than those of the edible pasta. They would have a length to diameter ratio typically of 10 000 to 1 rather than about 100 to 1 for spaghetti. It is quite remarkable that such a random mass could crystallize to give an ordered structure and clearly significant molecular rearrangement and cooperative movement is required for crystallization to occur.

Well-defined examples of polymer crystals can be found on crystallization from dilute polymer solutions. Small isolated lamellar crystals are obtained although detailed measurements have shown that they are still only semi-crystalline. It is thought that in this case the molecules are folded and the fold regions give rise to the non-crystalline component in the crystals. There is a higher degree of perfection in solution-grown polymer crystals than in the melt-crystallized counterparts. The polymer molecules are thought to be in the form of isolated coils in dilute solution and crystallization is not hindered by factors such as entanglements.

Single crystals of many materials occur naturally (e.g. quartz and diamond). For other materials such as metals and semi-conductors the growth of single crystals from the melt is now a routine matter. With polymers it is only recently that true single crystals have been prepared. This can only be done by using the process of solid-state polymerization outlined in Section 2.5.6. The most perfect crystals are obtained with certain substituted diacetylene monomers which polymerize topochemically to give polymer crystals which can be 100 percent crystalline and are of macroscopic dimensions.

4.1.2 *Determination of crystal structure*

Crystalline solids consist of regular three-dimensional arrays of atoms. In polymers, the atoms are joined together by covalent bonds along the macromolecular chains. These chains pack together side-by-side and lie along one particular direction in the crystals. It is possible to specify the structure of any crystalline solid by defining a regular pattern of atoms which is repeated in the structure. This repeating unit is known as the *unit cell* and the crystals are made up of stacks of the cells. In atomic solids such as metals the atoms lie in simple close-packed arrays (like billiard balls) and the structure can be specified by defining simple cubic or hexagonal unit cells containing only a few atoms. The packing units in molecular solids such as polymer crystals are more complicated than in atomic solids. In this case the unit cells are made up of the repeating segments of the polymer chains packed together, often with several segments in each unit cell. Depending upon the complexity of the polymer chain there can be tens or even hundreds of atoms in the unit cell. The spatial arrangement of the atoms is controlled by the covalent bonding within a particular molecular segment, with the polymer segments held together in the crystals by secondary van der Waals or hydrogen bonding. Since the polymer chains lie along one particular direction in the cell and there is only relatively weak secondary bonding between the molecules, the crystals have very anisotropic properties.

When the crystal structure of a molecular compound is analysed both the relative positions of the atoms on the molecular repeat units and the arrangement of these segments in the unit cell must be determined. This three-dimensional structure is normally determined using X-ray diffraction, involving measurement of the positions and intensities of all the X-ray maxima from a single crystal sample. Computer-controlled, four-circle diffractometers are now available which allow the relative positions of all the atoms in crystals with quite complex structures to be determined as a matter of routine in a period of a few days, but such machines required good, relatively large crystals.

For most conventional polymers, large single crystals are not available, but the complete crystal structures of several polydiacetylenes polymerized as macroscopic single crystals in the solid state (Section 2.5.6) have been determined recently. The degree of perfection shown by these polydiacetylene single crystals can be seen in Fig. 4.2. This is a single crystal rotation photograph obtained by using a cylindrical film with the chain direction as the oscillation axis. The diffraction pattern is obtained by rotating the crystal about one particular axis and is in the form of spots lying along parallel layer lines. The lattice parameter in the direction of the rotation axis of the crystal can be determined directly from the layer line spacing. Since the rotation axis coincides with the chain direction then this is, of course, the chain direction repeat conventionally indexed for polymers as c in the unit cell. If the layer lines are indexed $l = 0$ for the equatorial line and $l = 1$ for the first layer etc., then c is simply related to the height, h_l, of the lth layer line above the equatorial line. The crystal lattice can be considered as acting as a diffraction grating as shown in Fig. 4.3(a). The layer lines are then due to discrete differences in path length, $l\lambda$, where λ is the wavelength of the radiation. The angle of deviation of the beam, ϕ, is therefore given

$$\sin \phi = l\lambda/c \qquad (4.2)$$

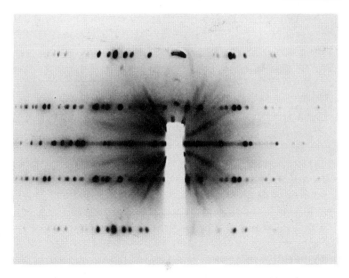

Fig. 4.2 *Single-crystal X-ray oscillation photograph obtained from a polydiacetylene single crystal. The chain direction is vertical.*

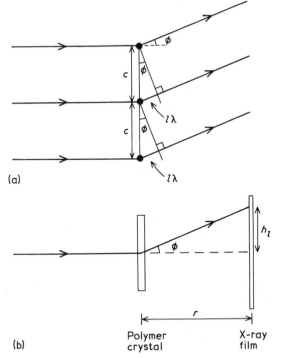

(a)

(b)

Polymer
crystal

X-ray
film

Fig. 4.3 *Schematic illustration of the formation of the layer lines on an oscillation photograph. (a) Diffraction by lattice in one dimension. (b) Experimental arrangement.*

The angle ϕ is related to h_l and the radius of the film, r, through the simple geometrical construction shown in Fig. 4.3(b). This leads to the relation

$$\tan \phi = h_l/r \qquad (4.3)$$

Combining Equations (4.2) and (4.3) gives the lattice parameter c as

$$c = \frac{l\lambda}{\sin \tan^{-1}(h_l/r)} \qquad (4.4)$$

Macroscopic single crystals are a special case and large single crystals of conventional crystalline polymers are not available. Crystal-structure determinations are less precise and the sophisticated methods of crystal-structure determination cannot be used without modification. Instead of using single crystals, samples are normally prepared in the form of highly oriented, drawn or rolled fibres. They can be analysed by obtaining rotation photographs, but flat film X-ray patterns are more usually obtained. A flat film pattern of oriented polypropylene is shown in Fig. 4.4. This type of X-ray photograph is often called a fibre pattern and they are characteristic of oriented fibres of semicrystalline polymers. They are analogous to rotation photographs, but the layer lines are in the form of hyperbolae because the film is flat rather than cylindrical. It is not necessary to rotate the polymer fibres because they are polycrystalline with a spread of crystal orientation about the fibre axis. In polymer fibres the polymer molecules are oriented approximately parallel to the fibre axis and since this is parallel to a crystal axis (normally defined as c), the fibre pattern is equivalent to a rotation photograph about this crystallographic axis for a single crystal sample. It is possible to analyse the layer lines in fibre patterns in a similar way to rotation photographs if certain precautions are taken. Equation (4.4) can only be applied when h_l is measured directly above and below the central spot and if r is taken as the specimen-to-film distance rather than the film radius. Since the chain direction normally coincides with the fibre axis in oriented polymer samples the dimensions of the c axis in the unit cell can be found directly from the height of the lth layer line.

In order to determine other unit cell dimensions it is necessary to assign the crystallographic indices (hkl) to each spot in the fibre pattern. This is often difficult to do for single crystal patterns and there are even greater problems with semi-crystalline polymers. The reflections tend to be in the form of arcs rather than spots because the orientation of the crystals is somewhat imperfect. Also they tend to be rather diffuse because the dimensions of the crystallites are usually small (a few hundred Å) and they contain many lattice imperfections. It is sometimes possible to improve the resolution of the diffraction patterns by using doubly oriented samples. These are fibres which have been rolled as well as stretched. They can have

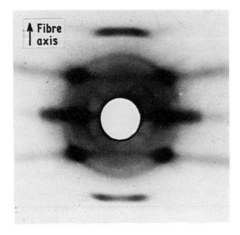

Fig. 4.4 *Flat-plate X-ray diffraction pattern obtained from a sample of oriented polypropylene.*

a three-dimensional texture rather than the normal cylindrical symmetry possessed by fibres.

Once the position and intensities of all the (*hkl*) reflections have been measured as accurately as sample perfection allows, the polymer crystallographer does not normally have sufficient information to determine the crystal structure of the polymer and so he has to resort to a series of educated guesses as to what this structure may be. There is now a considerable amount of experience in this area and some empirical rules have been noted to help with structure determination. The polymer chains are normally in their lowest energy conformations within the crystal as the bonding in the crystal is not sufficiently strong to perturb the nature of the chain. The chains pack into the crystal so as to fill space as efficiently as possible. The van der Waals' radii of the pendant atoms on the chain limit the distance of closest approach of each chain.

It is also possible to call upon other information which is generally available. The structures of chemically similar polymers are often known and these are often useful as a starting-point for structure determination. The stereochemical nature of the polymer chains depends upon the method of synthesis and knowledge of this is important. Spectroscopic techniques can be used to determine the detailed microstructure of the polymer molecules and can give information on the conformation and packing of the chains within the crystals. Also it is possible to determine the lowest energy crystal structure by taking into account inter- and intra-molecular interactions for different structural models.

Once a promising structure has been found it is necessary to compare the measured positions and intensities of the (*hkl*) reflections with those predicted for the structure that is proposed. The agreement that is obtained is never perfect and normally one has to refine the proposed structure until a structure with the best fit to the measured data is obtained. This means that the crystal structures that are quoted below tend to be somewhat idealized and in many cases better fits are almost certainly possible.

4.1.3 *Polymer crystal structures*

The crystal structures of several hundred polymer molecules have now been determined. The structure of a few common polymers are listed in Table 4.1. The name and chemical repeat unit of the polymer are given in column one. The unit cell of any crystal structure can be assigned to one of the seven basic crystal systems (triclinic, monoclinic, etc.) and this is given as the first entry in column two. The packing and symmetry of the polymer segments in the unit cell allows the structure to be assigned to one of the 230 space groups* which are given as the second entry in column two. Polymer molecules pack into crystals either in the form of zig-zags or helices and this is also indicated in column two. The nomenclature used to describe the zig-zag or helix is $A*u/t$, where A represents the class of the helix given by the number of skeletal atoms in the asymmetric unit of the chain. The parameter u represents the number of these units on the helix corresponding to the crystallographic (chain direction) repeat and t is the number of turns of the helix in this crystallographic repeat. This nomenclature can be used to describe both helices and zig-zags. For example, if polyethylene is considered as polymethylene ($-CH_2-)_n$ the $1*2/1$ helix represents a planar zig-zag. Alternatively it could be considered as ($-CH_2-CH_2-)_n$ and the zig-zag chain conformation would be designated $2*1/1$. The unit cell dimensions (Å) are listed in column three in the order a, b, c, and the chain axis is indicated by an asterisk. The angles between the axes of the unit cell α, β and γ are given in order in column four. The number of chain repeat units per unit cell are given in column five and the densities of the crystals determined from the crystal structure are given in the final column.

4.1.4 *Factors determining crystal structure*

There are certain structural requirements that are essential before a polymer molecule can crystallize. It is necessary that the polymer chains must be linear although a limited number of branches on a polymer chain tend to limit the extent of crystallization but do not stop it completely. In a

*See International Tables for X-ray crystallography, Vol. 1.

TABLE 4.1 *Details of the crystal structures of various common polymers (After Wunderlich, reproduced with permission.)*

Macromolecule	Crystal System Space group Mol. Helix	Unit cell axes	Unit cell angles	No. units	ρ_c (g cm^{-3})
Polyethylene I —CH$_2$—	Orthorhombic Pnam 1*2/1	7.418 4.946 2.546*	90° 90° 90°	4	0.9972
Polyethylene II —CH$_2$—	Monoclinic C2/m 1*2/1	8.09 2.53* 4.79	90° 107.9° 90°	4	0.998
Polytetrafluorethylene I —CF$_2$—	Triclinic P1 1*13/6	5.59 5.59 16.88*	90° 90° 119.3°	13	2.347
Polytetrafluoroethylene II —CF$_2$—	Trigonal P3$_1$ or P3$_2$ 1*15/7	5.66 5.66 19.50*	90° 90° 120°	15	2.302
Polypropylene (iso) —CH$_2$—CHCH$_3$—	Monoclinic P2$_1$/c 2*3/1	6.66 20.78 6.495*	90° 99.62° 90°	12	0.946
Polystyrene (iso) —CH$_2$—CHC$_6$H$_5$—	Trigonal R$\bar{3}$c 2*3/1	21.9 21.9 6.65*	90° 90° 120°	18	1.127
Polypropylene (Syndio) —CH$_2$—CHCH$_3$—	Orthorhombic C222$_1$ 4*2/1	14.50 5.60 7.40*	90° 90° 90°	8	0.930
Poly(vinyl chloride) (Syndio) —CH$_2$—CHCl—	Orthorhombic Pbcm 4*1/1	10.40 5.30 5.10*	90° 90° 90°	4	1.477
Poly(vinyl alcohol) (atac) —CH$_2$—CHOH—	Monoclinic P2/m 2*1/1	7.81 2.51* 5.51	90° 97.7° 90°	2	1.350
Poly(vinyl fluoride) (atac) —CH$_2$—CHF—	Orthorhombic Cm2m 2*1/1	8.57 4.95 2.52*	90° 90° 90°	2	1.430
Poly(4-methyl-1-penetene) (iso) —CH$_2$—CH— ‎⁣ CH$_2$—CH(CH$_3$)$_2$	Tetragonal P$\bar{4}$ 2*7/2	20.3 20.3 13.8*	90° 90° 90°	28	0.822
Poly(vinylidene chloride) —CH$_2$—CCl$_2$—	Monoclinic P2$_1$ 4*1/1	6.73 4.68* 12.54	90° 123.6° 90°	4	1.957
1,4-Polyisoprene (*cis*) —CH$_2$—CCH$_3$=CH—CH$_2$—	Orthorhombic Pbac 8*1/1	12.46 8.86 8.1*	90° 90° 90°	8	1.009
1,4-Polyisoprene (*trans*) —CH$_2$—CCH$_3$=CH—CH$_2$—	Orthorhombic P2$_1$2$_1$2$_1$ 4*1/1	7.83 11.87 4.75*	90° 90° 90°	4	1.025

Macromolecule	Crystal System Space group Mol. Helix	Unit cell axes	Unit cell angles	No. units	ρ_c (g cm^{-3})
Polyoxymethylene I	Trigonal	4.471	90°	9	1.491
—CH$_2$—O—	P3$_1$ or P3$_2$	4.471	90°		
	2∗9/5	17.39*	120°		
Polyoxymethylene II	Orthorhombic	4.767	90°	4	1.533
—CH$_2$—O—	P2$_1$2$_1$2$_1$	7.660	90°		
	2∗2/1	3.563*	90°		
Poly(ethylene terephthalate)	Triclinic	4.56	98.5°	1	1.457
—(CH$_2$)$_2$—O—CO—C$_6$H$_4$—	P$\bar{1}$	5.96	118°		
—CO—O—	12∗1/1	10.75*	112°		
Nylon 6, α	Monoclinic	9.56	90°	8	1.235
—(CH$_2$)$_5$—CO—NH—	P2$_1$	17.24*	67.5°		
	7∗2/1	8.01	90°		
Nylon 6, γ	Monoclinic	9.33	90°	4	1.163
—(CH$_2$)$_5$—CO—NH—	P2$_1$/a	16.88*	121°		
	7∗2/1	4.78	90°		
Nylon 6.6, α	Triclinic	4.9	48.5°	1	1.24
—(CH$_2$)$_6$NH—CO—(CH$_2$)$_4$—	P$\bar{1}$	5.4	77°		
—CO—NH—	14∗1/1	17.2*	63.5°		
Nylon 6.6, β	Triclinic	4.9	90°	2	1.25
—(CH$_2$)$_6$NH—CO—(CH$_2$)$_4$—	P$\bar{1}$	8.0	77°		
—CO—NH—	14∗1/1	17.2*	67°		

similar way copolymerization affects the perfection of a polymer chain, but it can be tolerated to a limited extent. One of the most important factors which determines the extent of crystallization and crystal structure is tacticity. In general, isotactic and syndiotactic polymers will crystallize whereas atactic polymers are non-crystalline, although there are notable exceptions to this rule. The most useful way of considering the factors controlling the structure of crystalline polymers is to look at specific examples.

The most widely studied crystalline polymer is *polyethylene*. The lowest-energy chain conformation of this macromolecule can be evaluated from considerations of the conformations of the n-butane molecule (Fig. 3.2). A series of all *trans* bonds produce the lowest-energy conformation in n-butane because of steric interactions and this is reflected in polyethylene where the most stable molecular conformation is the crystal with all *trans* bonds, i.e. the planar zig-zag. These molecules then pack into the orthorhombic unit cell given in Table 4.1. The arrangement of the molecules in the polyethylene crystal structure is illustrated in Fig. 4.5 with the polymer molecules lying parallel to the *c* axis. The molecules are held

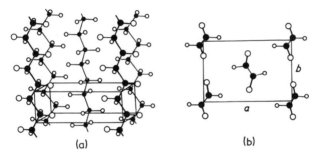

Fig. 4.5 *Crystal structure of orthorhombic polyethylene.* (a) *General view of unit cell.* (b) *Projection of unit cell parallel to the chain direction, c.* (● . . . *carbon atoms,* ○ . . . *hydrogen atoms*).

in position in the cell by the secondary van der Waals bonds between the chain segments. The interactions between the H atoms on the polymer chain determine the setting angle of the molecules in the cell. This is the angle that the molecular zig-zags make with the *a* or *b* axes. Orthorhombic polyethylene (I) is the most stable crystal structure but a monoclinic modification (II) can be formed by mechanical deformation of orthorhombic polyethylene. The molecules are still in the form of a planar zig-zag and the chain direction repeat and crystal density are virtually unchanged. The two structures differ just in the way the molecules are packed in the unit cells. Polymorphism, due to deformation or caused by different crystallization conditions, is not uncommon in polymers and other examples will be discussed below.

The copolymerization of ethylene with small amounts of propylene produces a branched form of polyethylene which is still capable of crystallization. The orthorhombic crystal structure is maintained and the *b* and *c* axes are almost unchanged. On the other hand, the *a* crystal axis undergoes an expansion which depends upon the proportion of —CH_3 groups on the polymer chain.

Polytetrafluoroethylene exists in two crystal modifications. The low-temperature form I is stable below 19°C and above this temperature the trigonal II form is found. The polymer can be considered as an analogue of polyethylene and it would undoubtedly have a similar crystal structure if the F atoms were not just too large to allow a planar zig-zag to be stable. The F atoms are significantly larger than H atoms and there is a considerable amount of steric repulsion when (—CF_2—)$_n$ is in the form of a planar zig-zag. Because of this the molecule adopts a helical conformation which allows the larger F atoms to be accommodated. Below 19°C the molecules are in the form of a 13/6 helix and at higher temperatures they untwist slightly into a 15/7 helix. The two types of polytetrafluoroethylene helices have very smooth molecular profiles and can be considered as

cylinders, as building molecular models will confirm. This smooth profile
has been suggested as being responsible for the good low-friction
properties of this polymer. The molecular cylinders pack very efficiently
into the crystal structures in a hexagonal array, rather like a stack of
pencils, and give rise to unit cells which have hexagonal dimensions (Table
4.1). But they are not assigned hexagonal crystal structures because the
twisted molecules do not have a sufficiently high degree of symmetry.
Poly(vinyl fluoride) is a related polymer which only has 25 percent of the F
atoms possessed by polytetrafluoroethylene. Consequently it does not
suffer from the steric repulsion of the pendant atoms and even the atactic
form adopts a crystal structure similar to polyethylene I (Table 4.1). In fact
the *b* and *c* unit cell dimensions are virtually identical to those of
polyethylene, but the *a* repeat is considerably larger, to accommodate the
bigger F atoms, as for the ethylene–propylene copolymers.

It is found that atactic vinyl polymers $(—CH_2—CHX—)_n$ will not
crystallize when X is large. We have just seen that when X = F
crystallization can still take place. This is also the case for atactic *poly(vinyl
alcohol)*. Since the —OH group is relatively small the molecules crystallize
in the form of a planar zig-zag into a distorted monoclinic version of the
polyethylene structure. In general it is found that vinyl polymers must be
either isotactic or syndiotactic before crystallization can occur. Isotactic
vinyl polymers invariably crystallize with the polymer molecules in a helical
conformation to accommodate the relatively large side groups. For
example, the molecules in the isotactic forms of both *polystyrene* and
polypropylene are in the form of a 3/1 helix as illustrated schematically in
Fig. 4.6. This helix is readily formed by having alternating *trans* and *gauche*
positions along the polymer chain. If the side groups are large and bulky
they require more space than is available in the 3/1 helix and the molecules
form looser helices.

The packing of the helices in isotactic vinyl polymers depends upon the
type of helix present and the nature and size of the side groups. The
packing is dictated by the intermeshing of the side groups, but it is also
complicated by the possibility of having two different kinds of helix both of
which can be either right- or left-handed. The trigonal crystal structure of
isotactic polystyrene (Table 4.1) is typical of many 2*3/1 helices whereas
the monoclinic structure of polypropylene is somewhat exceptional. In
syndiotactic vinyl polymers the conformation of the molecules is again
controlled by the size of the side groups. For example, the Cl atoms in
syndiotactic *poly(vinyl chloride)* are sufficiently small to allow the
backbone to remain in a planar zig-zag.

The final types of crystal structure that will be considered in detail are
those of *nylons* (polyamides) where hydrogen bonding, which can be many
times stronger than van der Waals bonding, controls chain packing. In both

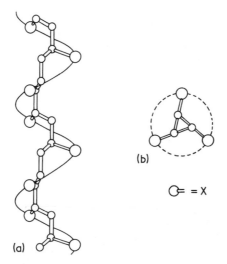

Fig. 4.6 *Schematic representation of the 3/1 helix found for isotactic vinyl polymers of the type* (—CH₂—CHX—)ₙ. (a) *Side view of helix.* (b) *View along helix. The X groups are* —CH₃, —⟨○⟩ , —CH=CH₂ *etc.*

forms of *nylon 6* and *nylon 66* the molecules are in the form of extended planar zig-zags joined together in hydrogen-bonded sheets. An example of this is given for nylon 66 in Fig. 4.7(a) where the hydrogen bonds between oxygen atoms and —NH groups in adjacent molecules are indicated by dashed lines. The α and β forms of this polymer differ in the way in which the hydrogen-bonded sheets are stacked as illustrated in Fig. 4.7. In the α form successive planes are displaced in the c direction whereas in the β form the hydrogen-bonded sheets have alternating up and down displacements. The situation is slightly different in nylon 6 where the α and γ forms differ in the orientation of the molecules in the hydrogen-bonded sheets. They are antiparallel in the α form and parallel to each other in the γ form. This difference in orientation leads to a slightly lower chain direction repeat in the γ form.

4.1.5 *Solution-grown single crystals*

It has been known for many years that crystallization can occur in polymers, but it was not until 1957 that the preparation of isolated polymer single crystals was first reported. They were grown by precipitation from dilute solution. The first crystals that were prepared were of polyethylene but, in the years since, single crystals of nearly every known soluble crystalline polymer have been prepared. Polyethylene crystals are by far

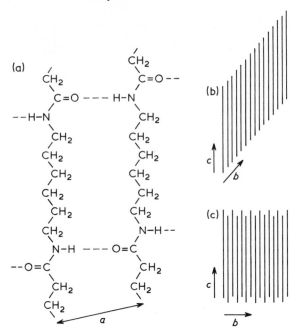

Fig. 4.7 *Schematic representation of the α and β forms of nylon 66. (a) Molecular arrangement of hydrogen-bonded sheets common to both forms. (b) Stacking of sheets in the α form, viewed parallel to the sheets. (c) Stacking of the sheets in the β form. (After Wunderlich).*

the most widely studied examples of polymer single crystals and most of the discussion here will be concerned with polyethylene crystals.

A characteristic feature of solution-grown polymer crystals is that they are small. They can usually just about be resolved in an optical microscope, but they are most readily examined in an electron microscope. Fig. 4.8(a) is an electron micrograph of a polyethylene single crystal which has been grown by precipitation from dilute solution. The crystals are normally precipitated either by cooling a hot solution, which is the most widely used method, or by the addition of a non-solvent. The polyethylene crystal in Fig. 4.8(a) is plate-like (lamellar) and of the order of 10 μm across. The crystal thickness can be measured by shadowing the crystal under vacuum with a heavy metal such as gold, at a known angle. A simple geometrical construction allows the crystal thickness to be calculated from the length of the shadow. For most solution-growth crystals the thickness is found to be of the order of 100 Å. It can also be seen from Fig. 4.8(a) that the lamellar crystals have straight edges which make angles which are characteristic for a given polymer. This regular shape reflects the underlying crystalline nature of the lamellae.

Fig. 4.8 *Solution-grown polyethylene single crystal.* (a) *Electron micrograph of a lamellar crystal (Courtesy of Dr P. Allan).* (b) *Selected area diffraction pattern (Courtesy of Dr I.G. Voigt–Martin).*

The crystals are too small to be studied with X-rays and the relative orientation of the polymer molecules within the lamellar crystals can only be determined by electron diffraction. This can conveniently be done in an electron microscope set up for selected area electron diffraction. An electron diffraction pattern from a polyethylene single crystal with the electron beam perpendicular to the crystal surface is shown in Fig. 4.8(b). The diffraction pattern consists of a regular array of discrete spots and there is no evidence of an amorphous halo. This means that the entity in Fig. 4.8(a) diffracts as good single crystals of non-polymeric solids such as metals. Because of this, the diffraction pattern can be analysed in the standard way and so it can be considered as corresponding to a section of the reciprocal lattice perpendicular to the crystallographic direction parallel to the electron beam. The diffraction pattern in Fig. 4.8(b) can be indexed as that expected for conventional orthorhombic polyethylene (Table 4.1) and it corresponds to a beam direction of [001]. Since the chain direction in orthorhombic polyethylene is c it means that the molecules in polyethylene single crystals are *perpendicular to the crystal surface*. This observation is somewhat surprising since as the polymer molecules are many thousands of Ångstroms long and the crystals only ~100 Å thick it must mean that the polymer molecules *fold back and forth* between the top and bottom surfaces of the lamellae. Although the example given in Fig. 4.8 is for polyethylene it has been found that solution-grown crystals of polymers are normally lamellar and that the molecules are always folded such that they are perpendicular (or nearly perpendicular) to the crystal surface.

A distinctive feature of polymer single crystals is that the crystal edges are normally straight and make well-defined angles with each other. It is also found from electron diffraction that the edge faces correspond to particular crystallographic planes. The crystal shown in Fig. 4.8(a) has four {110} type edge faces. Sometimes polyethylene crystals are truncated and can have {100} edges as well. The different types of polyethylene crystals are shown schematically in Fig. 4.9. It is also found that the crystals are divided into sectors. The crystal illustrated in Fig. 4.9(a) shows four {110}

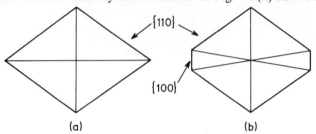

(a)   (b)

Fig. 4.9 *Schematic diagrams of two forms of lamellar single crystals of polyethylene.* (a) *Crystal with* {110} *sectors.* (b) *Crystal with* {100} *sectors as well.*

sectors and that in Fig. 4.9(b) has in addition two extra {100} sectors. This sectorization gives a clue concerning the nature of the folding which will be discussed later.

Measurement of the density and other properties of solution-grown polymer crystals has shown that they are not perfect single crystals. The density of the crystals is always less than the theoretical density which means that non-crystalline material must also be present in the crystal. It is generally thought that the bulk of this non-crystalline component resides in the *fold surfaces* of the crystals. Examination of molecular models of the folding molecule shows that if bonds are to remain unstrained, the folding can be achieved in polyethylene within the space of not less than five bonds with three of them in *gauche* conformations and it is obvious that the packing of folded molecules must be less efficient than that of extended molecules. The non-crystalline nature of the fold surface is also reflected in the observation that single crystals of polyethylene and other polymers are not always flat. Such crystals can often be seen in a collapsed form in the electron microscope containing pleats or corrugations. The true form of the crystals was only found by direct observation of single crystals suspended in a liquid using optical microscopy. It was shown that some polyethylene crystals were in the form of a hollow pyramid as shown schematically in Fig. 4.10. The fold surface in pyramidal polyethylene single crystals has been shown to be close to the {312} plane of the polyethylene unit cell. The exact form of the single crystals depends upon several factors such as the temperature at which the crystals were grown. Pyramidal crystals tend to grow at high temperatures whereas corrugated crystals form at lower crystallization temperatures.

There has been considerable argument over the years concerning the way in which folding occurs and the nature of the fold plane. The different models that have been suggested to account for folding in polymer crystals are illustrated schematically in Fig. 4.11. The models range from random re-entry ones, where a molecule leaves and re-enters a crystal randomly, to adjacent re-entry models, whereby molecules leave and re-enter the crystals in adjacent positions. Two particular adjacent re-entry models

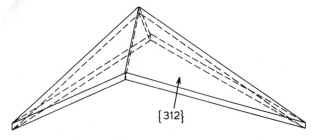

Fig. 4.10 *Schematic representation of a pyramidal polyethylene single crystal. (After Schultz).*

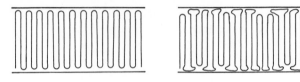

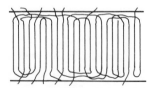

Fig. 4.11 *Schematic illustrations of the different types of folding suggested for polymer single crystals.* (a) *Adjacent re-entry with sharp folds.* (b) *Adjacent re-entry with loose folds.* (c) *Random re-entry or 'switch-board' model.*

have been suggested where the folds are either regular and tight or irregular and of variable length.

The main consensus of opinion appears to be that in single crystals grown from dilute solution the molecules undergo adjacent re-entry. This leads to there being a crystallographic plane on which the molecules fold, normally known as a *fold plane*. In the polyethylene crystals drawn in Fig. 4.9(a) it is thought that the fold planes are parallel to the crystal edges. This means that the fold planes must be of the type {110} and in truncated crystals (Fig. 4.9 (b)) there are thought to be sectors of {100} folding as well. The fold surface itself is thought to be fairly irregular, consisting of a mixture of loose and tight folds and chain ends.

4.1.6 *Solid-state polymerized single crystals*

It has recently become possible to prepare true 100 percent crystalline polymer crystals by the process of solid-state polymerization described in Section 2.5.6. Examples of solid-state polymerized polydiacetylene single crystals in the form of both lozenges and fibres are shown in Fig. 4.12. The crystals have regular facets and are of macroscopic dimensions and so similar to single crystals that can be prepared for non-polymeric materials such as metals or ceramics. The polymer molecules are parallel to the fibre axis in the fibrous crystals. A typical X-ray rotation photograph of a similar polydiacetylene crystal has been given in Fig. 4.2 and it indicates that in such crystals the molecules are in a chain-extended conformation. Density measurements have shown that the crystals have their theoretical densities calculated from a knowledge of the crystal structure and so it is concluded that the crystals are virtually defect-free with chain-folding being

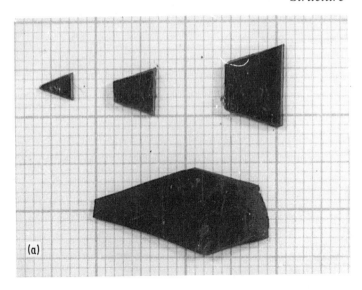

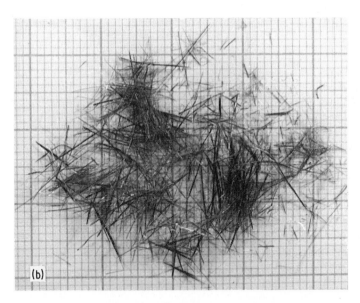

Fig. 4.12 *Examples of polydiacetylene single crystals in the form of* (a) *lozenges and* (b) *fibres. The crystals are placed on graph paper with mm spacing. (Crystals supplied by Dr D. Bloor).*

completely absent. Such crystals have allowed considerable advances to be made in the understanding of the fundamental physical properties of crystalline polymers. In particular, important new aspects of the deformation and fracture of polymer crystals have been revealed.

4.2 Semi-crystalline polymers

The isolated chain-folded lamellar single crystals described in the previous section are only obtained by crystallization from dilute polymer solutions. As the solutions are made more concentrated the crystal morphologies change and more complex crystal forms are obtained. In dilute solution the polymer coils are isolated from each other, but as it becomes more concentrated the molecules become entangled. Particular chains, therefore, have the probability of being incorporated in more than one crystal giving rise to inter-crystalline links. Crystals grown from concentrated solutions are often found either as lamellae with spiral overgrowth as shown in Fig. 4.13 or as aggregates of lamellar crystals. Under certain conditions the crystals can be even more complex showing either twinned or dendritic structures similar to the morphologies found in non-polymeric crystalline solids.

This particular section is concerned with the structures that are formed principally by crystallization from the melt. This is not fundamentally

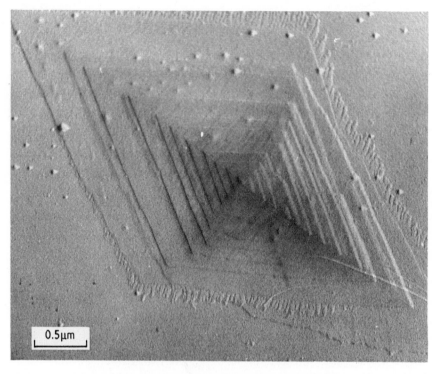

0.5μm

Fig. 4.13 *Electron micrograph of a polyethylene crystal grown from CCl₄ at 80°C showing a large growth spiral. The micrograph was obtained by replication and the replica has been shadowed with a heavy metal. (Courtesy of Dr G. Lieser).*

different from solution crystallization especially when the solutions are concentrated. In the melt chain entanglements are of extreme importance and consequently the crystals that form are more irregular than those obtained from dilute solution. A feature common to both melt and solution crystallization is that the solid polymer is only *semi-crystalline* and contains both crystalline and amorphous components.

4.2.1 *Spherulites*

If a melt-crystallized polymer is prepared in the form of a thin film, either by sectioning a bulk sample or casting the film directly, and then viewed in an optical microscope between crossed polars a characteristic structure is normally obtained. It consists of entities known as *spherulites* which show a characteristic Maltese cross pattern in cross-polarized light. A typical spherulite is shown in Fig. 4.14. They form by nucleation at different points in the sample and grow as spherical entities. The growth of the spherulites stops when impingment of adjacent spherulites occurs. At first sight they look very similar to grains in a metal being roughly of the same dimensions and growing in a similar way. However, the grains in a metal are individual single crystals whereas detailed investigation of polymeric spherulites shows that they consist of numerous crystals radiating from a central nucleus.

There are many examples of spherulites in other crystalline solids,

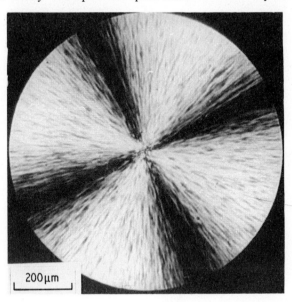

Fig. 4.14 *A single spherulite growing in isotactic polystyrene. Optical micrograph taken in polarized light. (Courtesy of Dr D. Tod).*

particularly in naturally occurring mineral rocks. The typical extinction patterns seen in polarized light are due to the orientation of the crystals, within the spherulites. Analysis of the Maltese cross patterns has indicated that the molecules are normally aligned tangentially in polymer spherulites. It has been shown that the *b* crystals axis is radial in polyethylene spherulites and the *a* and *c* crystallographic directions are tangential. In some polymer samples the spherulites are sometimes seen to be ringed. This has been attributed to a regular twist in the crystals.

The detailed structure of polymer spherulites can best be studied by electron microscopy because of the small size of the individual crystals. Fig. 4.15 gives an electron micrograph of a spherulite in polyisoprene. The non-crystalline areas of this polymer can be stained with osmium tetroxide and in the micrograph shown, the crystals appear light and the amorphous regions are dark. The spherulite can be seen to consist of crystals radiating out from a central nucleus. Close examination of the micrograph shows that the individual crystals are lamellar. When they are viewed edge-on they appear as fine lines of the order of 100 Å thick and appear as broad sheets when the lamellar surfaces are parallel to the plane of the polymer film. The dimensions of these crystal lamellae are, of course, similar to

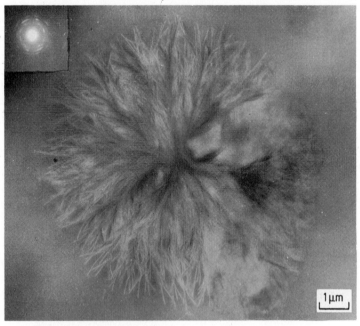

Fig. 4.15 *Electron micrograph of a thin film of polyisoprene. The film has been stained with osmium tetroxide which causes the non-crystalline areas to be dark whereas the crystals remain bright. (Courtesy of Dr C. K. L. Davies).*

those of solution-grown lamellar single crystals. Further evidence for the lamellar nature of the crystals in polymer spherulites has been obtained by examination of the debris left after the degradation of spherulitic polyethylene and polypropylene in nitric acid. The acid selectively oxidizes the inter-crystalline material and the crystal fragments that are left are found by electron microscopy to be lamellar. Selected area electron diffraction has shown that the polymer molecules are oriented approximately normal to the lamellar surfaces. This is further evidence of the similarity between melt-crystallized and solution-grown lamellae.

Electron microscopy has also revealed the twist of the crystals in the spherulites that was suspected from optical microscopy. Surface replicas taken from spherulitic polymers have clearly shown that the lamellae twist and this is also evident from the micrographs in Fig. 4.15, but the reason for this twisting is not yet understood. A possible structure of the twisted lamellae in polyethylene spherulites is illustrated schematically in Fig. 4.16.

An obvious question which arises is that of the nature of the conformation of the molecules within the crystals. There is accumulated evidence that inter-crystalline links must exist between the lamellae in spherulitic polymers. The similarity between the morphologies of and molecular orientation in these lamellae and solution-grown crystals clearly implies that a certain degree of chain folding must also take place. However, definitive experiments in this area have been extremely difficult to perform. The best evidence of the conformation of polymer molecules in melt-crystallized polymers has come from recent neutron-scattering studies of mixtures of a homopolymer and deuterated version of the same polymer. The results of such experiments are open to different interpretations but the consensus of opinion appears to be that the structure is similar to that shown schematically in Fig. 4.17. There are thought to be small segments of fairly sharp chain-folding and a large number of inter-crystalline links with a given molecule shared between at least two crystals.

4.2.2 *Degree of crystallinity*

Melt-crystallized polymers are never completely crystalline. This is because there are an enormous number of chain entanglements in the melt and it is impossible for the amount of organization required to form a 100 percent crystalline polymer to take place during crystallization. The degree of crystallinity is of great technological and practical importance and several methods, which do not always produce precisely the same results, have been devised to measure it.

The crystallization of a polymer from the melt is accompanied by a reduction in specimen volume due to an increase in density. This is because the crystals have a higher density than the molten or non-crystalline

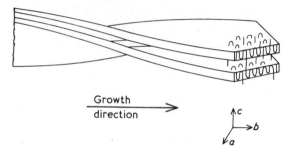

Growth direction →

↑c
↓a →b

Fig. 4.16 *Schematic representation of a possible model for twisted lamellae in spherulitic polyethylene showing chain-folds and intercrystalline links.*

polymer and this effect provides the basis of the *density* method for the determination of the degree of crystallinity. The technique relies upon the observation that there is a relatively large and measurable difference (up to 20 percent) between the densities of the crystalline and amorphous regions of the polymer. This method can yield both the volume fraction of crystals ϕ_c and the mass fraction x_c from measurement of sample density ρ.

If V_c is the volume of crystals and V_a the volume of amorphous material then the total specimen volume, V, is given by

$$V = V_c + V_a \tag{4.5}$$

Similarly the mass of the specimen W is given by

$$W = W_c + W_a \tag{4.6}$$

where W_c and W_a are the masses of crystalline and amorphous material in the sample respectively. Since density ρ is mass per volume then it follows from Equation (4.6) that

$$\rho V = \rho_c V_c + \rho_a V_a \tag{4.7}$$

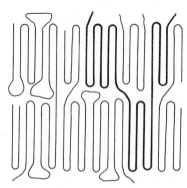

Fig. 4.17 *A schematic representation of one possible model for the conformation of a deuterated polymer molecule (heavy line) in a matrix of protonated molecules (light lines).*

substituting for V_a from Equation (4.5) into Equation (4.7) and rearranging leads to

$$\frac{V_c}{V} = \left(\frac{\rho - \rho_a}{\rho_c - \rho_a}\right) = \phi_c \qquad (4.8)$$

since ϕ_c is equal to the volume of crystals divided by the total specimen volume. The mass fraction x_c of crystals is similarly defined as

$$x_c = W_c/W = \rho_c V_c/\rho V \qquad (4.9)$$

and combining Equations (4.8) and (4.9) gives

$$x_c = \frac{\rho_c}{\rho}\left(\frac{\rho - \rho_a}{\rho_c - \rho_a}\right) \qquad (4.10)$$

This equation relates the degree of crystallinity x_c to the specimen density ρ and the densities of the crystalline and amorphous components.

The density of the polymer sample can be readily determined by *flotation* in a density-gradient column. This is a long vertical tube containing a mixture of liquids with different densities. The column is set up with a light liquid at the top and a steady increase in density towards the bottom. It is calibrated with a series of floats of known density. The density of a small piece of the polymer is determined from the position it adopts when it is dropped into the column. The density of the crystalline regions ρ_c can be calculated from knowledge of the crystal structure (Table 4.1). The term ρ_a can sometimes be measured directly if the polymer can be obtained in a completely amorphous form, for example by rapid cooling of a polymer melt. Otherwise it can be determined by extrapolating either the density of the melt to the temperature of interest or that of a series of semi-crystalline samples to zero crystallinity. Equations (4.8) and (4.10) are only valid if the sample contains no holes or voids, which are often present in moulded samples, and if ρ_a remains constant. In practice, since the packing of the molecules in the amorphous areas is random it is likely that ρ_a will be different in specimens which have had different thermal treatments.

A powerful method of determining the degree of crystallinity is *wide-angle X-ray scattering* (WAXS). A typical WAXS curve for a semi-crystalline polymer is given in Fig. 4.18 where the intensity of X-ray scattering is plotted against diffraction angle, 2Θ. The relatively sharp peaks are due to scattering from the crystalline regions and the broad underlying 'hump' is due to scattering from non-crystalline areas. In principle, it should be possible to determine the degree of crystallinity from the relative areas under the crystalline peaks and the amorphous hump. In practice it is often difficult to resolve the curve into areas due to each phase. The shape of the amorphous hump can be determined from the

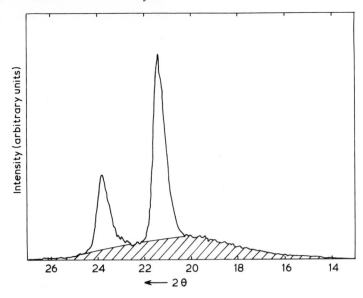

Fig. 4.18 *WAXS curves for a medium-density polyethylene. The intensity of scattering is plotted as a function of 2Θ. The amorphous hump is shaded.*

WAXS curve for a completely amorphous sample, obtained by rapidly cooling a molten sample. For certain polymers such as polyethylene this can be difficult or even impossible to do and the amorphous scattering can only be estimated. Also corrections should be made for disorder in the crystalline regions which can give rise to a reduction in area of the sharp peaks. Nevertheless, an approximate idea of the crystallinity can be obtained from the simple construction shown in Fig. 4.18. A horizontal base-line is drawn between the extremities of the scattering curve to remove the background scattering. The amorphous hump is then traced in either from a knowledge of the scattering from a purely amorphous sample or by estimating from experience with other polymers. The mass fraction of crystals x_c is then given to a first approximation by

$$x_c = A_c/(A_a + A_c) \tag{4.11}$$

where A_a is the area under the amorphous hump and A_c is area remaining under the crystalline peaks. A more sophisticated analysis involving resolution of the curve in Fig. 4.18 into three distinct components, two crystalline peaks and one amorphous hump, lying on the same baseline can produce slightly more accurate results.

Although the density and WAXS methods of determining the degree of crystallinity of semi-crystalline polymers are by far the most widely used techniques several others have been used with varying degrees of success.

They include physical methods such as the determination of electrical resistivity, enthalpy or specific heat capacity of the semi-crystalline polymer which require knowledge of the values of these different parameters for both the crystalline and amorphous phases. Spectroscopic methods such as n.m.r. and infrared spectroscopy which have been outlined in Section 3.6 have also been employed. In general there is found to be an approximate correlation between the different methods of measurement employed although the results often differ in detail.

4.2.3 *Crystal thickness and chain extension*

It is known that lamellar crystalline entities are obtained by the crystallization of polymers from both the melt and dilute solution. The most characteristic dimension of these lamellae is their thickness and this is known to vary with the crystallization conditions. The most important variable which controls the crystal thickness is the *crystallization temperature* and the most detailed measurements of this have been made for solution-crystallized lamellae. Several methods of crystal thickness determination have been used. They include shadowing lamellae deposited upon a substrate and measuring the length of the shadow by electron microscopy, measuring the thickness by interference microscopy and determining it by using small-angle X-ray scattering (SAXS) from a stack of deposited crystals. When the lamella single crystals are allowed to settle out from solution they form a solid mat of parallel crystals. The periodicity of the stack in the mat is related to the crystal thickness and the stack scatters X-rays in a similar way to the atoms in a crystal lattice. Bragg's law is obeyed

$$n\lambda = 2d \sin \Theta \tag{4.12}$$

where n is an integer, λ the wavelength of the radiation, d is the periodicity of the array and Θ the diffraction angle. Since d is the order of 100 Å then by using X-rays ($\lambda \sim 1$ Å) Θ will be very small and the diffraction maxima are very close to the main beam.

The variation of lamellar thickness with crystallization temperature for polyoxymethylene crystallized from a variety of solvents is shown in Fig. 4.19(a). For a given solvent the lamellar thickness is found to increase with increasing crystallization temperature. This behaviour is typical of many crystalline polymers such as polyethylene or polystyrene for which the length of the fold period increases as the crystallization temperature is raised. The different behaviour in the various solvents displayed in Fig. 4.19(a) is again typical of solution crystallization and Fig. 4.19(b) is a plot of the lamellar thickness against the reciprocal of the supercooling $\Delta T(= T_s - T_c)$ which gives a master-curve of all the data in Fig. 4.19(a). This

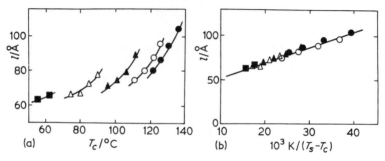

Fig. 4.19 (a) *Dependence of lamellar thickness, l, upon crystallization temperature, T_c for polyoxymethylene crystallized from different solvents. (b) Master curve for all the data in (a) plotted against the reciprocal of the supercooling. Solvents:* ■ . . . *phenol,* △ . . . *m-Cresol,* ▲ . . . *furfuryl alcohol,* ○ . . . *benzyl alcohol,* ● . . . *Acetophenone. (After Magill).*

indicates that the crystallization process is controlled by the difference between the crystallization temperature T_c and solution temperature T_s, rather than the actual temperature of crystallization. This has important implications for the theories of polymer crystallization which will be discussed later (Section 4.3.3).

The lamellar thickness for melt-crystallized polymers is rather more difficult to determine to any great accuracy. SAXS curves can be obtained from such samples but they tend to be much weaker and less well-defined than their counterparts from single crystal mats. If the lamellar thickness is sufficiently large (>500 Å) it is possible to measure it by electron microscopic examination of sections of bulk samples. The variation of lamellar thickness with crystallization temperature for melt-crystallized polyethylene is shown in Fig. 4.20. The behaviour is broadly similar to that of solution-crystallized polymers in that the lamellar thickness increases as the crystallization temperature is raised. In fact it is found that at high supercoolings the thickness of solution and melt-crystallized lamellae are comparable. However, at low supercoolings there is a rapid rise in lamellar thickness with crystallization temperature which is not encountered with solution crystallization. It is thought that this is due to an isothermal thickening process whereby the crystals become thicker with time when held at a constant temperature close to the melting temperature. On the other hand there is very little effect of crystallization time upon the thickness of solution-grown lamellae.

As might be expected, the *molar mass* of the polymer only has a strong effect upon lamellar thickness when the length of the molecules is comparable to the crystal thickness. Normally the molecular length is many times greater than the fold length and so even large increases in molar mass produce only correspondingly small increases in lamellar thickness. Another variable which has a profound effect upon the crystallization of

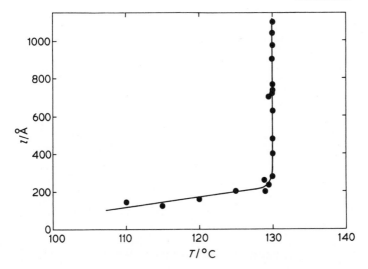

Fig. 4.20 *Lamellar thickness as a function of crystallization temperature for isothermally melt-crystallized polyethylene. (After Wunderlich).*

polyethylene and a few other polymers from the melt is *pressure*. The melting temperature of polymers increases rapidly with pressure and it is found that crystallization of polyethylene at low supercoolings at pressures above about 3 kbar can produce crystal lamellae in which some of the molecules are in fully extended conformations. Crystals of up to 10 μm thick have been reported for high molar mass polymer. This chain-extended type of morphology can also be found in polytetrafluoroethylene crystallized slowly at ambient pressure and a micrograph of a replica of a fracture surface of this polymer is given in Fig. 4.21. The micrograph shows long tapering lamellae containing striations approximately perpendicular to the lamellar surfaces. The striations are caused by steps on the fracture surface and define the chain direction within each lamella. In the chain-extended lamellae seen in Fig. 4.21 it is thought that the molecules fold backwards and forwards several times as it is known that the molecular length is somewhat greater than the average crystal thickness. Chain-extended morphologies can have extremely high degrees of crystallinity, sometimes in excess of 95 percent. However, their mechanical properties are extremely disappointing. Chain-extended polyethylene is relatively stiff but absence of inter-crystalline link molecules causes it to be very brittle and it crumbles like cheese when deformed mechanically.

The effect of the different variables upon the fold length and lamellar thickness of polyethylene is summarized in Table 4.2. The number of positive or negative signs indicates the magnitude of the effect in each case.

Fig. 4.21 *Electron micrograph of a replica of a fracture surface of melt-crystallized polytetrafluorethylene showing chain-extended crystals. The striations define the chain-direction in each crystal. (Young, Chapter 7 in 'Developments in Polymer Fracture' edited by E. H. Andrews, Applied Science Publishers Ltd, 1979, reproduced with permission).*

4.2.4 Crystallization with orientation

If polymer melts or solutions are crystallized under stress morphologies which are strikingly different from those obtained in the absence of stress are found. This observation has important technological consequences as many polymers are processed in the form of fibres, injection moulded or extruded and all of these processes involve the application of stress to the material during crystallization.

Some of the earliest observations on the effect of stress upon crystallization were made by Andrews upon thin stretched films of natural rubber. It was found that crystals formed transverse to the direction of the applied stress as shown in Fig. 4.22(a) for isotactic polystyrene. The

TABLE 4.2 *Effect of changing supercooling ΔT, pressure p, molar mass M and time t upon the lamellar thickness of polyethylene (After Wunderlich, reproduced with permission.)*

	ΔT	p	M	t
Solution crystallized	$--$	0	0	0
Melt crystallized (High T)	$--$	$+++$	0	$+$
Melt crystallized (Low T)	$---$	$++++$	$+$	$++$
Annealed	$--$	$++$	$(+)$	$++$

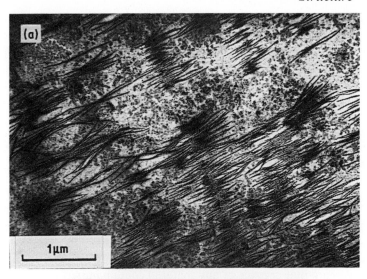

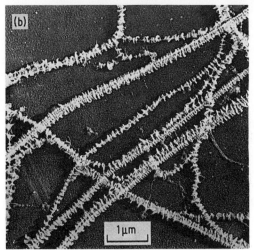

Fig. 4.22 *Electron micrographs showing the effect of applied stress upon the morphologies of crystalline polymers. (a) Row-nucleated structure in melt-crystallized isotactic polystyrene. (Courtesy of Dr J. Petermann.) (b) Shish-kebabs obtained from a stirred solution of polyethylene. (Courtesy of Dr A. Pennings).*

crystals are found to nucleate on a central backbone and grow in perpendicular directions. Similar row-nucleated structures have since been found on crystallizing oriented melts of many other polymers.

A related structure is found on cooling rapidly stirred polymer solutions. Fig. 4.22(b) shows the type of morphology which results from the rapid stirring of a polyethylene solution and such structures have been termed *shish-kebab morphologies*. They consist of a central backbone (shish) with lamellar overgrowth (kebabs). It has been found by electron diffraction that in both the backbone and the lamellar overgrowths the polymer molecules are parallel to the shish-kebab axis. It is thought that the molecules are folded in the overgrowth, but that in the backbone there is a considerable degree of chain extension, although some folds are also thought to be present.

Oriented polymer structures are extremely important as commercial materials since they allow the high unidirectional strength of the polymer molecules to be exploited.

4.2.5 *Defects in crystalline polymers*

It is clear that crystalline polymers are by no means perfect from a structural viewpoint. They contain crystalline and amorphous regions and probably also areas which are partially disordered. It has been recognized for many years that crystals of any material contain imperfections such as dislocations or point-defects and there is no fundamental reason why even the relatively well-ordered crystalline regions of crystalline polymers should not also contain such defects. In fact, it is now known that polymer crystals contain defects which are similar to those found in other crystalline solids, but in considering such imperfections it is essential to take into account the macromolecular structure of the crystals.

The presence of defects in a crystal will give rise to broadening of any diffraction maxima such as the rings seen in the X-ray diffraction pattern in Fig. 4.1. In principle, the degree of broadening can be used to determine the amount of disorder in the crystal caused by the presence of defects. However, broadening of the diffraction maxima can also be caused by the crystals having a finite size and since polymer crystals are normally relatively small most of the observed broadening is usually caused by the size effect. It is possible to estimate crystal size from the degree of broadening. If the crystals have a mean dimension normal to the (hkl) planes of L_{hkl} then the broadening of the X-ray diffraction peaks δs due to the size effect is given by*

$$\delta_s = \frac{1}{L_{hkl}} \tag{4.13}$$

The parameter δs is the breadth of the diffraction peak at half-height and is expressed in units of s, where s is defined through the equation

$$s = (2/\lambda) \sin \theta \tag{4.14}$$

---

*e.g. J.M. Schultz, *Polymer Materials Science*.

where Θ is the diffraction angle and λ is the wavelength of the X-ray used. It is possible to derive similar equations relating δs to the broadening due to defects but since it is quite difficult to measure δs with sufficient accuracy to usefully exploit the analysis, the reader is referred to more advanced texts for further details*.

In considering the types of defects that may be present in polymer crystals it is best to look at the different types separately. The defects which have been found to exist or postulated are as follows.

(1) *Point defects*

The presence of point defects such as vacancies or interstitial atoms or ions is well-established in atomic and ionic crystals. The situation is somewhat different in macromolecular crystals where the types of point defects are restricted by the long-chain nature of the polymer molecules. It is relatively easy to envisage the types of defects that may occur. They could include chain ends, short branches, folds or copolymer units. There is accumulated evidence that the majority of this chain disorder is excluded from the crystals and incorporated in non-crystalline regions. However, it is also clear that at least some of it must be present in the crystalline areas.

Another type of point defect which could occur and is unique for macromolecules is a molecular kink. For example, a kink could be generated in a planar zig-zag chain consisting of all-*trans* bonds by making two of the bonds *gauche*. The sequence of five consecutive bonds would be . . . tg^+tg^-t . . . rather than . . . $ttttt$ . . . Such a kinked chain could be incorporated into a crystal with only a relatively slight local distortion. One important aspect of such a kink defect is that in polyethylene it enables an extra —CH_2— group to be incorporated in the crystal and motion of the defect along the chain allows the transport of material across the crystal. It has been suggested that the motion of such defects may be the mechanism whereby crystals thicken during annealing (Section 4.3.4).

(2) *Dislocations*

It is found that in metal crystals slip can take place at stresses well below the theoretically calculated shear stress. The movement of line defects known as dislocations was invoked to account for this observation and with the advent of electron microscopy the presence of such defects was finally proved. The two basic types of dislocations found in crystals, screw and edge, are shown in Fig. 4.23. The dislocation is characterized by its line and Burgers vector. If the Burgers vector is parallel to the line it is termed a *screw dislocation* and if it is perpendicular it is called an *edge dislocation*. In general the Burgers vector and dislocation line may be at any angle as

*e.g. J.M. Schultz—*Polymer Materials Science*.

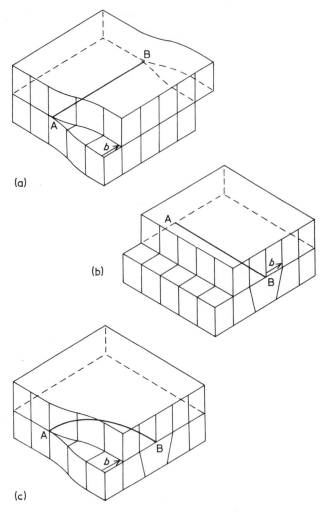

Fig. 4.23 *Schematic representation of dislocations in crystals. (a) Screw dislocation. (b) Edge dislocation. (c) Mixed screw and edge dislocation. The line of dislocation (AB) and the Burgers vector (b) are indicated in each case. (After Kelly and Groves).*

shown in Fig. 4.23(c). When this is the case the dislocation has both edge and screw components and it is said to be *mixed*. The dislocations are drawn in Fig. 4.23 for an atomic crystal but there is no *a priori* reason why they cannot exist in polymer crystals. However, there are several important differences between polymeric and atomic crystals which lead to restrictions upon the type of dislocations that may occur in polymer crystals. The differences are as follows. (a) There is a high degree of anisotropy in the bonding in polymer crystals. The covalent bonding in the chain direction is very much stronger than the relatively weak transverse secondary bonding.

(b) Polymer molecules tend to resist bond bending and dislocations which tend to do this are not favoured.

(c) The crystal morphology can also affect the stability and motion of the dislocations. The occurrence of chain folding and the fact that polymer crystals are often relatively thin may affect matters.

The possibility of dislocations being involved in the deformation of polymer crystals has received considerable attention over recent years. This will be considered later (Section 5.5.5) and this section will be concerned with dislocations that may be present in undeformed crystals. The most obvious example of the occurrence of dislocations in polymer crystals is seen in Fig. 4.13 where the crystals contain growth spirals. It is thought that there is a screw dislocation with a Burgers vector of the size of the fold length ($\sim$100 Å) at the centre of the spiral. Such a dislocation is illustrated schematically in Fig. 4.24 and the Burgers vector and dislocation line are parallel to each other and the chain direction.

Dislocations with Burgers vectors of a similar size to the unit cell dimensions have been deduced from Moiré patterns obtained from overlapping lamellar single crystals observed in the electron microscope. The Moiré pattern is caused by double diffraction when the two crystals are slightly misaligned and they allow the presence of defects in the crystals to be established. This technique has allowed edge dislocations with Burgers vectors perpendicular to the chain direction to be identified in several different types of polymer crystals. Such dislocations are thought to be due to the presence of a terminated fold plane in crystals containing molecules folding by adjacent re-entry. If this is so it seems likely that the dislocation will not be able to move without breaking bonds at the fold surface.

With recent advances and improvements in the instrumentation of electron microscopy there have been numerous reports of the observation of dislocations in atomic, ionic and molecular crystals through direct imaging of the crystal lattice. However, this technique is extremely difficult to apply to polymer crystals because of the general susceptibility of polymer crystals to radiation damage in the electron beam.

(3) *Other defects*
There are many other possible types of imperfections in crystalline polymers as well as point-defects and dislocations. The fold surface and chain folds can be considered as defects. It is also possible to consider the 'grain' boundaries between crystals as areas containing defects.

The effect of all the many types of defects discussed above is to introduce disorder into the polymer crystals. The result of their presence is to deform and distort the crystal lattice and produce broadening of the X-ray

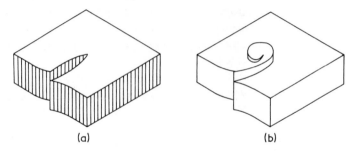

Fig. 4.24 *Schematic representation of screw dislocations with Burgers vectors parallel to the chain direction in lamellar polymer crystals. (a) Illustration of relation between the dislocation line and the chain direction (indicated by striations). (b) Screw dislocation leading to growth spiral.*

diffraction maxima (Fig. 4.1). This type of disorder in a crystalline structure has been termed *paracrystallinity*. The polymer paracrystal is modelled by allowing the unit cell dimensions to vary from one cell to another and this allows the broadening of the X-ray patterns, over and above that expected from crystal size effects, to be explained. The concept of paracrystallinity has proved extremely useful in characterizing the structures of semicrystalline polymers, but there is clearly more yet to be learned about the detailed structure of these materials.

4.3 Crystallization and melting

4.3.1 *General considerations*

Crystallization is the process whereby an ordered structure is produced from a disordered phase, usually a melt or dilute solution, and melting can be thought of as being essentially the opposite of this process. When the temperature of a polymer melt is reduced to the melting temperature there is a tendency for the random tangled molecules in the melt to become aligned and form small ordered regions. This process is known as *nucleation* and the ordered regions are called nuclei. These nuclei are only stable below the melting temperature of the polymer since they are disrupted by thermal motion above this temperature. The second step in the crystallization process is *growth* whereby the crystal nuclei grow by the addition of further chains. Crystallization is therefore a process which takes place by two distinct steps, nucleation and growth which may be considered separately.

Nucleation is classified as being either homogeneous or heterogeneous. During homogeneous nucleation in a polymer melt or solution it is envisaged that small nuclei form randomly throughout the melt. Although this process has been analysed in detail from a theoretical viewpoint it is thought that, in the majority of cases of crystallization from polymer melts and solutions, nucleation takes place heterogeneously on foreign bodies such as dust particles or the walls of the containing vessel. The number of nuclei formed depends, when all other factors are kept constant, upon the temperature of crystallization. At low undercoolings nucleation tends to be sporadic and during melt crystallization a relatively small number of large spherulites form. On the other hand when the undercooling is increased many more nuclei form and a large number of small spherulites are obtained.

The growth of a crystal nucleus can, in general, take place either one, two, or three dimensionally with the crystals in the form of rods, discs or spheres respectively. The growth of polymer crystals takes place by the incorporation of the macromolecular chains within crystals which are normally lamellar. Crystal growth will then be manifest as the change in the lateral dimensions of lamellae during crystallization from solution or the change of spherulite radius during melt crystallization. An important experimental observation which simplifies the theoretical analysis is that the change in linear dimensions of the growing entities at a given temperature of crystallization is usually linear with time. This means that the spherulite radius, r, will be related to time, t, through an equation of the form

$$r = \nu t \tag{4.15}$$

where ν is known as the growth rate. This equation is usually valid until the spherulites becomes so large that they touch each other. The change in the linear dimensions of lamellae tends to obey an equation of a similar form for solution crystallization. The growth rate ν is strongly dependent upon the crystallization temperature as shown in Fig. 4.25. It is found that the growth rate is relatively low at crystallization temperatures just below the melting temperature of the polymer, but as the supercooling is increased there is a rapid increase in ν. However, it is found that eventually there is a peak in ν and further lowering of the crystallization temperature produces a reduction in ν. Also, for a given polymer it is found that the growth rate at a particular temperature depends upon the molar mass with ν increasing as M is reduced (Fig. 4.25). The reason for the peak in ν is thought to be due to two competing effects. The thermodynamic driving force for crystallization will increase as the crystallization temperature is lowered. But as the temperature is reduced there will be an increase in viscosity and transport of material to the growth point will be more difficult and so ν peaks and eventually decreases as the temperature is reduced even though the driving force continues to increase.

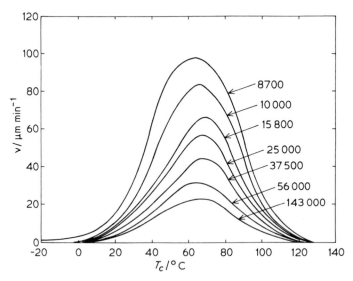

Fig. 4.25 *Dependence of crystal growth rate v upon crystallization temperature T_c, for different fractions of poly(tetramethyl-p-phenylene) siloxane (molar mass given in g mol^{-1}). (After Magill).*

4.3.2 *Overall crystallization kinetics*

The degree of crystallinity has an important effect upon the physical properties of a polymer. If a polymer melt is allowed to crystallize the crystallinity clearly increases as the spherulites form and grow and so the analysis of the crystallization of a polymer is of profound importance in understanding structure/property relationships in polymers. The nucleation and growth of spherulites in a polymer liquid can be readily analysed if a series of assumptions is made.

If a polymer melt of mass W_o is cooled below the crystallization temperature then spherulites will nucleate and grow over a period of time. It may be assumed that nucleation is homogeneous and that at a given temperature the number of nuclei formed per unit time per unit volume (i.e. the rate of nucleation) is a constant, N. The total number of nuclei formed in time interval, dt will then be $NW_o dt/\rho_L$ where ρ_L is the density of the liquid polymer. After a length of time, t these nuclei will have grown into spherulites of radius, r. The volume of each spherulite will be $4\pi r^3/3$ or using Equation (4.15), $4\pi v^3 t^3/3$. If the density of spherulitic material is ρ_s then the mass of each spherulite will be $4\pi v^3 t^3 \rho_s/3$. The total mass of spherulitic material, dW_s, present at time, t, and grown from the nuclei formed in the time interval, dt, will be given by

$$dW_s = \tfrac{4}{3}\pi v^3 t^3 \rho_s NW_o \frac{dt}{\rho_L} \qquad (4.16)$$

The total mass of spherulitic material formed after time, t, from all nuclei is then given by

$$W_s = \int_0^t \frac{4\pi\nu^3\rho_s NW_o t^3}{3\rho_L} dt \qquad (4.17)$$

which can be integrated to give

$$\frac{W_s}{W_o} = \frac{\pi N\nu^3\rho_s t^4}{3\rho_L} \qquad (4.18)$$

Alternatively this can be expressed in terms of the mass of liquid, W_L, remaining after time, t, since $W_s + W_L = W_o$

i.e. $$\frac{W_L}{W_o} = 1 - \frac{\pi N\nu^3\rho_s t^4}{3\rho_L} \qquad (4.19)$$

This analysis is highly simplified and the equations are only valid for the early stages of crystallization. However, the essential features of spherulitic crystallization are predicted. It is expected that the mass fraction of the crystals should depend initially upon t^4. It also follows that if the nuclei are formed instantaneously then a t^3 dependence would be expected as only the change in spherulite volume is time dependent.

Equation (4.19) is only valid in the initial stages of crystallization and must be modified to account for impingement of the spherulites. Also the analysis is slightly incorrect because, during crystallization, there is a reduction in the overall volume of the system and the centre of the spherulites move closer to each other. However, it can be shown that when impingement is taken into account W_L/W_o is related to t through an equation of the form

$$W_L/W_o = \exp(-zt^4) \qquad (4.20)$$

This type of equation is generally known as an Avrami equation and when t is small Equations (4.19) and (4.20) have the same form. If types of nucleation and growth other than those considered here are found the Avrami equation can be expressed as

$$W_L/W_o = \exp(-zt^n) \qquad (4.21)$$

where n is called the Avrami exponent.

From an experimental viewpoint it is much easier to follow the crystallization process by measuring the change in specimen volume rather than the mass of spherulitic material. If the initial and final specimen volumes are defined as V_o and V_∞ respectively and the specimen volume at time t is given by V_t then it follows that

$$V_t = \frac{W_L}{\rho_L} + \frac{W_s}{\rho_s} = \frac{W_o}{\rho_s} + W_L\left(\frac{1}{\rho_L} - \frac{1}{\rho_s}\right) \qquad (4.22)$$

and since

$$V_o = \frac{W_o}{\rho_L} \quad \text{and} \quad V_\infty = \frac{W_o}{\rho_s}$$

then Equation (4.22) becomes

$$V_t = V_\infty + W_L\left(\frac{V_o}{W_o} - \frac{V_\infty}{W_o}\right) \qquad (4.23)$$

Rearranging and combining Equations (4.21) and (4.23) gives

$$\frac{W_L}{W_o} = \frac{V_t - V_\infty}{V_o - V_\infty} = \exp(-zt^n) \qquad (4.24)$$

This equation then allows the crystallization process to be monitored by measuring how the specimen volume changes with time. In practice this is normally done by dilatometry. The crystallizing polymer sample is enclosed in a dilatometer and the change in volume is monitored from the change in height of a liquid which is proportional to the specimen volume. In terms of heights measured in the dilatometer Equation (4.24) can be expressed as

$$\left(\frac{V_t - V_\infty}{V_o - V_\infty}\right) = \left(\frac{h_t - h_\infty}{h_o - h_\infty}\right) = \exp(-zt^n) \qquad (4.25)$$

The Avrami exponent, n, can be determined from the slope of a plot of log $\{\ln[(h_t - h_\infty)/(h_o - h_\infty)]\}$ against log t. Fig. 4.26 shows an Avrami plot for polypropylene crystallizing at different temperatures. It is often difficult to estimate n from such plots because its value can vary with time. Also, non-integral values can be obtained and care must be exercised in using the Avrami analysis, as interpretation of the value of n in terms of specific nucleation and growth mechanisms can sometimes be ambiguous.

Serious deviations from the Avrami expression can be found particularly towards the later stages of crystallization because secondary crystallization often occurs and there is usually an increase in crystal perfection with time. They can take place relatively slowly and make estimation of V_∞ rather difficult.

4.3.3 *Molecular mechanisms of crystallization*

Although the Avrami analysis is fairly successful in explaining the phenomenology of crystallization it does not give any insight into the molecular process involved in the nucleation and growth of polymer crystals. There have been many attempts to develop theories to explain the important aspects of crystallization. Some of the theories are highly

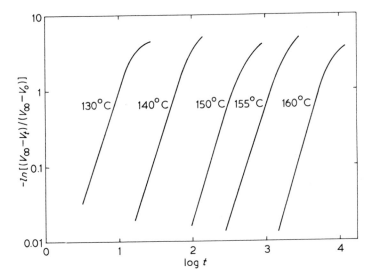

Fig. 4.26 *Avrami plot for polypropylene crystallizing from the melt at different indicated temperatures. The data points have been left off for clarity. (After Parrini and Corrieri, Makromol. Chem.* **62** *(1963), 83.)*

sophisticated and involve lengthy mathematical treatments. To date, none have been completely successful in explaining all the aspects of polymer crystallization. The important features that any theory must explain are the following characteristic experimental observations.

(a) Polymer crystals are usually thin and lamellar when crystallized from both dilute solution and the melt.

(b) A unique dependence is found between the lamellar thickness and crystallization temperature and, in particular, the lamellar thickness is found to be proportional to $1/\Delta T$ (Fig. 4.19).

(c) Chain folding is known to occur during crystallization from dilute solution and probably occurs to a certain extent during melt crystallization.

(d) The growth rates of polymer crystals are found to be highly dependent upon the crystallization temperature and molar mass of the polymer.

There are major difficulties in testing the various theories such as the lack of reliable experimental data on well-characterized materials and an absence of some important thermodynamic data on the crystallization process. Nevertheless, it is worth considering the predictions of the various theories.

The most widely accepted approach is the kinetic description due to Hoffman, Lauritzen and others which has been used to explain effects observed during polymer crystallization. It is essentially an extension of the approach used to explain the kinetics of crystallization of small molecules. The process is divided up into two stages, nucleation and growth and the main parameter that is used to characterize the process is the Gibbs free energy, G, which has been defined in Equation (4.1). It follows from this equation that the change in free energy, ΔG on crystallization at a constant temperature, T, is given by

$$\Delta G = \Delta H - T\Delta S \qquad (4.26)$$

where ΔH is the enthalpy change and ΔS is the change in entropy. It is envisaged that in the primary nucleation step a few molecules pack side-by-side to form a small cylindrical crystalline embryo. This process involves a change in free energy since the creation of a crystal surface, which has a surface energy will tend to cause G to increase whereas incorporation of molecules in a crystal causes a reduction in G (Section 4.1.1) which will depend upon the crystal volume. The result of these two competing effects upon G is illustrated schematically as a function of crystal size in Fig. 4.27. When the embryo is small the surface-to-volume ratio is high and so the overall value G increases because of the rapid increase in surface energy. However, as the embryo becomes larger the surface-to-volume ratio decreases and there will be a critical size above which G starts to decrease and eventually the free energy will be less than that of the original melt. This concept of primary nucleation has been widely applied to many crystallizing systems. The main difference between polymers and small molecules or atoms is in the geometry of the embryos. It is expected that the form of the free-energy change in all cases will be as

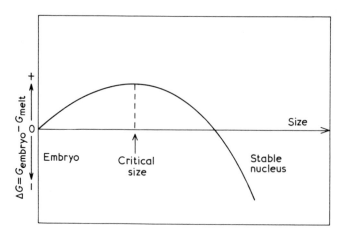

Fig. 4.27 *Schematic representation of change in free energy for the nucleation process during polymer crystallization.*

shown in Fig. 4.27. The peak in the curve may be regarded as an energy barrier and it is envisaged that at the crystallization temperature there will be sufficient thermal fluctuations to allow it to be overcome. Once the nucleus is greater than the critical size it will grow spontaneously as this will cause G to decrease.

The kinetic theories envisage the growth of the polymer crystals as taking place by a process of secondary nucleation on a pre-existing crystal surface. This process is illustrated schematically in Fig. 4.28, whereby molecules are added to a molecularly smooth crystal surface. This process is similar to primary nucleation but differs somewhat because less new surface per unit volume of crystal is created than in the primary case and so the activation energy barrier is lower. The first step in the secondary nucleation process is the laying down of a molecular strand on an otherwise smooth crystal surface. This is followed by the subsequent addition of further segments through a chain-folding process. It is found that chain folding only occurs for flexible polymer molecules. Extended-chain crystals are obtained from more rigid molecules. It is thought that the folding only takes place when there is relatively free rotation about the polymer backbone. The basic energetics of the crystallization process can be developed if it is assumed that the polymer lamellae have a fold surface energy of γ_e, a lateral surface energy of γ_s and that the free-energy change on crystallization is ΔG_v per unit volume. If the secondary nucleation process illustrated in Fig. 4.28 is then considered the increase in surface

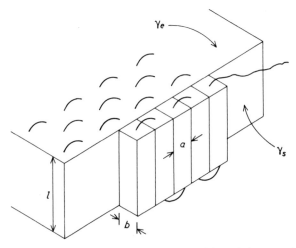

Fig. 4.28 *Model of the growth of a lamellar polymer crystal through the successive laying down of adjacent molecular strands. The different parameters are defined in the text. (Adapted from Magill).*

free energy involved in laying down n adjacent molecular strands of length l will be

$$\Delta G_n(\text{surface}) = 2bl\gamma_s + 2nab\gamma_e$$

when each strand has a cross-sectional area of ab. There will be a reduction in free energy because of the incorporation of molecular strands in a crystal which will be given by

$$\Delta G_n(\text{crystal}) = -n\,abl\,\Delta G_v$$

The overall change in free energy when n strands are laid down will then be

$$\Delta G_n = 2bl\gamma_s + 2nab\gamma_e - nabl\,\Delta G_v \qquad (4.27)$$

The value of ΔG_v can be estimated from a simple calculation. It is assumed that polymer crystals have an 'equilibrium melting temperature', T_m^0, which is the temperature at which a crystal without any surface would melt. If unit volume of crystal is considered then at this temperature it follows from Equation (4.26) that

$$\Delta G_v = \Delta H_v - T_m^0\Delta S_v \qquad (4.28)$$

where ΔH_v and ΔS_v are the enthalpies and entropies of fusion per unit volume respectively. However, at T_m^0 there is no change in free energy for the idealized boundaryless crystal since melting and crystallization are equally probable and so $\Delta G_v = 0$ at this temperature which means that

$$\Delta S_v = \Delta H_v/T_m^0 \qquad (4.29)$$

For crystallization below this temperature ΔG_v will be finite and it is envisaged that ΔS_v will not be very temperature dependent and so at a temperature, $T\ (<T_m^0)$ ΔG_v can be approximated to

$$\Delta G_v = \Delta H_v - T\Delta H_v/T_m^0 \qquad (4.30)$$

Since the degree of undercooling ΔT is given by $(T_m^0 - T)$ then rearranging Equation (4.30) gives

$$\Delta G_v = \Delta H_v\Delta T/T_m^0 \qquad (4.31)$$

Inspection of Equation (4.27) shows that for a given value of n, ΔG_n has a maximum value when l is small and decreases as l increases. Eventually a critical length of strand, l^0, will be achieved and the secondary nucleus will be stable ($\Delta G_n = 0$). Also, normally, n is large and so the term $2bl\gamma_s$ is negligible. A final approximate equation relating l^0 to ΔT can be obtained by combining Equations (4.27) and (4.31) which gives

$$l^0 \sim \frac{2\gamma_e T_m^0}{\Delta H_v\Delta T} \qquad (4.32)$$

The proportionality between l and ΔT which is observed experimentally is therefore predicted theoretically. The analysis outlined above is highly simplified, but it serves to illustrate the important aspects of the kinetic approach to polymer crystallization.

The kinetic approach can also be used to explain why chain folding is found for solution-grown crystals. The separation between individual molecules is relatively high in dilute solutions and so growth can take place most rapidly by the successive deposition of chain-folded strands of the same molecule once it starts to be incorporated in a particular crystal.

4.3.4 *Melting*

The melting of polymer crystals is essentially the reverse of crystallization, but it is more complicated than the melting of low molar mass crystals. There are several characteristics of the melting behaviour of polymers which distinguishs them from other materials. They can be summarized as follows.

(a) It is not possible to define a single melting temperature for a polymer sample as the melting generally takes place over a range of temperature.

(b) The melting behaviour depends upon the specimen history and in particular upon the temperature of crystallization.

(c) The melting behaviour also depends upon the rate at which the specimen is heated.

These observations are a reflection of the peculiar morphologies that polymer crystals can possess and in particular the fact that polymer crystals are normally thin. The concept of an equilibrium melting temperature T_m^0 (Section 4.3.3) is introduced because of the variability in the melting behaviour. This corresponds to the melting temperature of an infinitely large crystal. However, there is still a good deal of disagreement over the values of T_m^0 even for widely studied polymers such as polyethylene. This is particularly disturbing because accurate values of T_m^0 are required in order to test the theories of crystallization and melting quantitatively. The value of T_m^0 can be estimated by a simple extrapolation procedure. It is found that the observed melting temperature, T_m, for a polymer sample is always greater than the crystallization temperature, T_c and a plot of T_m versus T_c is usually linear as shown in Fig. 4.29. Since T_m can never be lower than T_c the line $T_m = T_c$ will represent the lower limit of the melting behaviour. The point at which the extrapolation of the upper line meets the $T_m = T_c$ line then represents the melting temperature of a polymer crystallized infinitely slowly and for which crystallization and melting would take place at the same temperature. This intercept therefore gives T_m^0.

There is found to be a strong dependence of the observed melting

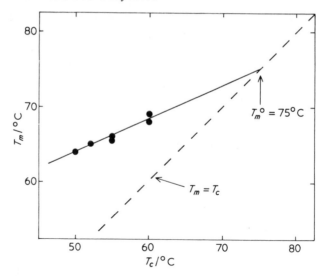

Fig. 4.29 *Plot of melting temperature,* T_m, *against crystallization temperature,* T_c, *for poly(dl-propylene oxide) (After Magill).*

temperature of a polymer crystal, T_m, upon the crystal thickness, l. This can be explained by considering the thermodynamics of melting a rectangular lamellar crystal with lateral dimensions x and y. If it is assumed that the crystal has side surface energy of γ_s and top and bottom surface energies of γ_e then melting causes a decrease in surface free energy of $(2xl\gamma_s + 2yl\gamma_s + 2xy\gamma_e)$. This is compensated by an increase in free energy of ΔG_v per unit volume due to molecules being incorporated in the melt rather than in a crystal. The overall change in free energy on melting the lamellar crystal is given by

$$\Delta G = xyl\Delta G_v - 2l(x + y)\gamma_s - 2xy\gamma_e \tag{4.33}$$

The value of ΔG_v is given by Equation (4.31). For lamellar crystals the area of the top and bottom surfaces will be very much larger than the sides and so the term $2l(x + y)\gamma_s$ can be neglected. At the melting point of the crystal $\Delta G = 0$ and so it follows from Equations (4.31) and (4.33) that

$$T_m = T_m^0 - \frac{2\gamma_e T_m^0}{l\Delta H_v} \tag{4.34}$$

where ΔH_v is the enthalpy of fusion per unit volume of the crystals. Inspection of this equation shows that for finite size crystals T_m will always be less than T_m^0. Also it allows T_m^0 and γ_e to be calculated if T_m is determined as a function of l. A plot of T_m against $1/l$ is predicted to have a slope of $-2\gamma_e T_m^0/\Delta H_v$ with T_m^0 as the intercept. The value of ΔH_v can be measured by calorimetry in a separate experiment.

A process which affects the melting behaviour of crystalline polymers and is of interest in its own right is *annealing*. This is a term which is normally used to describe the heat-treatment of metals and the annealing of polymers bears a similarity to that of metals. It is found that when crystalline polymers are heated to temperatures just below the melting temperature there is an increase in lamellar crystal thickness. The driving force is the reduction in free energy gained by lowering the surface area of a lamellar crystal when it becomes thicker and less wide. The lamellar thickening only happens at relatively high temperatures when there is sufficient thermal energy available to allow the necessary molecular motion to take place.

A certain amount of annealing usually takes place when a crystalline polymer sample is heated and melted. The increase in lamellar thickness, l, causes an increase in T_m (Equation 4.35). This means that the measured melting temperature will depend upon the heating rate because annealing effects will be lower for more rapid rates of heating. However, high heating rates can give rise to other problems such as with thermal conductivity and so great care is necessary when measuring the melting temperatures of polymer samples especially when the melting behaviour is being related to the as-crystallized structure which could change during heating.

4.3.5 *Factors affecting T_m*

The use of polymers in many practical applications is often limited by their relatively low melting temperatures. Because of this there has been considerable interest in determining the factors which control the value of T_m and in synthesizing polymers which have high melting temperatures. For a particular type of polymer the value of T_m depends upon the molar mass and degree of chain branching. This is because there is a higher proportion of chain ends in low-molar-mass polymers and the branches in non-linear polymers both have the effect of introducing defects into the crystals and so lower their T_m. If the molar mass of the polymer is sufficiently high that the polymer has useful mechanical properties the effect of varying M upon T_m is not strong. In contrast, the presence of branches in a high-molar-mass sample of polyethylene can reduce T_m by 30°C.

The over-riding factor which determines the melting points of different polymers is their chemical structure. The melting points of several polymers are listed in Table 4.3. It is most convenient to consider the melting points of different polymers using polyethylene ($-CH_2-CH_2-)_n$ as a reference. The first factor which must be considered is the stiffness of the main polymer chain. This is controlled by the ease at which rotation can take place about the chemical bonds along the chain. In general,

TABLE 4.3 *Approximate values of melting temperature, T_m, for various polymers.*

Repeat unit		T_m/K
—CH$_2$—CH$_2$—		410–419
—CH$_2$—CH$_2$—O—		340
—CH$_2$—CH$_2$—CO—O—		395
—CH$_2$—⟨O⟩—CH$_2$—		670
—CH$_2$—CH$_2$—CO—NH—		603
—CH$_2$—CH$_2$—CH$_2$—CO—NH—		533
—CH$_2$—CH$_2$—CH$_2$—CH$_2$—CO—NH—		531
	Side group (X)	
—CH$_2$—CHX—	—CH$_3$	460
	—CH$_2$—CH$_3$	398
	—CH$_2$—CH$_2$—CH$_3$	351
	—CH$_2$—CH(CH$_3$)$_2$	508
	⟨O⟩	450

incorporation of groups such as —O—, —O—O— or —CO—O— in the main chain increases flexibility and so lowers T_m (Table 4.3). On the other hand the presence of a phenyl group in the main chain increases the stiffness and causes a large increase in T_m.

Another important factor which causes an increase in T_m is the presence of polar groups such as the amide linkage —CONH— which allows intermolecular hydrogen-bonding to take place within the crystals. The presence of the hydrogen-bonding tends to stabilize the crystals and so raise their T_m. The melting points of different polyamides are very sensitive to the degree of intermolecular bonding and the value of the T_m is reduced as the number of —CH$_2$— groups between the amide linkages is increased (Table 4.3).

A third factor which governs the value of T_m in different polymers is the type and size of any side-groups present on the polymer backbone. This is most easily shown when the effect of having different aliphatic side-groups in vinyl polymers of the type (—CH$_2$—CHX—)$_n$ is considered. The presence of a —CH$_3$ side-group regularly placed along the polyethylene chain leads to a reduction in chain flexibility and means that polypropylene has a higher melting point than polyethylene. However, if the side-group is long and flexible the T_m is lowered as the length is increased. On the other hand, an increase in the bulkiness of the side-group restricts rotation about bonds in the main chain and so has the effect of raising T_m (Table 4.3).

It can be seen from the considerations outlined above that it is possible to exert a good deal of control upon the melting temperatures of different polymers. It must be borne in mind that, generally, factors which affect the T_m of a polymer also change the glass transition temperature (Section

4.4.3) and in general these two parameters cannot be varied independently of each other. Also, changing the structure of the polymer may affect the ease of crystallization and although the potential melting point of a crystalline phase may be high the amount of this phase that may form could be low.

4.4 **Amorphous polymers**

We have so far been concerned principally with the structure of crystalline polymers which can readily be studied by using standard X-ray and electron diffraction methods. However, there is an important category of polymers which have not yet been considered which can be completely non-crystalline. They are generally termed *amorphous* and include the well-known polymer glasses and rubbers. Although the properties of these materials have been studied at length, very little is known about their structure. This is because there is no well-defined order in the structure of amorphous polymers and so they cannot be analysed very easily using standard diffraction techniques.

4.4.1 *Structure in amorphous polymers*

Amorphous polymers can be thought of simply as frozen polymer liquids. Over the years there have been many attempts to analyse the structure of liquids of small molecules and they have met with a varied success. As may be expected, the necessity of having long chains makes the problem more difficult to solve in the case of polymer melts and amorphous polymers. Investigation of the structure of amorphous polymers has yielded a limited amount of information. Fig. 4.30 shows an X-ray diffraction pattern for an amorphous polymer. It has only one diffuse ring and can be compared with the corresponding pattern for a semi-crystalline polymer shown in Fig. 4.1 which consists of several relatively well-defined rings. In crystalline materials the discrete rings or spots on X-ray-diffraction patterns correspond to Bragg reflections from particular crystallographic planes within the crystals. The lack of such discrete reflection in non-crystalline materials is taken as an indication of the lack of crystalline order. The very diffuse scattering observed in Fig. 4.30 corresponds simply to the regular spacing of the atoms along the polymer chain or in side-groups and the variable separation of the atoms in adjacent molecules.

Information on the conformations of polymer chains in amorphous polymers has been obtained by using small-angle neutron diffraction. The technique involves measurement of the coherent scattering of neutrons by mixtures of deuterated and protonated polymer molecules. Analysis of the scattering data leads to an estimation of the radius of gyration (Section

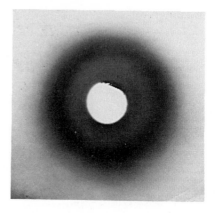

Fig. 4.30 *Flat-plate wide-angle X-ray diffraction pattern from atactic polystyrene. The polymer does not crystallize and shows only diffuse rings (cf. Fig. 4.1).*

3.1.2) of the polymer molecules. It has been clearly demonstrated that in both polymer glasses and polymer liquids the polymer chains, to a first approximation, have their unperturbed dimensions. This means that the molecules have the same conformations in the bulk as they have in a Theta solvent. This behaviour was predicted many years ago by Flory* and the simple reason for this is that a molecule in the bulk state will interfere with itself, but if it expands to decrease this interaction there is an increase in the interaction with its neighbours and so it adopts the unperturbed conformation.

There have been many suggestions over the years that both polymer melts and polymer glasses have domains of order. These are microscopic regions in which there is a certain amount of molecular order. This idea is not unique for polymers and similar ideas have been put forward to explain the structure of liquids and inorganic glasses. The main evidence for ordered regions in glassy polymers has come from examination of the structure of these materials by electron microscopy either by taking replicas of fracture surfaces or looking directly at the structure of thin films. Fig. 4.31 shows a micrograph of a fracture surface replica of an epoxy resin. The nodular structure on a scale of ~1000 Å can be clearly seen. Similar structures have been seen in many glassy polymers and the nodules have been interpreted as domains of order. However, the presence of order has not been confirmed by X-ray diffraction which shows the structures to be random and some of the electron micrographic evidence has been criticized.

There are some polymers, such as poly(ethylene terephthalate) and

*P.J. Flory. (1953). *Principles of Polymer Chemistry*, p. 602.

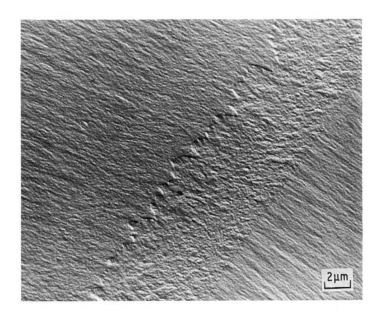

Fig. 4.31 *Electron micrograph of a replica of a fracture surface of an epoxy resin.*

isotactic polypropylene which normally crystallize rather slowly, that can be obtained in an apparently amorphous state by rapidly quenching the melt which does not allow sufficient time for crystals to develop by the normal processes of nucleation and growth. Crystallization can be induced by annealing the quenched polymer at an elevated temperature. The X-ray-diffraction patterns which are obtained from the quenched polymer tend to be diffuse and ill-defined. This could be either because the samples are truly non-crystalline or due to the presence of very small and imperfect crystals. Crystallization is manifest during annealing by the diffraction patterns becoming more well-defined. There must clearly be a certain amount of molecular rearrangement during annealing, but this cannot take place through large-scale molecular diffusion as the process occurs in the solid state. The mechanism of crystallization is therefore unclear and it is very difficult to decide whether the quenched polymers are truly amorphous or just contain very small crystals.

4.4.2 *The glass transition*

If the melt of a non-crystallizable polymer is cooled it becomes more viscous and flows less readily. If the temperature is reduced low enough it becomes rubbery and then as the temperature is reduced further it becomes a relatively hard and elastic polymer glass. The temperature at which the polymer undergoes the transformation from a rubber to a glass is known as the *glass transition temperature*, T_g. The ability to form glasses is not confined to non-crystallizable polymers. Any material which can be cooled sufficiently below its melting temperature without crystallizing will undergo a glass transition.

There is a dramatic change in the properties of a polymer at the glass-transition temperature. For example, there is a sharp increase in the stiffness of an amorphous polymer when its temperature is reduced below T_g. This will be dealt with further in the section on mechanical properties (Section 5.2.6). There are also abrupt changes in other physical properties such as heat capacity and thermal expansion coefficient. One of the most widely used methods of demonstrating the glass transition and determining T_g is by measuring the specific volume of a polymer sample as a function of the temperature as shown in Fig. 4.32. In the regimes above and below the glass transition temperature there is a linear variation in specific volume with temperature, but in the vicinity of the T_g there is a change in slope of the curve which occurs over several degrees. The T_g is normally taken as the point at which the extrapolations of the two lines meet. Another characteristic of the T_g is that the exact temperature depends upon the rate at which the temperature is changed. It is found that the lower the cooling rate the lower the value of T_g that is obtained. It is still a matter of some debate as to whether a limiting value of T_g would eventually be reached if the cooling rate were low enough. It is also possible to detect a glass transition in a semi-crystalline polymer, but the change in properties at T_g is usually less marked than for a fully amorphous polymer.

There have been attempts to analyse the glass transition from a thermodynamic viewpoint. Thermodynamic transitions are classified as being either first- or second-order. In the first-order transition there is an abrupt change in a fundamental thermodynamic property such as enthalpy, H or volume, V, whereas in a second-order transition only the first derivative of such properties changes. This means that during a first-order transition, such as melting, H and V will change abruptly whereas for a second-order transition changes will only be detected in properties such as heat capacity, C_p or volume thermal expansion coefficient, α which are defined as

$$C_p = \left(\frac{\partial H}{\partial T}\right)_p \quad \text{and} \quad \alpha = \frac{1}{V}\left(\frac{\partial V}{\partial T}\right)_p$$

respectively. As both of these parameters are found to change abruptly at the glass transition temperature it would appear that it may be possible to consider the glass transition as a second-order thermodynamic transition. However, more detailed thermodynamic measurements have shown that

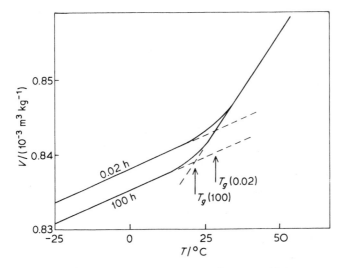

Fig. 4.32 *Variation of the specific volume of poly(vinyl acetate) with temperature taken after two different times (After Kovacs, (1958) J. Polym. Sci.* **30** *131).*

although the glass transition has the appearance of a thermodynamic second-order transition it cannot strictly be considered as one.

One of the most useful approaches to analysing the glass transition is to use the concept of *free volume*. This concept has been used in the analysis of liquids and it can be readily extended to the consideration of the glass transition in polymers. The free volume is the space in a solid or liquid sample which is not occupied by polymer molecules, i.e. the 'empty-space' between molecules. In the liquid state it is supposed that the free volume is high and so molecular motion is able to take place relatively easily because the unoccupied volume allows the molecules space to move and so change their conformations freely. It is envisaged that the free volume will be sensitive to the change in temperature and that most of the thermal expansion of the polymer rubber or melt can be accounted for by a change in the free volume. As the temperature of the melt is lowered the free volume will be reduced until eventually there will not be enough free volume to allow molecular rotation or translation to take place. The temperature at which this happens corresponds to T_g as below this temperature the polymer glass is effectively 'frozen'. The situation is represented schematically in Fig. 4.33. The free volume is represented as a shaded area. It is assumed to be constant at V_f^* below T_g and to increase as the temperature is raised above T_g. The total sample volume V therefore consists of volume occupied by molecules V_o and free volume V_f such that

$$V = V_o + V_f \tag{4.35}$$

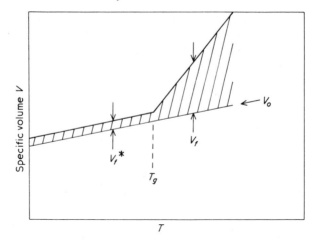

Fig. 4.33 *Schematic illustration of the variation of the specific volume, V, of a polymer with temperature, T. The free volume is represented by the shaded area.*

It is more convenient to talk in terms of fractional free volume, f, which is defined as $f = V_f/V$. At and below the T_g, f is given by $f_g = V_f^*/V$ and can be considered as being effectively constant. Above the T_g there will be an important contribution to V_f from the expansion of the melt. The free volume above T_g is then given by

$$V_f = V_f^* + (T - T_g)\left(\frac{\partial V}{\partial T}\right)$$

(4.36)

Dividing through by V gives

$$f = f_g + (T - T_g)\alpha_f$$

(4.37)

where α_f is the thermal expansion coefficient of the free volume which will be given, close to T_g, by the difference between the thermal expansion coefficients of the rubbery and glassy polymer. The equation for the fractional free volume will only be strictly valid over small increments of temperature above T_g. It is found for a whole range of different glassy polymers that f_g is remarkably constant and this concept of free volume has found important use in the analysis of the rate and temperature dependence of the viscoelastic behaviour of polymers between T_g and $T_g +$ 100 K (Section 5.2.7).

4.4.3 *Dependence of T_g upon chemical structure*

The onset of the glass transition marks a significant change in the physical properties of the polymer. A glassy polymer will lose its stiffness and have a tendency to flow above T_g. As many polymers are used in practical applications in the glassy state it is vital to know what factors control the T_g. At the glass transition the molecules which are effectively frozen in position in the polymer glass become free to rotate and translate and so it is not surprising that the value of the T_g will depend upon the physical and chemical nature of the polymer molecules. The effect of factors such as molar mass and branching will be considered in the next section and in this section the way in which the chemical nature of the polymer chain affects T_g will be examined.

The effect of the chemical nature of the polymer chain upon T_g is similar to the effect it has upon T_m (Section 4.3.5). The most important factor is chain flexibility which is governed by the nature of the chemical groups which constitute the main chain. Polymers such as polyethylene ($-CH_2-CH_2-)_n$ and polyoxyethylene ($-CH_2-CH_2-O-)_n$ which have relatively flexible chains because of easy rotation about main chain bonds, tend to have low values of T_g The value of the T_g of polyethylene is a matter of some dispute because samples are normally highly crystalline which makes measurement of T_g rather difficult. Values have been quoted between 140 K and 270 K. However, the incorporation of units which impede rotation and stiffen the chain clearly cause a large increase in T_g. For example, the presence of a p-phenylene ring in the polyethylene chain causes poly(p-xylylene) to have a T_g of the order of 80°C.

In vinyl polymers of the type ($-CH_2-CHX-)_n$ the nature of the side group has a profound effect upon T_g as can be seen in Table 4.4. The presence of side-groups on the main-chain has the effect of increasing T_g through a restriction of bond rotation. Large and bulky side-groups tend to cause the greatest stiffening and if the side-group itself is flexible the effect is not so marked. The presence of polar groups such as $-Cl$, $-OH$ or $-CN$ tends to raise T_g more than non-polar groups of equivalent size. This is because the polar interactions will restrict rotation further and so poly(vinyl chloride) ($-CH_2-CHCl-)_n$ has a higher T_g than polypropylene ($-CH_2-CHCH_3-)_n$.4.4

It is clear that the chemical factors outlined above as controlling T_g are also ones which have an effect upon T_m (Section 4.3.5). In fact, the same factors tend to raise or lower both T_g and T_m as both are controlled principally by main-chain stiffness. It is not surprising, therefore, that a correlation is found between T_g and T_m for polymers which exhibit both types of transition. It is found that when they are expressed in Kelvins the value of T_g is generally between 0.5 and 0.8 T_m. This behaviour is shown schematically in Fig. 4.34. It demonstrates that it is not possible to control T_g and T_m independently for homopolymers. There are possibilities of having more control over T_g and T_m separately by making copolymers. For example, random copolymers of nylon 66 and nylon 610 can be made for

TABLE 4.4 *Approximate values of glass transition temperature, T_g, for various polymers.*

Repeat unit	T_g/K
—CH$_2$—CH$_2$—	140–270
—CH$_2$—CH$_2$—O—	206
—⟨O⟩—O—	357
—CH$_2$—⟨O⟩—CH$_2$—	353

	Side group (X)	
—CH$_2$—CHX—	—CH$_3$	250
	—CH$_2$—CH$_3$	249
	—CH$_2$—CH$_2$—CH$_3$	233
	—CH$_2$—CH(CH$_3$)$_2$	323
	—⟨O⟩	373
	—Cl	354
	—OH	358
	—CN	370

which the T_g is very little different from that of the homopolymers, because the stiffness of the main chain will be virtually unchanged. On the other hand, the irregularity introduced into the main chain reduces the ability of the molecules to crystallize and so lowers T_m.

4.4.4 *Dependence of T_g upon molecular architecture*

We have, so far, only been concerned with the way in which the chemical nature of the polymer molecules affects T_g. It is also known that the physical characteristics of the molecules such as molar mass, branching and cross-linking affect the temperature of the glass transition. The value of the T_g is found to increase as the molar mass of the polymer is increased and the behaviour can be approximated to an equation of the form

$$T_g^\infty = T_g + K/M \qquad (4.38)$$

where T_g^∞ is the value of T_g for a polymer sample of infinite molar mass and K is a constant. The form of this equation can be justified using the concept of free volume if it assumed that there is more free volume involved with each chain end than a corresponding segment in the middle part of the chain. In a polymer sample of density, ρ and molar mass, $\bar{M}_n$, the number of chains per unit volume is given by $\rho N_A/\bar{M}_n$ and so the number of chain ends per unit volume is $2\rho N_A/\bar{M}_n$ where N_A is the Avagadro number. If Θ is the contribution of one chain end to the free volume then the total fractional free volume due to chain ends f_c is given by

$$f_c = 2\rho N_A \Theta/\bar{M}_n \qquad (4.39)$$

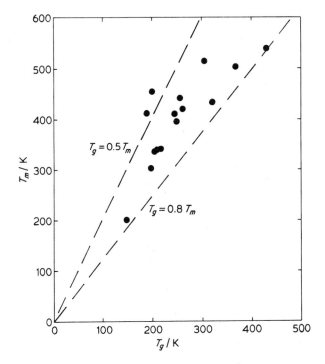

Fig. 4.34 *Plot of T_m against T_g for a variety of common polymers such as polyethylene, polypropylene, polystyrene, poly(ethylene oxide) etc. (Data taken from Boyer, (1963) Rubber Chem. Tech.* **36**, *1303).*

It can be reasoned that if a polymer with this value of f_c has a glass transition temperature of T_g then f_c will be equivalent to the increase in the free volume on expanding the polymer thermally between T_g and T_g^∞. This means that

$$f_c = \alpha_f(T_g^\infty - T_g) \tag{4.40}$$

where α_f is the thermal expansion coefficient of the free volume (Section 4.4.2). Combining Equations (4.39) and (4.40) and rearranging leads to the final result

$$T_g = T_g^\infty - \frac{2\rho N_A \Theta}{\alpha_f \overline{M}_n} \tag{4.41}$$

which is of exactly the same form as Equation (4.38) when $K = 2\rho N_A \Theta / \alpha_f$. The observed dependence of the T_g upon molar mass is therefore predicted from considerations of free volume.

A small number of branches on a polymer chain are found to reduce the value of T_g. This can again be analysed using the free volume concept. It follows from the above analysis that the glass transition temperature of a chain possessing a total number of x ends per chain is given by an equation of the form

$$T_g = T_g^x - \frac{\rho x N_A \Theta}{\alpha_f \bar{M}_n} \tag{4.42}$$

where T_g^x is again the glass transition temperature of a linear chain of infinite molar mass. Since a linear chain has two ends the number of branches per chain is $(x - 2)$. Equation (4.42) is only valid when the number of branches is low. A high density of branching will have the same effect as side groups in restricting chain mobility and hence raising T_g.

The presence of chemical cross-links in a polymer sample has the effect of increasing T_g although when the density of cross-links is high the range of the transition region is broadened and the glass transition may not occur at all in highly cross-linked materials. Cross-linking tends to reduce the specific volume of the polymer which means that the free volume is reduced and so the T_g is raised because molecular motion is made more difficult.

4.4.5 *Elastomers*

A non-crystalline linear polymer above its T_g behaves like a viscous liquid and will normally flow under its own weight. This viscous flow takes place by the sliding of the molecular chains past each other and it can be stopped by cross-linking the polymer. In this state the polymer exhibits *elastomeric* behaviour. It can be extended to many times its initial dimensions and will spring back rapidly to take up its original shape. The elastomeric properties of natural rubber, which is a naturally occurring form of *cis*-polyisoprene, have been known for many years. It is rather tacky and tends to flow in its natural state, but on vulcanization (i.e. cross-linking) it exhibits useful elastomeric behaviour. The vulcanization process was first discovered in the USA in 1839 by Goodyear who found that the properties of the rubber were greatly improved by heating with sulphur. This process introduces intermolecular cross-links, but as the reaction with sulphur is rather difficult to control, more sophisticated peroxide cross-linking agents are also employed these days.

The two naturally occurring isomers of polyisoprene have quite different physical properties. The chemical structures of *cis* and *trans* polydienes have been discussed in Section 2.5.4. When the chains have a *cis* configuration crystallization is relatively difficult. The degree of crystallinity in natural rubber is fairly low and the melting-point is just above ambient temperature ($\sim 35°C$). The chains in *cis*-polyisoprene can also coil relatively easily by rotation about single bonds and when the polymer is cross-linked it has good elastomeric properties. On the other hand in gutta percha or balata which are two forms of *trans*-polyisoprene the degree of crystallinity is higher because crystallization takes place more readily and the crystals are more stable (T_m ~75°C). The chains in the *trans* form are also less flexible. These factors make the material relatively hard and rigid with no useful elastomeric properties.

The properties of a particular elastomer are controlled by the nature of the cross-linked network. Ideally this can be envisaged as an amorphous network of polymer chains with branching-points from which four chains emanate. Every length of polymer chain can be thought of as being anchored at two separate sites. In reality this is a great oversimplification and the likely situation is sketched in Fig. 4.35. In a polymer of finite molar mass there will be some loose chain ends. Also it is possible that intramolecular cross-linking can take place leading to the presence of loops. Neither of these two structural irregularities will contribute to the strength of the network. On the other hand, entanglements can act as effective cross-links, even in an unvulcanized rubber, at least during short-term loading.

Crystallization takes place only very slowly in natural rubber at ambient temperature, but when a sample is stretched crystallization takes place relatively rapidly. The rate and degree of crystallization depends upon the extension of the sample and the length of time the extension is maintained. It is very rapid at strains of above about 300–400 percent and degrees of crystallinity of over 30 percent have been reported at the highest extensions. Crystallization is manifest by the appearance of arcs in the X-ray diffraction patterns. The unstretched material is non-crystalline and the diffraction pattern shows only diffuse rings. On deforming a rubber the molecules tend to become aligned parallel to the stretching direction and since polymer molecules always pack side-by-side into crystals this

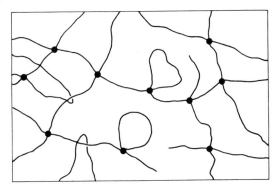

Fig. 4.35 *Schematic diagram of the molecules in an elastomer showing normal cross-linking, chain ends, intra-molecular loops and entanglements.*

alignment reduces the entropy change on crystallization (Section 4.1.1). The stretching process, therefore, aids the crystallization process and the observed arcing in the diffraction patterns follows from the molecular alignment which causes the crystallites to form with their c-axes approximately parallel to the stretching direction. On the other hand, when crystallization takes place in an unstretched rubber the diffraction pattern contains only rings showing that the crystallites have random orientations.

Although natural rubber has satisfactory properties in many applications there are now many synthetic alternatives available. Natural rubber is rather prone to chemical degradation by ozone and has poor resistance to solvents. Synthetic rubbers are usually better in these respects. The most important use of rubbers is in motor vehicle tyres in which styrene–butadiene rubber (a random copolymer, SBR), is mainly used along with a variety of additives such as carbon black. This reinforces the rubber and improves the strength, stiffness and abrasion resistance.

Further reading

Alexander, L.E. (1969), *X-ray Diffraction Methods in Polymer Science*, Wiley-Interscience, New York.

Haward, R.N. (ed.) (1973), *The Physics of Glassy Polymers*, Applied Science Publishers, London.

Kelly, A. and Groves, G.W. (1970), *Crystallography and Crystal Defects*, Longman, London.

Magill, J.H. (1977), 'Morphogenesis of Solid Polymers' in *Treatise on Materials Science and Technology*, Vol. 10A (ed. J.M. Schultz), Academic Press, New York.

Schultz, J.M. (1974), *Polymer Materials Science*, Prentice-Hall Inc., New Jersey.

Treloar, L.R.G. (1958), *The Physics of Rubber Elasticity*, Clarendon Press, Oxford.

Wunderlich, B. (1973), *Macromolecular Physics, Vol. 1: Crystal structure, morphology, defects*, Academic Press, London.

Wunderlich, B. (1976), *Macromolecular Physics Vol. 2: Crystal nucleation, growth and annealing*, Academic Press, London.

Problems

Unit cell dimensions for polyethylene may be assumed to be as follows:

$a = 7.41$ Å, $b = 4.96$ Å, $c = 2.54$ Å, $\alpha = \beta = \gamma = 90°$
The wavelength of CuKα radiation is 1.542 Å

4.1 A flat-plate X-ray diffraction pattern was obtained from an unoriented sample of polyethylene using CuKα radiation. It consisted of three rings of radius 22.2 mm, 36.6 mm and 19.7 mm. The specimen-to-film distance was 50 mm. Calculate the d-spacing of the planes giving rise to these reflections. It is thought that these reflections are from (hk0) type planes. Draw a scale diagram of the unit cell of polyethylene projected along the c-axis. Measure the spacing of the low-index (hk0) type planes and hence identify the planes giving rise to the observed reflections.

4.2 A flat-plate X-ray diffraction pattern was obtained from an oriented polymer fibre using CuKα radiation. It consisted of a series of spots arranged along a horizontal line containing the central spot and first-order layer lines above and below the central line. It was found that with a specimen-to-film distance of 30 mm the first order lines were 22.9 mm above and below the central spot.

(a) Calculate the spacing of the chain repeat c of the polymer crystals.

(b) Explain why only first-order layer lines are observed for this polymer and suggest how higher order lines may be obtained.

(c) Determine the angle through which the specimen would have to be tilted in order to obtain the (001) reflection on the diffraction pattern.

4.3 The crystalline density of polyethylene is 1000 kg m^{-3} and the density of amorphous polyethylene is 865 kg m^{-3}. Calculate the mass fraction crystallinity in a sample of linear polyethylene of density 970 kg m^{-3} and in a sample of branched polyethylene of density 917 kg m^{-3}. Why do the two samples have considerably different crystallinities?

4.4 A linear homopolymer was crystallized from the melt at crystallization temperatures (T_c) within the range 270 K to 330 K. Following complete crystallization the melting temperatures (T_m) were measured by differential scanning calorimetry.

T_c/K	T_m/K
270	300.0
280	306.5
290	312.5
300	319.0
310	325.0
320	331.0
330	337.5

Determine graphically the equilibrium melting temperature, T_m°. Small-angle X-ray scattering experiments using CuKα radiation gave the positions of the first maxima as:

T_c/K	Θ/degrees
270	0.44
280	0.39
290	0.33
300	0.27
310	0.22
320	0.17
330	0.11

Using the Bragg equation calculate the values of the long period for each crystallization temperature.

The degree of crystallinity was measured to be 45 percent for all samples. Calculate the lamellar thickness in each case and hence determine graphically the fold surface energy, γ_e, if the enthalpy of fusion of the polymer per unit volume ΔH_V is $1.5 \times 10^8 \mathrm{J\ m^{-3}}$.

4.5 The refractive index, $\bar{n}$, of a polymer was determined as a function of temperature. The results are given below:

$T/^\circ C$	$\bar{n}$
20	1.5913
30	1.5898
40	1.5883
50	1.5868
60	1.5853
70	1.5838
80	1.5822
90	1.5801
100	1.5766
110	1.5725
120	1.5684
130	1.5643

Determine the glass transition temperature of the polymer.

4.6 Discuss the reasons for the differences in glass transition temperature for the following pairs of polymers with similar chemical structures.

(a) Polyethylene (~150 K) and polypropylene (~250 K).
(b) Poly(methyl acrylate) (283 K) and poly(vinyl acetate) (305 K).
(c) Poly(1-butene) (249 K) and poly(2-butene) (200 K).
(d) Poly(ethylene oxide) (232 K) and poly(vinyl alcohol) (358 K).
(e) Poly(ethyl acrylate) (249 K) and poly(methyl methacrylate) (378 K).

4.7 The glass transition temperature, T_g, of a linear polymer with number average molar mass $\bar{M}_n = 2300$ was found to be 121°C. The value of T_g increased to 153°C for a sample of the same linear polymer with $\bar{M}_n = 9000$. A branched version of the same polymer with $\bar{M}_n = 5200$ was found to have a T_g of 115°C. Determine the average number of branches on the molecules of the branched polymer.

5 Mechanical Properties

5.1 General Considerations

Polymers are often thought of as being mechanically weak and their mechanical properties have consequently been somewhat ignored in the past. These days many polymers are used in structural engineering applications and are subjected to appreciable stresses. This increase in use has been due to several factors. One of the most important being that, although on an absolute basis their mechanical strength and stiffness may be relatively low compared with metals and ceramics, when the low density of polymers is taken into account their specific strength and stiffness become more comparable with those of conventional materials. Also the fabrication costs of polymeric components are usually considerably lower than for other materials. Polymers melt at relatively low temperature and can be readily moulded into quite intricate components using a single moulding.

Perhaps the most over-riding reason for using polymers is that they can display unique mechanical properties. One polymer that has been used for many years is natural rubber which is elastomeric. It can be stretched to very high extensions and will snap back immediately the stress is removed. Polyolefins are widely used in domestic containers and in packaging because of their remarkable toughness and flexibility and ease of fabrication. Other polymers are employed as adhesives. Epoxy resins can produce bonds in metals which can be as strong as those made by conventional joining methods such as welding or rivetting.

With the now increasingly more widespread use of polymers an understanding of their mechanical properties is becoming essential. In this chapter the mechanical properties have been considered from a phenomenological viewpoint and in terms of the molecular deformation processes which occur. This latter approach has been facilitated by a better understanding of polymer structure which was outlined in the previous section.

5.1.1 Stress

Force can be exerted externally on a body in two particular ways. Gravity and inertia can be thought of as *body forces* since they act directly on all the individual particles in the body. The other type are *surface* or *contact forces* which act only on the surface of the body, but their effect is transmitted to the particles inside the body through the atomic and molecular bonds. In consideration of the mechanical properties of polymers we are mainly interested in the effect of applying surface forces such as stress or pressure to the material but it must be remembered that certain polymers such as non-cross-linked rubbers will flow under their own weight, a simple response to the body force of gravity.

   Although the response of these forces reflects the displacement of the individual particles within the body the system is normally considered from a macroscopic viewpoint and it is regarded as a continuum. In order to define the state of stress at a point within the body we consider the surface forces acting on a small cube of material around that point as shown in Fig. 5.1. Each of these surface forces is divided by the area upon which it acts and then it is resolved into components which are parallel to the three co-ordinate axes. In total there will be nine stress components given as

$$\begin{matrix} \sigma_{11} & \sigma_{12} & \sigma_{13} \\ \sigma_{21} & \sigma_{22} & \sigma_{23} \\ \sigma_{31} & \sigma_{32} & \sigma_{33} \end{matrix}$$

where the first subscript gives the normal to the plane on which the stress acts and the second subscript defines the direction of the stress. The components σ_{11}, σ_{22} and σ_{33} are known as normal stresses as they are perpendicular to the plane on which they act and the others are shear stresses. The nine stress components are not independent of each other. If the cube is in equilibrium and not rotating it is necessary that for the shear stresses

$$\sigma_{12} = \sigma_{21}, \sigma_{23} = \sigma_{32}, \sigma_{31} = \sigma_{13}$$

and so the state of stress at a point in a body in equilibrium can be defined by a stress tensor σ_{ij} which has six independent components

$$\sigma_{ij} = \begin{pmatrix} \sigma_{11} & \sigma_{12} & \sigma_{13} \\ \sigma_{21} & \sigma_{22} & \sigma_{23} \\ \sigma_{31} & \sigma_{32} & \sigma_{33} \end{pmatrix}$$

The knowledge of the state of stress at a point provides enough information to calculate the stress acting on any plane within the body.

   It is possible to express the stress acting at a point in terms of three *principal stresses* acting along principal axes. In this case the shear stresses ($i \neq j$) are all zero and the only terms remaining in the stress tensor are σ_{11}, σ_{22} and σ_{33}. It is often possible to determine the principal axes from simple inspection of the body and the state of stress. For example, two principal axes always lie in the plane of a free surface. In the consideration of the

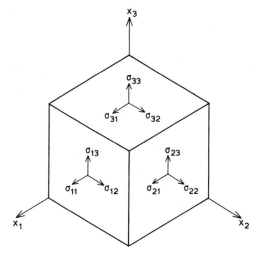

Fig. 5.1 *Schematic illustration of the nine stress components for a rectangular coordinate system.*

mechanical properties of polymers it is often useful to divide the stress tensor into its hydrostatic and deviatoric components. The hydrostatic pressure p is given by

$$p = \tfrac{1}{3}(\sigma_{11} + \sigma_{22} + \sigma_{33}) \tag{5.2}$$

and the deviatoric stress tensor σ'_{ij} is found by subtracting the hydrostatic stress components from the overall tensor such that

$$\sigma'_{ij} = \begin{pmatrix} (\sigma_{11} - p) & \sigma_{12} & \sigma_{13} \\ \sigma_{21} & (\sigma_{22} - p) & \sigma_{23} \\ \sigma_{31} & \sigma_{32} & (\sigma_{33} - p) \end{pmatrix} \tag{5.3}$$

5.1.2 *Strain*

When forces are applied to a material the atoms change position in response to the force and this change is known as *strain*. A simple example is that of a thin rod of material of length, l, which is extended a small amount δl by an externally applied stress. In this case the strain e could be represented by

$$e = \delta l / l \tag{5.4}$$

For a general type of deformation consisting of extension and compression in different directions and shears the situation is very much more complicated. However, the analysis can be greatly simplified by assuming that any strains are small and terms in e^2 can be neglected. In practice the analysis can only normally be strictly applied to the elastic deformation of crystals and a rather different approach has to be applied in the case of, for example, rubber elasticity where very high strains are involved.

In the analysis of *infinitesimal strains* the displacements of particular points in a body are considered. It is most convenient to use a rectangular co-ordinate system with axes x_1, x_2 and x_3 meeting at the origin 0. The analysis is most easily explained by looking at the two-dimensional case as shown in Fig. 5.2. If P is a point in the body with co-ordinates (x_1, x_2) then when the body is deformed it will move to a point P' given by co-ordinates $(x_1 + u_1, x_2 + u_2)$. However, another point Q which is infinitesimally close to P will not be displaced by the same amount. If Q originally has co-ordinates of $(x_1 + dx_1, x_2 + dx_2)$ its displacement to Q' will have components $(u_1 + du_1, u_2 + du_2)$. Since dx_1 and dx_2 are infinitesimal we can write

$$du_1 = \frac{\partial u_1}{\partial x_1}dx_1 + \frac{\partial u_1}{\partial x_2}dx_2 \qquad (5.5)$$

and

$$du_2 = \frac{\partial u_2}{\partial x_1}dx_1 + \frac{\partial u_2}{\partial x_2}dx_2 \qquad (5.6)$$

It is necessary to define the partial differentials and this can be done by writing them as

$$e_{11} = \frac{\partial u_1}{\partial x_1} \qquad e_{12} = \frac{\partial u_1}{\partial x_2}$$

$$e_{21} = \frac{\partial u_2}{\partial x_1} \qquad e_{22} = \frac{\partial u_2}{\partial x_2}$$

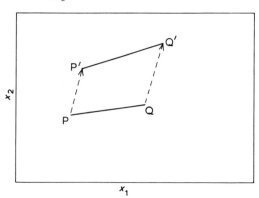

Fig. 5.2 *General displacement in two dimensions of points P and Q in a body (After Kelly and Groves).*

Equations (5.5) and (5.6) can be written more neatly as

$$du_i = e_{ij}\,dx_j \qquad (j = 1,2)$$

Since du_i and dx_j are vectors then e_{ij} must be a tensor and the significance of the various e_{ij} can be seen by taking particular examples of states of strain.

The simplest case is where the strains are small and the vector PQ can be taken as first parallel to the x_1 axis and then parallel to the x_2 axis as shown in Fig. 5.3(a). This enables the distortion of a rectangular element to be determined by considering the displacement of each vector separately. Since PQ_1 is parallel to x_1 then $dx_2 = 0$ and so

$$du_1 = \frac{\partial u_1}{\partial x_1}dx_1 = e_{11}dx_1 \tag{5.7}$$

$$du_2 = \frac{\partial u_2}{\partial x_1}dx_1 = e_{21}dx_1 \tag{5.8}$$

The significance of these equations can be seen from Fig. 5.3(a). When e_{11} and e_{21} are small the term e_{11} corresponds to the extensional strain of PQ_1 resolved along x_1 and e_{21} gives the anticlockwise rotation of PQ_1. It also follows that e_{22} will correspond to the extensional strain of PQ_2 resolved along x_2 and e_{12} the clockwise rotation of PQ_2.

The simple tensor notation of strain outlined above is not an entirely satisfactory representation of the strain in a body as it is possible to have certain components of $e_{ij} \neq 0$ without the body being distorted. This situation is illustrated in Fig. 5.3(b). In this case $e_{11} = e_{22} = 0$ but a small rotation, ω of the body can be given by $e_{12} = -\omega$ and $e_{21} = \omega$. This problem can be overcome by expressing the general strain tensor e_{ij} as a sum of an anti-symmetrical tensor and a symmetrical tensor such that

$$\left. \begin{array}{l} e_{ij} = \tfrac{1}{2}(e_{ij} - e_{ji}) + \tfrac{1}{2}(e_{ij} + e_{ji}) \\ \text{or} \quad e_{ij} = \qquad \omega_{ij} \qquad + \qquad \varepsilon_{ij} \end{array} \right\} \tag{5.9}$$

The term ω_{ij} then gives the rotation and ε_{ij} represents the pure strain.

This analysis can be readily extended to three dimensions and the strain tensor e_{ij} has nine rather than four components. The tensor ε_{ij} which represents the pure strain is again given by

$$\varepsilon_{ij} = \tfrac{1}{2}(e_{ij} + e_{ji})$$

and it can be written more fully as

$$\begin{pmatrix} \varepsilon_{11} & \varepsilon_{12} & \varepsilon_{13} \\ \varepsilon_{21} & \varepsilon_{22} & \varepsilon_{23} \\ \varepsilon_{31} & \varepsilon_{32} & \varepsilon_{33} \end{pmatrix} = \begin{pmatrix} e_{11} & \tfrac{1}{2}(e_{12} + e_{21}) & \tfrac{1}{2}(e_{13} + e_{31}) \\ \tfrac{1}{2}(e_{21} + e_{12}) & e_{22} & \tfrac{1}{2}(e_{23} + e_{32}) \\ \tfrac{1}{2}(e_{31} + e_{13}) & \tfrac{1}{2}(e_{32} + e_{23}) & e_{33} \end{pmatrix} \tag{5.10}$$

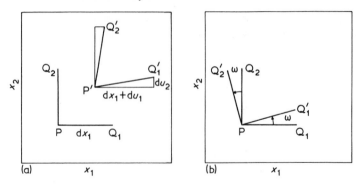

Fig. 5.3 *Two-dimensional displacements used in the definition of strain.* (a) *General displacements.* (b) *Rotation when* $e_{11} = e_{22} = 0$ *but* $e_{12} = -\omega$ *and* $e_{21} = \omega$. *(After Kelly and Groves).*

Inspection of this equation shows that $\varepsilon_{12} = \varepsilon_{21}$, $\varepsilon_{23} = \varepsilon_{32}$, $\varepsilon_{31} = \varepsilon_{13}$ which means that the strain tensor ε_{ij} has only six independent components (cf. stress tensor σ_{ij} in the last section). The diagonal components of ε_{ij} correspond to the tensile strains parallel to the axes and the off-diagonal ones are then shear strains.

As with stresses it is possible to choose a set of axes such that the shear strains are all zero and the only terms remaining are the *principal strains*, ε_{11}, ε_{22} and ε_{33}. In fact, for an isotropic body the axes of principal stress and principal strain coincide. This type of deformation can be readily demonstrated by considering what happens to a unit cube with the principal axes parallel to its edges. On deformation it transforms into a rectangular brick of dimensions $(1 + \varepsilon_{11})$, $(1 + \varepsilon_{22})$ and $(1 + \varepsilon_{33})$.

5.1.3 *Relationship between stress and strain*

For many materials the relationship between stress and strain can be expressed, at least at low strains, by *Hooke's law* which says that stress is proportional to strain. It must be borne in mind that this is not a fundamental law of nature. It was discovered by empirical observation and certain materials, particularly polymers, tend not to obey it. Hooke's law enables us to define the Young's modulus, E of a material which for simple uniaxial extension or compression is given by

$$E = \text{stress/strain}$$

It will be readily appreciated that this simple relationship cannot be applied without modification to complex systems of stress. In order to do this Hooke's law must be generalized and the relationship between σ_{ij} and ε_{ij} is obtained in its most general form by assuming that every stress component is a linear function of every strain component.

i.e. $\quad \sigma_{11} = A\varepsilon_{11} + B\varepsilon_{22} + C\varepsilon_{33} + D\varepsilon_{12} \ldots$

and $\quad \sigma_{22} = A'\varepsilon_{11} + B'\varepsilon_{22} + C'\varepsilon_{33} + D'\varepsilon_{12} \ldots$

This can be expressed most neatly using tensor notation as

$$\sigma_{ij} = c_{ijkl}\varepsilon_{ij} \tag{5.11}$$

where c_{ijkl} is a fourth-rank tensor containing all the stiffness constants. In principle, there are $9 \times 9 = 81$ stiffness constants but the symmetry of σ_{ij} and ε_{ij} reduces them to $6 \times 6 = 36$.

The notation used is often simplified to a contracted form

$$\sigma_{11} \rightarrow \sigma_1, \quad \sigma_{22} \rightarrow \sigma_2, \quad \sigma_{33} \rightarrow \sigma_3$$
$$\sigma_{23} \rightarrow \sigma_4, \quad \sigma_{13} \rightarrow \sigma_5, \quad \sigma_{12} \rightarrow \sigma_6$$
$$\varepsilon_{11} \rightarrow \varepsilon_1, \quad \varepsilon_{22} \rightarrow \varepsilon_2, \quad \varepsilon_{33} \rightarrow \varepsilon_3$$
$$2\varepsilon_{23} \rightarrow \gamma_4, \, 2\varepsilon_{13} \rightarrow \gamma_5, \, 2\varepsilon_{12} \rightarrow \gamma_6$$

It is conventional to replace the tensor shear strains by the angles of shear γ, which are twice the corresponding shear component in the strain tensor. There is also a corresponding contraction in c_{ijkl} which can then be written as

$$c_{ij} = \begin{pmatrix} c_{11} & c_{12} & c_{13} & c_{14} & c_{15} & c_{16} \\ c_{21} & c_{22} & c_{23} & c_{24} & c_{25} & c_{26} \\ c_{31} & c_{32} & c_{33} & c_{34} & c_{35} & c_{36} \\ c_{41} & c_{42} & c_{43} & c_{44} & c_{45} & c_{46} \\ c_{51} & c_{52} & c_{53} & c_{54} & c_{55} & c_{56} \\ c_{61} & c_{62} & c_{63} & c_{64} & c_{65} & c_{66} \end{pmatrix} \tag{5.12}$$

Only 21 of the 36 components of the matrix are independent since strain energy considerations show that $c_{ij} = c_{ji}$. Also if the strain axes are chosen to coincide with the symmetry axes in the material there may be a further reduction in the number of independent components. For example, in a cubic crystal the x_1, x_2 and x_3 axes are equivalent and so

$$c_{11} = c_{22} = c_{33}$$

$$c_{12} = c_{23} = c_{31}$$

$$c_{44} = c_{55} = c_{66}$$

and it can be proved further that all the other stiffness constants are zero. This means that there are only three independent elastic constants for cubic crystals, c_{11} c_{12} and c_{44}. Polymer crystals are always of lower symmetry and consequently there are more independent constants. Orthorhombic polyethylene requires nine stiffness constants to fully characterize its elastic behaviour.

In the case of elastically isotropic solids there are only two independent elastic constants, c_{11} and c_{12} because $2c_{44} = c_{11} - c_{12}$. Since glassy polymers and randomly oriented semi-crystalline polymers fall into this category it is worth considering how the stiffness constants can be related to quantities such as Young's modulus, E, Poisson's ratio ν, shear modulus G and bulk modulus K which are measured directly. The *shear modulus* G relates the shear stress σ_4 to the angle of shear γ_4 through the equation

$$\sigma_4 = G\gamma_4 \tag{5.13}$$

and since $c_{44} = \gamma_4/\sigma_4$ then we have the simple result that the shear modulus is given by

$$G = c_{44} \tag{5.14}$$

The situation is not so simple in the case of *uniaxial tension* where all stresses other than σ_1 are equal to zero. The generalized Hooke's law can be expressed as

$$\begin{aligned} \sigma_1 &= c_{11}\varepsilon_1 + c_{12}\varepsilon_2 + c_{13}\varepsilon_3 \\ &= c_{12}(\varepsilon_1 + \varepsilon_2 + \varepsilon_3) + (c_{11} - c_{12})\varepsilon_1 \\ &= c_{12}\Delta + 2c_{44}\varepsilon_1 \end{aligned} \tag{5.15}$$

where $\Delta = \varepsilon_1 + \varepsilon_2 + \varepsilon_3$ represents, at low strains, the fractional volume change or dilatation. The *bulk modulus* K relates the hydrostatic pressure p to the dilatation Δ through the equation

$$p = K\Delta \tag{5.16}$$

Since $p = \frac{1}{3}(\sigma_1 + \sigma_2 + \sigma_3)$ we can write using Equation (5.15)

$$p = (c_{12} + \tfrac{2}{3}c_{44})\Delta \tag{5.17}$$

Comparing Equations (5.16) and (5.17) enables K to be expressed as

$$K = c_{12} + \tfrac{2}{3}c_{44} \tag{5.18}$$

This equation then allows an expression for the Young's modulus E to be obtained. The Young's modulus relates the stress and strain in uniaxial tension through

$$\sigma_1 = E\varepsilon_1 \tag{5.19}$$

In uniaxial tension $\sigma_2 = \sigma_3 = 0$ and so

$$\sigma_1 = 3p = (3c_{12} + 2c_{44})\Delta \tag{5.20}$$

Substituting for σ_1 from Equation (5.15) and rearranging gives

$$\varepsilon_1 = \frac{(c_{12} + c_{44})\Delta}{c_{44}} \tag{5.21}$$

and so the modulus is given by

$$E = \frac{\sigma_1}{\varepsilon_1} = \frac{c_{44}(3c_{12} + 2c_{44})}{(c_{12} + c_{44})} \tag{5.22}$$

The final constant that can be derived is *Poisson's ratio* ν which is the ratio of the lateral strain to the longitudinal extension measured in a uniaxial tensile test. It is then given by

$$\nu = -\frac{\varepsilon_2}{\varepsilon_1} = -\frac{\varepsilon_3}{\varepsilon_1} \tag{5.23}$$

This can be related to Δ through

$$\nu = \tfrac{1}{2}(1 - \Delta/\varepsilon_1) \tag{5.24}$$

which using Equation (5.21) gives

$$\nu = \frac{c_{12}}{2(c_{12} + c_{44})} \tag{5.25}$$

It is clear that since the four constants, G, K, E and ν can be expressed in terms of three stiffness constants, only two of which are independent, then G, K, E and ν must all be related to each other. It is a matter of simple algebraic manipulation to show that, for example

$$G = \frac{E}{2(1 + \nu)} \quad \text{and} \quad K = \frac{E}{3(1 - 2\nu)} \quad \text{etc.}$$

The final point of this exercise is to show how an isotropic body will respond to a general system of stresses. A stress of σ_1 will cause a strain of ε_1/E along the x_1 direction, but will also lead to strains of $-\nu\sigma_1/E$ along the x_2 and x_3 directions. Similarly a stress of σ_2 will cause strains of σ_2/E along x_2 and of $-\nu\sigma_2/E$ in the other directions. If it is assumed that the strains due to the different stresses are additive then using the original notation it follows that

$$\varepsilon_{11} = \frac{\sigma_1}{E} - \frac{\nu}{E}(\sigma_{22} + \sigma_{33}) = e_{11} \tag{5.26}$$

$$\varepsilon_{22} = \frac{\sigma_2}{E} - \frac{\nu}{E}(\sigma_{33} + \sigma_{11}) = e_{22} \tag{5.27}$$

$$\varepsilon_{33} = \frac{\sigma_3}{E} - \frac{\nu}{E}(\sigma_{11} + \sigma_{22}) = e_{33} \tag{5.28}$$

The shear strains can be obtained directly from Equation (5.13) and are $\varepsilon_{12} = \sigma_{12}/2G$, $\varepsilon_{23} = \sigma_{23}/2G$, $\varepsilon_{31} = \sigma_{31}/2G$.

5.1.4 *Theoretical shear stress*

In the consideration of the resistance of materials to plastic flow it is useful to estimate theoretically the stress that may be required to shear the structure. It is necessary to do this before a sensible comparison can be made of the relative strengths of different types of materials such as metals and polymers. The general situation to be considered is shown in Fig. 5.4 where one sheet of atoms or molecules is allowed to slip over another under the action of a shear stress σ. The separation of the sheets is h and the repeat distance in the direction of shear is b. It is envisaged that the shear stress will vary periodically with the shear displacement x in an approximately sinusoidal manner and that the maximum resistance to shear will be when $x = b/4$. The variation of σ with the displacement x can then be written as

$$\sigma = \sigma_u \sin(2\pi x/b) \tag{5.29}$$

The initial slope of the curve of σ as a function of x will be equal to the shear modulus G of the material and so at small strains

$$\sigma = \sigma_u 2\pi x/b = Gx/h \tag{5.30}$$

and so the ideal shear strength is given by

$$\sigma_u = \frac{Gb}{2\pi h} \tag{5.31}$$

In a face-centred cubic metal $h = a/\sqrt{3}$ and $b = a/\sqrt{6}$ where a is the lattice parameter and so the theoretical shear strength is predicted to be $\sigma_u \approx G/9$. For chain direction slip on the (020) planes in the polyethylene crystal $b = 2.54$ Å and h is equal to the separation of the (020) planes which is 2.47 Å. The theoretical shear stress would be expected to be of the order of $G/6$. However, more sophisticated calculations of Equation (5.31) lead to lower estimates of σ_u which comes out to be of the order of $G/30$ for most materials. Even so, this estimate of the stress required to shear the structure is very much higher than the values that are normally measured and the discrepancy is due to the presence of defects such as *dislocations* within the crystals. The high values are only realized for certain crystal whiskers and other perfect crystals.

5.1.5 *Theoretical tensile strength*

Another parameter of interest in the consideration of the mechanical behaviour of materials is the stress that is expected to be required to cause

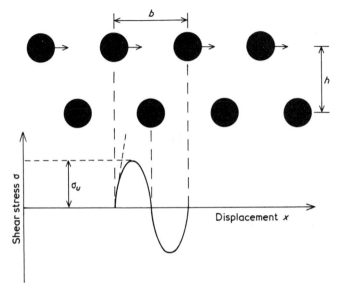

Fig. 5.4 *Schematic representation of the shear of layers of atoms in crystal and the expected variation of shear stress with displacement.*

cleavage through the breaking of bonds. The situation is visualized in Fig. 5.5 in which a bar of unit cross-sectional area, made of a perfectly elastic material is pulled in tension. Work is done in stressing the material and since fracture involves the creation of two new surfaces thermodynamic considerations demand that the work cannot be less than the surface energy of the new surfaces. The expected dependence of the stress σ upon the displacement x is shown in Fig. 5.5 and it is envisaged that this can be approximated to a sine function of wavelength λ and of the form

$$\sigma = \sigma_t \sin(2\pi x/\lambda) \tag{5.32}$$

where σ_t is the maximum tensile stress. This can be written for low strains as

$$\sigma = \sigma_t 2\pi x/\lambda \tag{5.33}$$

but another equation can be derived relating σ to x since the material is perfectly elastic. If the equilibrium separation of the atomic planes perpendicular to the tensile axis is h then for small displacements Hooke's law gives

$$\sigma = Ex/h \tag{5.34}$$

and comparing Equations (5.33) and (5.34) leads to

$$\sigma_t = \frac{\lambda E}{2\pi h} \tag{5.35}$$

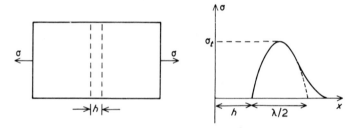

Fig. 5.5 *Model used to calculate the theoretical cleavage stress of a crystal.*

The work done when the bar is fractured is given approximately by the area under the curve and equating this to the energy required to create the two new surfaces leads to

$$\int_0^{\lambda/2} \sigma_t \sin(2\pi x/\lambda)\mathrm{d}x = 2\gamma$$

or $$\gamma = \lambda\sigma_t/2\pi$$

This then gives to the result that

$$\sigma_t = \sqrt{\frac{E\gamma}{h}} \tag{5.36}$$

Putting typical values of E, γ and h into this equation for a variety of materials shows that it is generally expected that

$$\sigma_t \approx E/10 \tag{5.37}$$

This result is particularly interesting in the context of polymers because although their moduli are generally low there are certain circumstances in which E can be made large. Oriented amorphous and semi-crystalline polymers have relatively high moduli and the moduli of polymer crystals in the chain direction are comparable to those of metals. High tensile strengths might therefore be expected to be displayed when polymers are deformed in ways which exploit these high moduli.

In general, it is found that the tensile strengths of materials are very much lower than $E/10$ because brittle materials fracture prematurely due to the presence of flaws (Section 5.6.2) and ductile materials undergo plastic deformation through the motion of dislocations. However, fine whiskers of glass, silica and certain polymer crystals which do not contain any flaws and are not capable of plastic deformation have values of fracture strength which are close to the theoretically predicted ones.

5.2 Viscoelasticity

A distinctive feature of the mechanical behaviour of polymers is the way in which their response to an applied stress or strain depends upon the rate or time period of loading. This dependence upon rate and time is in marked contrast to the behaviour of elastic solids such as metals and ceramics which, at least at low strains, obey *Hooke's law* and the stress is proportional to the strain and independent of loading rate. On the other hand the mechanical behaviour of liquids is completely time dependent. It is possible to represent their behaviour at low rates of strain by *Newton's law* whereby the stress is proportional to the strain-rate and independent of the strain. The behaviour of most polymers can be thought of as being somewhere between that of elastic solids and liquids. At low temperatures and high rates of strain they display elastic behaviour whereas at high temperatures and low rates of strain they behave in a viscous manner, flowing like a liquid. Polymers are therefore termed *viscoelastic* as they display aspects of both viscous and elastic types of behaviour.

Polymers used in engineering applications are often subjected to stress for prolonged periods of time, but it is not possible to know how a polymer will respond to a particular load without a detailed knowledge of its viscoelastic properties. There have been many studies of the viscoelastic properties of polymers and several textbooks written specifically about the subject of viscoelasticity. It is intended to give a brief introduction to what is really a vast subject and to leave the reader to consult more advanced texts for further details.

5.2.1 *General time-dependent behaviour*

The exact nature of the time dependence of the mechanical properties of a polymer sample depends upon the type of stress or straining cycle employed. The variation of the stress σ and strain e with time t is illustrated schematically in Fig. 5.6 for a simple polymer tensile specimen subjected to four different deformation histories. In each case the behaviour of an elastic material is also given as a dashed line for comparison. During *creep* loading a constant stress is applied to the specimen at $t = 0$ and the strain increases rapidly at first, slowing down over longer time periods. In an elastic solid the strain stays constant with time. The behaviour of a viscoelastic material during *stress relaxation* is shown in Fig. 5.6(b). In this case the strain is held constant and the stress decays slowly with time whereas in an elastic solid it would remain constant. The effect of deforming a viscoelastic material at a *constant stress-rate* is shown in Fig. 5.6(c). The increase in strain is not linear and the curve becomes steeper with time and also as the stress-rate is increased. If different *constant strain*

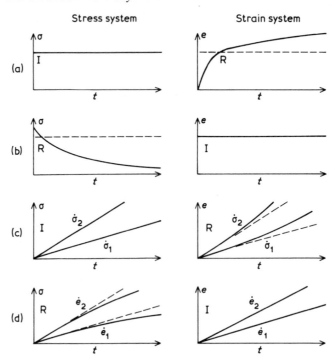

Fig. 5.6 *Schematic representation of the variation of stress and strain with time indicating the input (I) and responses (R) for different types of loading. (a) Creep, (b) Relaxation, (c) Constant stressing rate, (d) Constant straining rate (After Williams).*

rates are used the variation of stress with time is not linear. The slope of the curve tends to decrease with time, but it is steeper for higher strain-rates. The variation of both strain and stress with time is linear for constant stress- and strain-rate tests upon elastic materials.

5.2.2 Viscoelastic mechanical models

It is clear that any theoretical explanations of the above phenomena should be able to account for the dependence of the stress and strain upon time. Ideally it should be possible to predict, for example, the stress relaxation behaviour from knowledge of the creep curve. In practice, with real polymers this is somewhat difficult to do but the situation is often simplified by assuming that the polymer behaves as a *linear viscoelastic* material. It can be assumed that the deformation of the polymer is divided into an elastic component and a viscous component and that the deformation of the polymer can be described by a combination of Hooke's law and Newton's law. The linear elastic behaviour is given by Hooke's law as

$$\sigma = Ee \tag{5.38}$$

$$\text{or} \quad d\sigma/dt = E\,de/dt \tag{5.39}$$

where E is the elastic modulus and Newton's law describes the linear viscous behaviour through the equation

$$\sigma = \eta\frac{de}{dt} \tag{5.40}$$

where η is the viscosity and de/dt the strain rate. It should be noted that these equations only apply at small strains. A particularly useful method of formulating the combination of elastic and viscous behaviour is through the use of mechanical models. The two basic components used in the models are an elastic spring of modulus E which obeys Hooke's law (Equation 5.38) and a viscous dashpot of viscosity, η, which obeys Newton's law (Equation 5.40). The various models that have been proposed involve different combinations of these two basic elements.

(i) *Maxwell model*
This model was proposed in the last century by Maxwell to explain the time-dependent mechanical behaviour of viscous materials, such as tar or pitch. It consists of a spring and dashpot in series as shown in Fig. 5.7(a). Under the action of an overall stress σ there will be an overall strain e in the system which is given by

$$e = e_1 + e_2 \tag{5.41}$$

where e_1 is the strain in the spring and e_2 the strain in the dashpot. Since the elements are in series the stress will be identical in each one and so

$$\sigma_1 = \sigma_2 = \sigma \tag{5.42}$$

Equations (5.39) and (5.40) can be written therefore as

$$\frac{d\sigma}{dt} = E\frac{de_1}{dt} \quad \text{and} \quad \sigma = \eta\frac{de_2}{dt}$$

for the spring and dashpot respectively. Differentiation of Equation (5.41) gives

$$\frac{de}{dt} = \frac{de_1}{dt} + \frac{de_2}{dt} \tag{5.43}$$

and so for the Maxwell model

$$\frac{de}{dt} = \frac{1}{E}\frac{d\sigma}{dt} + \frac{\sigma}{\eta} \tag{5.44}$$

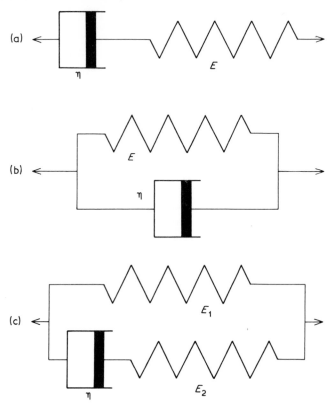

Fig. 5.7 *Mechanical models used to represent the viscoelastic behaviour of polymers.* (a) *Maxwell model,* (b) *Voigt model,* (c) *Standard linear solid.*

At this stage it is worth considering how closely the Maxwell model predicts the mechanical behaviour of a polymer. In the case of creep the stress is held constant at $\sigma = \sigma_0$ and so $d\sigma/dt = 0$. Equation (5.44) can therefore be written as

$$\frac{de}{dt} = \frac{\sigma_0}{\eta} \tag{5.45}$$

Thus, the Maxwell model predicts Newtonian flow as shown in Fig. 5.8(a). The strain is expected to increase linearly with time which is clearly not the case for a viscoelastic polymer (Fig. 5.6) where de/dt decreases with time.

The model is perhaps of more use in predicting the response of a polymer during stress relaxation. In this case a constant strain $e = e_0$ is imposed on the system and so $de/dt = 0$. Equation (5.44) then becomes

$$0 = \frac{1}{E}\frac{d\sigma}{dt} + \frac{\sigma}{\eta} \tag{5.46}$$

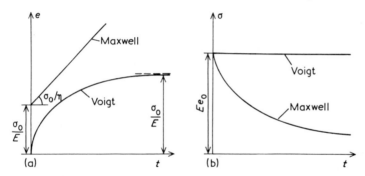

Fig. 5.8 *The behaviour of the Maxwell and Voigt models during different types of loading.* (a) *Creep (constant stress σ_0)*, (b) *Relaxation (constant strain e_0) (After Williams)*.

and hence

$$\frac{d\sigma}{\sigma} = -\frac{E}{\eta}dt \tag{5.47}$$

This can be readily integrated if at $t = 0$, $\sigma = \sigma_0$ and so

$$\sigma = \sigma_0 \exp\left(-\frac{Et}{\eta}\right) \tag{5.48}$$

where σ_0 is the initial stress. The term η/E will be constant for a given Maxwell model and it is sometimes referred to as a 'relaxation time', τ_0. The equation can then be written as

$$\sigma = \sigma_0 \exp(-t/\tau_0) \tag{5.49}$$

and it predicts an exponential decay of stress as shown in Fig. 5.8(b). This is a rather better representation of polymer behaviour than the Maxwell model gave for creep. However, Equation (5.49) predicts that the stress relaxes completely over long a period of time which is not normally the case for a real polymer.

(ii) *Voigt model*
This is also known as the Kelvin model. It consists of the same elements as are in the Maxwell model but in this case they are in parallel (Fig. 5.7b) rather than in series. The parallel arrangement of the spring and the dashpot means that the strains are uniform

i.e. $e = e_1 = e_2$ (5.50)

and the stresses in each component will add to make an overall stress of σ such that

$$\sigma = \sigma_1 + \sigma_2 \tag{5.51}$$

The individual stresses σ_1 and σ_2 can be obtained from Equations (5.38) and (5.40) and are

$$\sigma_1 = Ee \quad \text{and} \quad \sigma_2 = \eta\frac{de}{dt}$$

for the spring and the dashpot respectively. Putting these expressions into Equation (5.51) and rearranging gives

$$\frac{de}{dt} = \frac{\sigma}{\eta} - \frac{Ee}{\eta} \tag{5.52}$$

The usefulness or otherwise of the model can again be determined by examining its response to particular types of loading. The Voigt or Kelvin model is particularly useful in describing the behaviour during creep where the stress is held constant at $\sigma = \sigma_0$. Equation (5.52) then becomes

$$\frac{de}{dt} + \frac{Ee}{\eta} = \frac{\sigma_0}{\eta} \tag{5.53}$$

This simple differential equation has the solution

$$e = \frac{\sigma_0}{E}\left[1 - \exp\left(-\frac{Et}{\eta}\right)\right] \tag{5.54}$$

The constant ratio η/E can again be replaced by τ_0, the relaxation time and so the variation of strain with time for a Voigt model undergoing creep loading is given by

$$e = \frac{\sigma_0}{E}[1 - \exp(-t/\tau_0)] \tag{5.55}$$

This behaviour is shown in Fig. 5.8(a) and it clearly represents the correct form of behaviour for a polymer undergoing creep. The strain rate decreases with time and $e \to \sigma_0/E$ and $t \to \infty$. On the other hand, the Voigt model is unsuccessful in predicting the stress relaxation behaviour of a polymer. In this case the strain is held constant at e_0 and since $de/dt = 0$ then Equation (5.52) becomes

$$\frac{\sigma}{\eta} = \frac{Ee_0}{\eta} \tag{5.56}$$

$$\text{or} \quad \sigma = Ee_0 \tag{5.57}$$

which is the linear elastic response indicated in Fig. 5.8(b).

(iii) *Standard linear solid.*
It has been demonstrated that the Maxwell model describes the stress relaxation of a polymer to a first approximation whereas the Voigt model

similarly describes creep. A logical step forward is therefore to find some combination of these two basic models which can account for both phenomena. A simple model which does this is known as the standard linear solid, one example of which is shown in Fig. 5.7(c). In this case a Maxwell element and spring are in parallel. The presence of the second spring will stop the tendency of the Maxwell element undergoing simple viscous flow during creep loading, but will still allow the stress relaxation to occur.

There have been many attempts at devising more complex models which can give a better representation of the viscoelastic behaviour of polymers. As the number of elements increases the mathematics becomes more complex. It must be stressed that the mechanical models only give a mathematical representation of the mechanical behaviour and as such do not give much help in interpreting the viscoelastic properties on a molecular level.

5.2.3 *Boltzmann superposition principle*

A corner-stone of the theory of linear viscoelasticity is the *Boltzmann superposition principle*. It allows the state of stress or strain in a viscoelastic body to be determined from knowledge of its entire deformation history. The basic assumption is that during viscoelastic deformation in which the applied stress is varied, the overall deformation can be determined from the algebraic sum of strains due to each loading step. Before the use of the principle can be demonstrated it is necessary, first of all, to define a parameter known as the *creep compliance J(t)* which is a function only of time. It allows the strain after a given time $e(t)$ to be related to the applied stress σ for a linear viscoelastic material since

$$e(t) = J(t)\sigma \tag{5.58}$$

The use of the superposition principle can be demonstrated by considering the creep deformation caused by a series of step loads as shown in Fig. 5.9. If there is an increment of stress $\Delta\sigma_1$ applied at a time τ_1 then the strain due this increment at time, t is given by

$$e_1(t) = \Delta\sigma_1 J(t - \tau_1) \tag{5.59}$$

If further increments of stress, which may be either positive or negative, are applied then the principle assumes that the contributions to the overall strain, $e(t)$ from each increment are additive so that

$$
\begin{aligned}
e(t) &= e_1(t) + e_2(t) + \dots \\
&= \Delta\sigma_1 J(t - \tau_1) + \Delta\sigma_2 J(t - \tau_2) + \dots \\
&= \sum_{n=0}^{n} J(t - \tau_n)\Delta\sigma_n
\end{aligned}
\tag{5.60}
$$

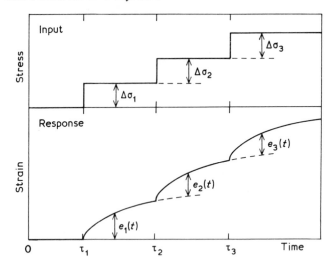

Fig. 5.9 *The response of a linear viscoelastic material to a series of loading steps (After Ward).*

It is possible to represent this summation in an integral form such that

$$e(t) = \int_{-\infty}^{t} J(t - \tau)d\sigma(t) \tag{5.61}$$

The integral is taken from $-\infty$ to t because it is necessary to take into account the entire deformation history of the viscoelastic sample as this will affect the subsequent behaviour. This equation can be used to determine the strain after any general loading history. It is normally expressed as a function of τ which gives

$$e(t) = \int_{-\infty}^{t} J(t - \tau)\frac{d\sigma(\tau)}{d\tau}d\tau \tag{5.62}$$

The use of the Boltzmann superposition principle can be seen by considering specific simple examples. A trivial case is for a single loading step of σ_0 at a time $\tau = 0$. In this case $J(t - \tau) = J(t)$ and so $e(t) = \sigma_0 J(t)$. A more useful application is demonstrated in Fig. 5.10 whereby a stress of σ_0 is applied at time $\tau = 0$ and taken off at time $\tau = t_1$. The strain due to loading, e_1, is given by

$$e_1 = \sigma_0 J(t) \tag{5.63}$$

and that due to unloading is similarly

$$e_2 = -\sigma_0 J(t - t_1) \tag{5.64}$$

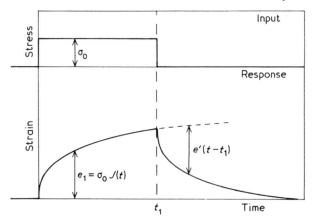

Fig. 5.10 *The response of a linear viscoelastic material to loading followed by unloading, illustrating the Boltzmann superposition principle.*

The total strain after time $t(>t_1)$, $e(t)$ is given by

$$e(t) = \sigma_0 J(t) - \sigma_0 J(t - t_1) \tag{5.65}$$

In this region $e(t)$ is decreasing and this process is known as recovery. If the recovered strain $e'(t - t_1)$ is defined as the difference between the strain that would have occurred if the initial stress had been maintained and the actual strain then

$$e'(t - t_1) = \sigma_0 J(t) - e(t) \tag{5.66}$$

Combining Equations (5.65) and (5.66) gives the recovered strain as

$$e'(t - t_1) = \sigma_0 J(t - t_1) \tag{5.67}$$

which is identical to the creep strain expected for a stress of $+\sigma_0$ applied after a time t_1. It is clear, therefore, that the extra extension or recovery that results from each loading or unloading event is independent of the previous loading history and is considered by the superposition principle as a series of separate events which add up to give the total specimen strain.

It is possible to analyse stress relaxation in a similar way using the Boltzmann superposition principle. In this case it is necessary to define a stress relaxation modulus $G(t)$ which relates the time-dependent stress $\sigma(t)$ to the strain e through the relationship

$$\sigma(t) = G(t)e \tag{5.68}$$

If a series of incremental strains Δe_1, Δe_2 etc. are applied to the specimen at times, τ_1, τ_2 etc., the total stress will be given by

$$\sigma(t) = \Delta e_1 G(t - \tau_1) + \Delta e_2 G(t - \tau_2) + \ldots \tag{5.69}$$

after time t. This equation is analogous to Equation (5.60) and can be given in a similar way as an integral (cf. Equation 5.62)

$$\sigma(t) = \int_{-\infty}^{t} G(t - \tau)\frac{de(\tau)}{d\tau}d\tau \tag{5.70}$$

This equation can then be used to predict the overall stress after any general straining programme.

5.2.4 *Dynamic mechanical testing*

So far only the response of polymers during creep and stress relaxation has been considered. Often, they are subjected to variable loading at a moderately high frequency and so it is pertinent to consider their behaviour during this type of deformation. The situation is most easily analysed when an oscillating sinusoidal load is applied to a specimen at a particular frequency. If the applied stress varies as a function of time according to

$$\sigma = \sigma_0 \sin \omega t \tag{5.71}$$

where ω is the angular frequency (2π times the frequency in Hz), the strain for an elastic material obeying Hooke's law would vary in a similar manner as

$$e = e_0 \sin \omega t \tag{5.72}$$

However, for a viscoelastic material the strain lags somewhat behind the stress (e.g. during creep). This can be considered as a damping process and the result is that when a stress defined by Equation (5.71) is applied to the sample the strain varies in a similar sinusoidal manner, but out of phase with the applied stress. Thus, the variation of stress and strain with time can be given by expressions of the type

$$\left.\begin{array}{l} e = e_0 \sin \omega t \\ \sigma = \sigma_0 \sin(\omega t + \delta) \end{array}\right\} \tag{5.73}$$

where δ is the 'phase angle' or 'phase lag' i.e. the relative angular displacement of the stress and strain (Fig. 5.11). The equation for stress can be expanded to give

$$\sigma = \sigma_0 \sin \omega t \cos \delta + \sigma_0 \cos \omega t \sin \delta \tag{5.74}$$

The stress can therefore be considered as being resolved into two components; one of $\sigma_0 \cos \delta$ which is in phase with the strain and another $\sigma_0 \sin \delta$ which is $\pi/2$ out of phase with the strain. Hence it is possible to define two dynamic moduli; E_1 which is in phase with the strain and E_2,

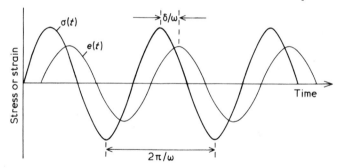

Fig. 5.11 *The variation of stress and strain with time for a viscoelastic material.*

$\pi/2$ out of phase with the strain. Since $E_1 = (\sigma_0/e_0)\cos\delta$ and $E_2 = (\sigma_0/e_0)\sin\delta$ Equation (5.74) becomes

$$\sigma = e_0 E_1 \sin\omega t + e_0 E_2 \cos\omega t \qquad (5.75)$$

The phase angle δ is then given by

$$\tan\delta = E_2/E_1 \qquad (5.76)$$

A complex notation is often favoured for the representation of the dynamic mechanical properties of viscoelastic materials. The stress and strain are given as

$$e = e_0 \exp i\omega t$$

$$\text{and} \quad \sigma = \sigma_0 \exp i(\omega t + \delta) \qquad (5.77)$$

where $i = (-1)^{1/2}$. The overall complex modulus $E^* = \sigma/e$ is then given by

$$E^* = \frac{\sigma_0}{e_0}\exp i\delta = \frac{\sigma_0}{e_0}(\cos\delta + i\sin\delta) \qquad (5.78)$$

From the definitions of E_1 and E_2 it follows that

$$E^* = E_1 + i E_2 \qquad (5.79)$$

and because of this, E_1 and E_2 are sometimes called the *real* and *imaginary* parts of the modulus respectively.

At this stage it is worth considering how the dynamic mechanical properties of a polymer can be measured experimentally. This can be done quite simply by using a *torsion pendulum* similar to the one illustrated schematically in Fig. 5.12. It consists of a cylindrical rod of polymer connected to a large inertial disc. The apparatus is normally arranged such that the weight of the disc is counter-balanced so that the specimen is not under any uniaxial tension or compression. The system can be set into oscillation by twisting and releasing the inertial disc and it undergoes sinusoidal oscillations at a frequency ω which depends upon the dimensions

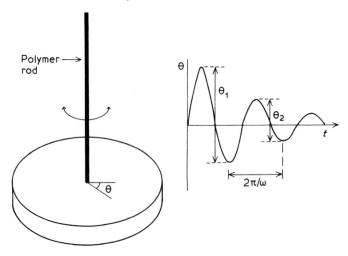

Fig. 5.12 *Schematic diagram of a torsion pendulum and the damping of the oscillations for a viscoelastic material.*

of the system and the properties of the rod. If the system is completely frictionless it will oscillate indefinitely when the rod is perfectly elastic. For a viscoelastic material such as a polymer the oscillations are damped and their amplitude decreases with time as shown in Fig. 5.12. The apparatus is sufficiently compact to allow measurements to be made over a wide range of temperature (typically $-200°C$—$+200°C$) but since the dimensions of the specimen cannot be varied greatly ω is limited to a relatively narrow range (typically 0.01–10Hz). Where a large range of frequency is required measurements of viscoelastic properties are usually made using forced-vibration techniques.

The motion of a torsion pendulum can be analysed relatively simply. If a circular cross-sectioned rod of radius a and length l is used then the torque T in the rod when one end is twisted by an angle Θ relative to the other is given by

$$T = \frac{G\pi a^4}{2l}\Theta \tag{5.80}$$

where G is the shear modulus of the rod. If the material is perfectly elastic then the equation of motion of the oscillating system is

$$I\ddot{\Theta} + T = 0 \tag{5.81}$$

where I is the moment of inertia of the disc. Hence

$$I\ddot{\Theta} + \frac{G\pi a^4}{2l}\Theta = 0 \tag{5.82}$$

describes the motion of the system. An equation of this form has the standard solution

$$\Theta = \Theta_0 \cos(\omega t + \alpha)$$

where

$$\omega = \left(\frac{G\pi a^4}{2lI}\right)^{1/2} \tag{5.83}$$

The shear modulus for an elastic rod is therefore given by

$$G = \frac{2lI\omega^2}{\pi a^4} \tag{5.84}$$

It was shown earlier that the Young's modulus of a viscoelastic material can be considered as a complex quantity E^*. It is possible to define the complex shear modulus in a similar way as

$$G^* = G_1 + iG_2 \tag{5.85}$$

where G_1 and G_2 are the real and imaginary parts of the shear modulus. If a viscoelastic material is used in the rod the equation of motion (Equation 5.82) then becomes

$$I\ddot{\Theta} + \frac{\pi a^4}{2l}(G_1 + iG_2)\Theta = 0 \tag{5.86}$$

If this equation is assumed to have a solution

$$\Theta = \Theta_0 e^{-\lambda t} e^{i\omega t} \tag{5.87}$$

then equating real and imaginary parts gives

$$G_1 = \frac{2lI}{\pi a^4}(\omega^2 - \lambda^2) \tag{5.88}$$

and

$$G_2 = \frac{2lI}{\pi a^4} 2\omega\lambda \tag{5.89}$$

Also the phase angle δ is given by

$$\tan \delta = E_2/E_1 = G_2/G_1 = 2\omega\lambda/(\omega^2 - \lambda^2) \tag{5.90}$$

A quantity that is normally measured for a viscoelastic material in the torsion pendulum is the logarithmic decrement Λ. This is defined as the natural logarithm of the ratio of the amplitude of two successive cycles.

i.e. $\Lambda = \ln(\Theta_n/\Theta_{n+1})$ (5.91)

The determination of Λ is shown in Fig. 5.12. Since the time period between two successive cycles is $2\pi/\omega$ then combining Equations (5.87) and (5.91) gives

$$\Lambda = 2\pi\lambda/\omega \tag{5.92}$$

Hence Equation (5.88) becomes

$$G_1 = \frac{2ll\omega^2}{\pi a^4}\left(1 - \frac{\Lambda^2}{4\pi^2}\right) \tag{5.93}$$

and Equation (5.89)

$$G_2 = \frac{2ll\omega^2}{\pi a^4}\frac{\Lambda}{\pi} \tag{5.94}$$

and δ is given by

$$\tan\delta = \frac{\Lambda/\pi}{1 - \Lambda^2/4\pi^2} \tag{5.95}$$

Usually Λ is small ($\ll 1$) and so the term $\Lambda^2/4\pi^2$ is normally neglected making Equation (5.93) become

$$G_1 \simeq \frac{2ll\omega^2}{\pi a^4} = G \tag{5.96}$$

and Equation (5.95) approximates to

$$\tan\delta \simeq \Lambda/\pi \tag{5.97}$$

when the damping is small.

The real part of the complex modulus G_1 is often called the *storage modulus* because it can be identified with the in-phase elastic component of the deformation. Elastic materials 'store' energy during deformation and release it on unloading. The imaginary part G_2 is sometimes called the *loss modulus* since it gives a measure of the energy dissipated during each cycle through its relation with Λ (Equation 5.94) and tan δ. The real and imaginary parts of E^* can be treated in an analogous way.

5.2.5 *Frequency dependence of viscoelastic behaviour*

It is found experimentally that when the values of storage modulus E_1, loss modulus E_2 and tan δ are measured for a polymer at a fixed temperature their values depend upon the testing frequency or rate. Generally it is found that tan δ and E_2 are usually small at very low and very high frequencies and their values peak at some intermediate frequency. On the other hand, E_1 is high at high frequencies when the polymer is displaying glassy behaviour and low at low frequencies when the polymer is rubbery. The value of E_1 changes rapidly at intermediate frequencies in the viscoelastic region where the damping is high and tan δ and E_2 peak (normally at slightly different frequencies).

The frequency dependence of the viscoelastic properties of a polymer can be demonstrated very simply using the Maxwell model (Section 5.2.2). The mechanical behaviour of the model is represented by Equation (5.44) as

$$\frac{de}{dt} = \frac{1}{E}\frac{d\sigma}{dt} + \frac{\sigma}{\eta} \tag{5.44}$$

where E is the modulus of the spring and η is the viscosity of the dashpot (Fig. 5.7a). It was shown earlier that the model has a characteristic relaxation time $\tau_0 = \eta/E$ and so Equation (5.44) becomes

$$E\tau_0\frac{de}{dt} = \tau_0\frac{d\sigma}{dt} + \sigma \tag{5.98}$$

It was shown in Section 5.2.4 that if the stress on a viscoelastic material is varied sinusoidally at a frequency ω then the variation of stress and strain with time can be represented by complex equations of the form

$$\left. \begin{array}{l} e = e_0\exp i\omega t \\ \sigma = \sigma_0\exp i(\omega t + \delta) \end{array} \right\} \tag{5.77}$$

where δ is the 'phase lag'. Substituting the relationships for σ and e given in Equation (5.77) into Equation (5.98) gives

$$\frac{E\tau_0 i\omega}{(\tau_0 i\omega + 1)} = \frac{\sigma_0\exp i(\omega t + \delta)}{e_0\exp i\omega t} = \frac{\sigma}{e} \tag{5.99}$$

The ratio σ/e is the complex modulus E^* and so

$$E_1 + iE_2 = E\tau_0 i\omega/(\tau_0 i\omega + 1) \tag{5.100}$$

Equating the real and imaginary parts then gives

$$E_1 = \frac{E\tau_0^2\omega^2}{(\tau_0^2\omega^2 + 1)} \quad \text{and} \quad E_2 = \frac{E\tau_0\omega}{(\tau_0^2\omega^2 + 1)} \tag{5.101}$$

and since tan $\delta = E_2/E_1$ it follows that

$$\tan \delta = 1/\tau_0\omega \tag{5.102}$$

The variation of E_1, E_2 and tan δ with frequency for a Maxwell model with a fixed relaxation time is sketched in Fig. 5.13. The behaviour of E_1 and E_2 is predicted quite well by the analysis with a peak in E_2 at $\tau_0\omega = 1$, but tan δ

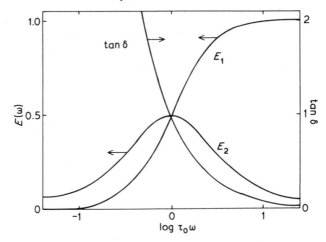

Fig. 5.13 *Variation of E_1, E_2 and tan δ with frequency ω for a Maxwell model with a spring modulus $E = 1$ and a fixed relaxation time τ_0.*

drops continuously as ω increases and the peak that is found experimentally does not occur. Better predictions can be obtained by using more complex models, such as the standard linear solid, and combinations of several models which have a range of relaxation times.

One of the main reasons for studying the frequency dependence of the dynamic mechanical properties is that it is often possible to relate peaks in E_2 and tan δ to particular types of molecular motion in the polymer. The peaks can be regarded as a 'damping' effect and occur at the frequency of some molecular motion in the polymer structure. There is particularly strong damping at the *glass transition*. We have, so far, considered this a purely static phenomenon (Section 4.4.2) being observed through changes in properties such as specific volume or heat capacity. The large increase in free volume which occurs above the T_g allows space for molecular motion to occur more easily. A major resonance occurs at T_g when the applied test frequency equals the natural frequency for main chain rotation. At higher frequencies there will be insufficient time for chain uncoiling to occur and the material will be relatively stiff. On the other hand at lower frequencies the chains will have more time to move and the polymer will appear to be soft and rubbery.

Peaks in E_2 can also be found at the *melting temperature*, T_m, in semi-crystalline polymers. This is due to the greater freedom of molecular motion possible when the molecules are no longer packed regularly into crystals.

It is possible that other types of molecular motion can occur such as the rotation of side-groups as these are often reflected in other damping occurring at somewhat different frequencies. The magnitude of the peaks in E_2 and tan δ in this case are usually smaller than those obtained at T_g and such phenomena are known as 'secondary transitions'. The fact that dynamic mechanical testing can be used to follow main chain and side-group motion in polymers makes it a powerful technique for the characterization of polymer structures.

The exact frequency at which the T_g and secondary transitions occur depends upon the temperature of testing as well as the structure of the polymer. In general, the frequency at which the transition takes place increases as the testing temperature is increased and for a given testing frequency it is possible to induce a transition by changing the testing temperature. From a practical viewpoint, it is normally much easier to keep the frequency fixed and vary the testing temperature rather than do measurement at a variable frequency and so most of the data available on the dynamic mechanical behaviour of polymers has been obtained at a fixed frequency and over a range of temperature. For a complete picture of the dynamic mechanical behaviour of a polymer it is desirable that measurements should be made over as wide a range of frequency and temperature as possible and it is sometimes possible to relate data obtained at different frequencies and temperatures through a procedure known as time–temperature superposition (see Section 5.2.7).

5.2.6 *Transitions and polymer structure*

The transition behaviour of a very large number of polymers has been widely studied as a function of testing temperature using dynamic mechanical testing methods. The types of behaviour observed are found to depend principally upon whether the polymers are amorphous or crystalline. Fig. 5.14 shows the variation of the shear modulus G_1 and tan δ with temperature for atactic polystyrene which is typical for amorphous polymers. The shear modulus decreases as the testing temperature is increased and drops sharply at the glass transition where there is a corresponding large peak in tan δ. Fig. 5.14(b) shows the variation of tan δ with temperature and it is possible to see minor peaks at low temperatures. These correspond to secondary transitions. The peak which occurs at the highest temperature is normally labelled α and the subsequent ones are called β, γ etc. In an amorphous polymer the α-transition or α-relaxation corresponds to the glass transition which can be detected by other physical testing methods (Section 4.4.2) and so the temperature of the α-relaxation depends upon the chemical structure of the polymer. In general, it is found that the transition temperature is increased as the main chain is made stiffer. Side-groups raise the T_g or α-relaxation temperature if they are polar or large and bulky and reduce it if they are long and flexible.

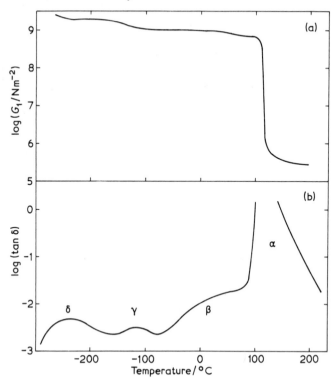

Fig. 5.14 *Variation of shear modulus* G_1 *and* tan δ *with temperature for polystyrene (After Arridge).*

The assignment of particular mechanisms to the secondary β, γ etc. transitions can sometimes be difficult since their position and occurrence depends upon which polymer is being studied. At least three secondary transitions have been reported for atactic polystyrene (Fig. 5.14b). The position of the β-peak at about 50°C is rather sensitive to testing frequency and it merges with the α-relaxation at high frequencies. It has been assigned to rotation of phenyl groups around the main chain or alternatively the co-operative motion of segments of the main chain containing several atoms. It is thought that the γ-peak may be due to the occurrence of head-to-head rather than head-to-tail polymerization (Section 2.2.4) and the δ-peak has been assigned to rotation of the phenyl group around its link with the main chain.

The interpretation of the relaxation behaviour of semi-crystalline polymers can be extremely difficult, because of their complex two-phase structure (Section 4.2). It is sometimes possible to consider the crystalline and amorphous regions as separate entities with the amorphous areas having a glass transition. It is clear, however, that as particular molecules can traverse both the amorphous and crystalline regions, this picture may be too simplistic. The relaxation behaviour of polyethylene has probably been more widely studied than that of any other semi-crystalline polymer and so this will be taken as an example, even though it is not completely understood. Fig. 5.15 shows the variation of tan δ with temperature for two samples of polyethylene, one high density (linear) and the other low density (branched). In polyethylene there are four transition regions designated α', α, β and γ. The γ transition is very similar in both the high- and low-density samples whereas the β-relaxation is virtually absent in the high density polymer. The α and α' peaks are also somewhat different in the pre-melting region.

The presence of branches upon the polyethylene molecules produces a difference in polymer morphology as well as in the molecules themselves. The degree of crystallinity and crystal size and perfection are all reduced by the branching and this can lead to complications in interpretation of the relaxation behaviour. The intensities of the α'- and α-relaxation decrease as the degree of crystallinity is reduced, implying that they are associated with motion within the crystalline regions. On the other hand, the intensity of the γ-relaxation increases with a reduction in crystallinity indicating that it is associated with the amorphous material and it has been tentatively assigned to a glass transition in the non-crystalline regions. The disappearance of the β-transition with the absence of branching has been taken as a strong indication that it is associated with relaxations at the branch-points. In general, the assignment of the peaks in crystalline polymers to particular types of molecular motion is rather difficult and often a matter for some debate.

5.2.7 *Time–temperature superposition*

It has been suggested earlier that it may be possible to interrelate the time and temperature dependence of the viscoelastic properties of polymers. It is thought that there is a general equivalence between time and temperature. For instance, a polymer which displays rubbery characteristics under a given set of testing conditions can be induced to show glassy behaviour by either reducing the temperature or increasing the testing rate or frequency. This type of behaviour is shown in Fig. 5.16 for the variation of the shear compliance J of a polymer with testing frequency measured at different temperatures in the region of the T_g. The material is rubber-like with a high compliance at high temperatures and low frequencies and becomes glassy with a low compliance as the temperature is reduced and the frequency increased.

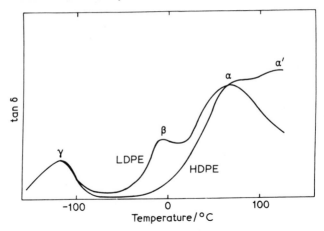

Fig. 5.15 *Variation of tan δ with temperature for high- and low-density polyethylene (Adapted from Ward).*

It was found empirically that all the curves in Fig. 5.16 could be superposed by keeping one fixed and shifting all the others by different amounts horizontally parallel to the logarithmic frequency axis. If an arbitrary reference temperature T_s is taken to fix one curve then if ω_s is the frequency of a point on the curve at T_s with a particular compliance and ω is the frequency of a point with the same compliance on a curve at a different temperature then the amount of shift required to superpose the two curves is a displacement of $(\log \omega_s - \log \omega)$ along the log frequency axis. The 'shift factor' a_T is defined by

$$\log a_T = \log \omega_s - \log \omega = \log(\omega_s/\omega) \tag{5.103}$$

and this parameter is a function only of temperature. The values of $\log a_T$ which must be used to superpose the curves in Fig. 5.16 are given in the insert. The density of the polymer changes with temperature and so this will also affect the compliance of the polymer. It is normally unnecessary to take this into account unless a high degree of accuracy is required. Much of the early work upon time–temperature superposition was done by Williams, Landel and Ferry and they proposed that a_T could be given by an equation of the form

$$\log a_T = \frac{-C_1(T - T_s)}{C_2 + (T - T_s)} \tag{5.104}$$

where C_1 and C_2 are constants and T_s is the reference temperature. This is normally called the WLF equation and was originally developed empirically. It holds extremely well for a wide range of polymers in the vicinity of

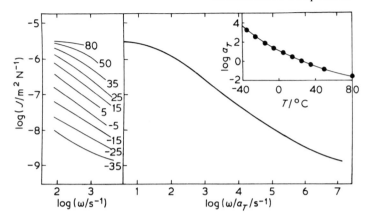

Fig. 5.16 *Example of the time–temperature superposition principle using shear compliance, J, data for polyisobutylene. The curves of J as a function of frequency obtained at different temperatures can all be shifted to lie on a master curve using the shift factor, a_T. (Using data of Fitzgerald, Grandine and Ferry, J. Appl. Phys. **24** (1953) 650.)*

the glass transition and if T_s is taken as T_g, measured by a static method such as dilatometry then

$$\log a_T = \frac{-C_1^g(T - T_g)}{C_2^g + (T - T_g)} \tag{5.105}$$

and the new constants C_1^g and C_2^g become 'universal' with values of 17.4 and 51.6 K respectively. In fact, the constants vary somewhat from polymer to polymer, but it is often quite safe to assume the universal values as they usually give shift factors which are close to measured values.

The WLF master curve produced by superposing data obtained at different temperatures is a very useful way of presenting the mechanical behaviour of a polymer. The WLF equation is also useful in predicting the mechanical behaviour of a polymer outside the range of temperature and frequency (or time) for which experimental data are available.

Although the WLF equation was originally developed by curve fitting it is possible to justify it theoretically from considerations of free volume. The concept of free volume was introduced in Section 4.4 in the context of the variation of the specific volume of a polymer with temperature and the abrupt change in slope of the $V - T$ curve at T_g. The fraction free volume f in a polymer is given as (Equation 4.37)

$$f = f_g + (T - T_g)\alpha_f \tag{5.106}$$

where f_g is the fractional free volume at T_g and α_f is the thermal expansion coefficient of the free volume. There are several ways of deriving a relationship of the form of the WLF equation, but perhaps the simplest is by assuming that the polymer behaves as a viscoelastic model having a relaxation time, τ_0 (see Section 5.2.2). For the Maxwell model $\tau_o = \eta/E$ where η is the viscosity of the dashpot and E is modulus of the spring. It can be assumed that E is independent of temperature and only η varies with temperature. If the shifts in the time–temperature superposition are thought of as a process of matching the relaxation times then the shift factor is given by

$$a_T = \frac{\tau_0(T)}{\tau_0(T_g)} = \frac{\eta(T)}{\eta(T_g)} \tag{5.107}$$

when the T_g is used as the reference temperature. It is possible to relate the viscosity to the free volume through a semi-empirical equation developed by Doolittle from the study of the viscosities of liquids. He was able to show that for a liquid η was related to the free volume V_f through an equation of the form

$$\ln \eta = \ln A + B(V - V_f)/V_f \tag{5.108}$$

where V is the total volume and A and B are constants. This equation can be rearranged to give

$$\ln \eta(T) = \ln A + B(1/f - 1) \tag{5.109}$$

and at the T_g

$$\ln \eta(T_g) = \ln A + B(1/f_g - 1) \tag{5.110}$$

Substitution for f in Equation (5.109) and subtracting Equation (5.110) from (5.109) gives

$$\ln \frac{\eta(T)}{\eta(T_g)} = B\left(\frac{1}{f_g + \alpha_f(T - T_g)} - \frac{1}{f_g}\right) \tag{5.111}$$

which can be rearranged to give

$$\log \frac{\eta(T)}{\eta(T_g)} = \log a_T = \frac{-(B/2.303 f_g)(T - T_g)}{f_g/\alpha_f + (T - T_g)} \tag{5.112}$$

which has an identical form as the WLF equation. The universal nature of C_1^g and C_2^g in the WLF equation implies that f_g and α_f should also be the same for different polymers. It is found that f_g is of the order of 0.025 and α_f approximately 4.8×10^{-4} K^{-1} for most amorphous polymers.

5.2.8 Non-linear viscoelasticity

An unfortunate aspect of the analysis of the viscoelastic behaviour of polymers is that for most polymers deformed in practical situations, the theory of linear viscoelasticity does not apply. It was emphasized earlier (Section 5.2.2) that Hooke's law and Newton's law are only obeyed at low strains. However, many polymers and especially semi-crystalline ones do not obey the Boltzmann superposition principle even at low strains and it is found that the exact loading route affects the final state of stress and strain. In this case empirical approaches are normally used and the mechanical properties are evaluated for the range of conditions of service expected (e.g. stress, strain, time and temperature) and are tabulated or displayed graphically for use by design engineers and anyone who requires the information.

5.3 Rubber Elasticity

Rubbers possess the remarkable ability of being able to stretch to five to ten times their original length and then retract rapidly to near their original dimensions when the stress is removed. This *elastomeric* behaviour sets them apart from materials such as metals and ceramics where the maximum reversible strains that can be tolerated are usually less than one percent. Most non-crystallizable polymers are capable of being obtained in the rubber-like state. Fig. 5.17 shows the variation of the Young's modulus E of an amorphous polymer with temperature for a given rate of frequency testing. At low temperatures the polymer is glassy with a high modulus ($\sim 10^9 \text{Nm}^{-2}$). The modulus then falls through the glass transition region where the polymer is viscoelastic and the modulus becomes very rate- and temperature-dependent and at a sufficiently high temperature the polymer becomes rubbery. However, all rubbery polymers do not show useful elastomeric properties as they will flow irreversibly on loading unless they are cross-linked. The modulus of a linear polymer drops effectively to zero at a sufficiently high temperature when it behaves like a liquid, but if a polymer is cross-linked it will remain approximately constant at $\sim 10^6 \text{Nm}^{-2}$ as the temperature is increased.

The properties of elastomeric materials are controlled by their molecular structure which has been discussed earlier (Section 4.4.5). They are basically all amorphous polymers above their glass transition and normally cross-linked. Their unique deformation behaviour has fascinated scientists for many years and there are even reports of investigations into the deformation of natural rubber from the beginning of the last century. Rubber elasticity is particularly amenable to analysis using thermodynamics, as an elastomer behaves essentially as an 'entropy spring'. It is even possible to derive the form of the basic stress–strain relationship from first principles by considering the statistical thermodynamic behaviour of the molecular network.

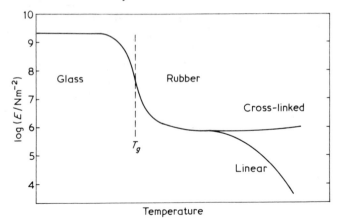

Fig. 5.17 *Typical variation of Young's modulus, E, with temperature for a polymer showing the effect of cross-linking upon E in the rubbery state.*

5.3.1 *Thermodynamics of rubber-elasticity*

As well as having the ability to sustain large reversible amounts of extension there are several other more subtle aspects of the deformation of rubbers which are rather unusual and merit further consideration. A well-known property is that when a sample of rubber is stretched rapidly it becomes warmer and conversely when it is allowed to equilibrate for a while at constant extension and then allowed to contract rapidly it cools down. This can be readily demonstrated by putting an elastic band to ones lips, which are quite sensitive to changes in temperature, and stretching it quickly. Another property is that the length of a sample of rubber held at constant stress decreases as the temperature is increased whereas an unstressed rubber exhibits conventional thermal expansion behaviour. It is possible to rationalize these aspects of the thermomechanical behaviour by considering the deformation of rubbers in thermodynamic terms. The use of thermodynamics allows the basic relationship between force, length and temperature to be established for an elastomeric specimen in terms of parameters such as internal energy and entropy. One important experimental observation which allows the analysis to be simplified is that the deformation of rubbers takes place approximately at constant volume.

The first law of thermodynamics establishes the relationship between the change in internal energy of a system dU and the heat $\bar{d}Q$ absorbed by the system and work $\bar{d}W$ done by the system as

$$dU = \bar{d}Q - \bar{d}W \tag{5.113}$$

The bars indicate that $\bar{d}Q$ and $\bar{d}W$ are inexact differentials because Q and W, unlike U, are not macroscopic functions of the system. If the length of an elastic specimen is increased a small amount dl by a tensile force f then an amount of work f dl will be done *on* the system. There is also a change in volume dV during elastic deformation and work p dV is done against the pressure, p. Since rubbers deform at roughly constant volume the contribution of p dV to dW at ambient pressure will be small and so the work done *by* the system (i.e. the specimen) when it is extended is given by

$$dW = -f\,dl \tag{5.114}$$

The deformation of elastomers can be considered as a reversible process and so dQ can be evaluated from the second law of thermodynamics which states that for a reversible process

$$dQ = T\,dS \tag{5.115}$$

where T is the thermodynamic temperature and dS is the change in entropy of the system. Combining Equations (5.113), (5.114) and (5.115) gives

$$f\,dl = dU - T\,dS \tag{5.116}$$

Most of the experimental investigation on elastomers have been done under conditions of constant pressure. The thermodynamic function which can be used to describe equilibrium under these conditions is the Gibbs free energy (Equation 4.1), but since elastomers tend to deform at constant volume it is possible to use the Helmholtz free energy, A in the consideration of equilibrium. It is defined as

$$A = U - TS \tag{5.117}$$

and for a change taking place under conditions of constant temperature

$$dA = dU - TdS \tag{5.118}$$

Comparing Equations (5.116) and (5.118) gives

$$fdl = dA \quad (const.\ T)$$

and so

$$f = (\partial A/\partial l)_T \tag{5.119}$$

Combining equations 5.118 and 5.119 enables the force to be expressed as

$$f = \left(\frac{\partial U}{\partial l}\right)_T - T\left(\frac{\partial S}{\partial l}\right)_T \tag{5.120}$$

The first term refers to the change in internal energy with extension and the second to the change in entropy with extension. An expression of this form could be used approximately to describe the response of any solid to an applied force. For most materials the internal energy term is dominant, but in the case of rubbers the change in entropy gives the largest contribution to the force. This can be demonstrated after modifying the entropy term in Equation (5.120). The Helmholtz free energy can be expressed in a differential form for any general change as

$$dA = dU - TdS - SdT \qquad (5.121)$$

and combining this equation with Equation (5.116) gives

$$dA = fdl - SdT \qquad (5.122)$$

and partial differentiation under conditions first of all of constant temperature and then of constant length gives

$$\left.\begin{array}{c} (\partial A/\partial l)_T = f \\ \text{and} \quad (\partial A/\partial T)_l = -S \end{array}\right\} \qquad (5.123)$$

But we have the standard relation for partial differentiation

$$\frac{\partial}{\partial l}\left(\frac{\partial A}{\partial T}\right)_l = \frac{\partial}{\partial T}\left(\frac{\partial A}{\partial l}\right)_T \qquad (5.124)$$

and applying this to Equations (5.123) gives

$$(\partial S/\partial l)_T = -(\partial f/\partial T)_l \qquad (5.125)$$

Equation (5.120) then becomes

$$f = \left(\frac{\partial U}{\partial l}\right)_T + T\left(\frac{\partial f}{\partial T}\right)_l \qquad (5.126)$$

These last two equations contain parameters which can be measured experimentally and so allow the entropy and internal energy terms to be calculated. This can be done by measuring how the force required to hold a piece of rubber at constant length varies with the absolute temperature. Fig. 5.18 shows the results of an early experiment of this type. The stress in the rubber drops as the temperature is reduced until about 220 K at which point it starts to rise again. This change in slope corresponds to the transition to a glassy state when the material is no longer elastomeric. The curve in the rubbery region is virtually linear and can be extrapolated approximately to zero tension at absolute zero. The curve in Fig. 5.18 is of great significance and can be related directly to Equation (5.126). The slope of the curve at any point $(\partial f/\partial T)_l$ gives the variation of entropy with extension at that temperature. Since the curve is linear it suggests that $(\partial S/\partial l)_T$ is fairly temperature independent and as the intercept is approximately zero it also implies that $(\partial U/\partial l)_T$ is small. This means that there is very little change in internal energy during the extension of a rubber and that the deformation is controlled by the change in entropy.

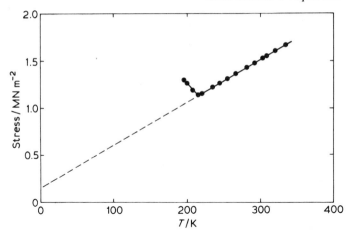

Fig. 5.18 *Variation of the stress in a rubber, as a function of temperature, held at a fixed extension of 350 percent (Data of Meyer and Ferri reported by Treloar.)*

The data in Fig. 5.18 were obtained at a fairly high extension (350 percent). When the experiment is repeated at lower extensions the slope of the curve decreases and below about 10 percent extension it becomes negative. This is caused by a reduction in stress due to thermal expansion of the rubber as the temperature is increased and it is known as the *thermo-elastic inversion effect*. If the effective change in extension due to thermal expansion is allowed for then the thermoelastic inversion effect disappears and the stress increases proportionately with temperature at low extensions as well as high extensions.

All the considerations of thermoelastic effects have so far been concerned with changes in stress at constant extension, but they can be extended directly to tests done at constant tension. In this case the tendency of the stress to increase as the temperature is raised will cause a specimen subjected to a constant stress to reduce its length as the temperature is increased. This can be demonstrated readily by hanging a weight on a length of rubber and changing the temperature. The weight moves upwards when the rubber is heated and it goes downwards when the rubber is cooled.

Another thermoelastic effect that was mentioned earlier is the tendency of a rubber to become warm when stretched rapidly. This type of behaviour is illustrated in Fig. 5.19 for the adiabatic extension of a rubber. Some of the heating can be explained by crystallization which can also occur when a rubber is stretched (Section 4.4.5) but a temperature rise is found even in non-crystallizable rubbers. This can be explained by consideration of the thermodynamic equilibrium. The rise in temperature

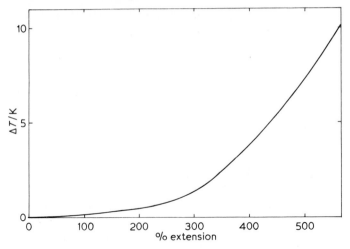

Fig. 5.19 *Increase in temperature, ΔT, upon the adiabatic extension of a rubber. (Data of Dart, Anthony and Guth reported by Treloar).*

as the length is increased adiabatically (isentropically) is characterized by $(\partial T/\partial l)_S$ which is given by the standard relation

$$\left(\frac{\partial T}{\partial l}\right)_S = -\left(\frac{\partial T}{\partial S}\right)_l\left(\frac{\partial S}{\partial l}\right)_T \qquad (5.127)$$

The first factor on the right-hand side can be expressed as

$$\left(\frac{\partial T}{\partial S}\right)_l = \left(\frac{\partial T}{\partial H}\right)_l\left(\frac{\partial H}{\partial S}\right)_l \qquad (5.128)$$

But

$$\left(\frac{\partial T}{\partial H}\right)_l = \frac{1}{C_l} \quad \text{and} \quad \left(\frac{\partial H}{\partial S}\right)_l = T$$

and so Equation (5.127) becomes

$$\left(\frac{\partial T}{\partial l}\right)_S = -\frac{T}{C_l}\left(\frac{\partial S}{\partial l}\right)_T \qquad (5.129)$$

where C_l is the heat capacity of the rubber at constant length. Combining Equations (5.125) and (5.129) gives

$$\left(\frac{\partial T}{\partial l}\right)_S = \frac{T}{C_l}\left(\frac{\partial f}{\partial T}\right)_l \qquad (5.130)$$

This equation shows clearly that the temperature of a rubber will rise when it is stretched adiabatically as long as $(\partial f/\partial T)_l$ is positive as is normally the case (Fig. 5.18). It is a direct consequence of the fact that when a rubber is stretched, work is transformed reversibly into heat.

5.3.2 *Statistical theory of rubber elasticity*

Although the thermodynamic approach shows clearly that when a rubber is deformed it behaves like an entropy spring, no specific deformation mechanisms are implied. In fact, none of the thermodynamic equations depend upon the macromolecular nature of the rubber although the behaviour obviously stems from the presence of polymer molecules in the rubber. The statistical approach looks directly at how the molecular structure changes during deformation and allows the change in entropy during deformation to be calculated and the stress–strain curve derived from first principles. It is similar to the calculation of the pressure of an ideal gas using statistical thermodynamics and it makes use of the calculations of the conformations of freely-jointed chains outlined in Section 3.1.2.

(i) *Entropy of an individual chain*

The rubber is assumed to be made up of a network of cross-linked polymer chains with the individual lengths of chain in their most random conformations. When the network is extended the molecules become uncoiled and their entropy is reduced. The force required to deform the rubber can therefore be related directly to this change in entropy. The first step in the analysis involves the calculation of the entropy of an individual polymer molecule. This can be done with the help of Equation (3.4) which gives the probability per unit volume $W(x, y, z)$ of finding one end a freely-jointed chain at a point (x, y, z) a distance r from the other end which is fixed at the origin. This probability can be expressed as

$$W(x,y,z) = (\beta/\pi^{1/2})^3 \exp(-\beta^2 r^2) \qquad (3.4)$$

and the function has been plotted graphically in Fig. 3.7(a). The parameter β is defined as $\beta = (3/(2nl^2))^{1/2}$ where n is the number of links of length l in the chain and β can be considered to be characteristic for a particular chain. Equation (3.4) allows the entropy of a single chain to be calculated using the Boltzmann relationship from statistical thermodynamics which is

$$S = k \ln \Omega \qquad (5.131)$$

where k is Boltzmann's constant and Ω is the number of possible conformations the chain can adopt. It can be assumed that Ω will be proportional to $W(x, y, z)$. This is clearly a reasonable assumption since when r is small $W(x, y, z)$ is large (Fig. 3.7a) and there will be many conformations available to the chain. $W(x, y, z)$ drops as r is increased and the number of possible conformations is reduced as the molecule becomes extended. The entropy, S, of a single chain can then be expressed by an equation of the form

$$S = c - k\beta^2 r^2 \qquad (5.132)$$

where c is a constant.

(ii) *Deformation of the polymer network*

Having obtained an expression for the entropy of a single chain it is a relatively simple matter to determine the change in entropy on deforming an elastomeric polymer network. It is assumed that when the rubber is in either the strained or unstrained state the junction points can be regarded as being fixed at their mean positions and that the lengths of chain between the points behave as freely jointed chains so that Equation (5.132) can be applied. It is also assumed that when the rubber is deformed the change in the components of the displacement vector, r, of each chain are proportional to the corresponding change in specimen dimensions. In the consideration of the deformation of rubbers large strains are encountered and it is more convenient to use *extension ratios* λ_1, λ_2, λ_3 rather than infinitessimal strains. The extension ratio, λ, in a particular direction is defined as the deformed length of the specimen in that direction divided by the original length. The axes of these ratios are chosen to coincide with the rectangular coordinate system used to characterize the conformations of the original lengths of chain. If an individual chain is considered, first of all, then if one junction point is fixed at the origin, O, a general deformation of the rubber will displace the other junction point from (x, y, z) to (x', y', z') as shown in Fig. 5.20. The assumption that the change in the components of the displacement vector will be proportional to the overall change in specimen dimensions means that

$$x' = \lambda_1 x, \quad y' = \lambda_2 y, \quad z' = \lambda_3 z \qquad (5.133)$$

Since r^2 is equal to $(x^2 + y^2 + z^2)$ then it follows from Equation (5.132) that the entropies of the individual chain will be

$$S = c - k\beta^2(x^2 + y^2 + z^2) \qquad \text{(before deformation)}$$
$$S' = c - k\beta^2(\lambda_1^2 x^2 + \lambda_2^2 y^2 + \lambda_3^2 z^2) \quad \text{(after deformation)}$$

and so the change in entropy of an individual chain which has its end displaced from (x, y, z) to (x', y', z') will be given by

$$\Delta S_i = S' - S = -k\beta^2[(\lambda_1^2 - 1)x^2 + (\lambda_2^2 - 1)y^2 + (\lambda_3^2 - 1)z^2] \qquad (5.134)$$

The polymer consists of a network of many such chains with a range of displacement vectors. If the number of these chains per unit volume is

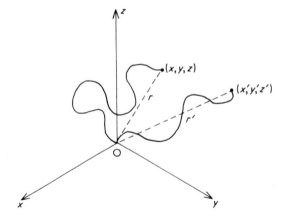

Fig. 5.20 *Schematic representation of the displacement of a junction point in a rubber from* (x, y, z) *to* (x', y', z') *during a general deformation (After Treloar).*

defined as N the number of dN which have ends initially in the small volume element $dx\, dy\, dz$ at the point (x, y, z) can be determined from the Gaussian distribution function as

$$dN = NW(x,y,z)dx\, dy\, dz$$
$$\text{or}\quad dN = N(\beta/\pi^{1/2})^3\exp[-\beta^2(x^2 + y^2 + z^2)]dx\, dy\, dz \qquad (5.135)$$

On deformation the change in entropy of these chains is given by $\Delta S_i dN$ and the total entropy change ΔS per unit volume of the sample during deformation is given by the sum of the entropy changes for all chains. This can be expressed in an integral form as

$$\Delta S = \int \Delta S_i\, dN$$

$$\text{or}\quad \Delta S = \int_{-\infty}^{\infty}\int_{-\infty}^{\infty}\int_{-\infty}^{\infty} -\frac{Nk\beta^5}{\pi^{3/2}}[(\lambda_1^2 - 1)x^2 + (\lambda_2^2 - 1)y^2$$
$$+ (\lambda_3^2 - 1)z^2]\exp[-\beta^2(x^2 + y^2 + z^2)]dx\, dy\, dz \qquad (5.136)$$

which when evaluated* gives the simple result

$$\Delta S = -\tfrac{1}{2}Nk(\lambda_1^2 + \lambda_2^2 + \lambda_3^2 - 3)$$

*The integrals in Equation (5.136) are of the types

$$\int_{-\infty}^{\infty} \exp(-\beta^2 x^2)x^2 dx = \pi^{1/2}/2\beta^3$$

and $\quad \displaystyle\int_{-\infty}^{\infty} \exp(-\beta^2 x^2)dx = \pi^{1/2}/\beta$

This equation relates the change in entropy to the extension ratios and the number of chains between cross-links per unit volume. It was shown earlier (Section 5.3.1) that there is ideally no change in internal energy U when a rubber is deformed and since the deformation takes place at constant volume then the change in the Helmholtz free energy per unit volume ΔA is

$$\Delta A = -T\Delta S = \tfrac{1}{2}NkT(\lambda_1^2 + \lambda_2^2 + \lambda_3^2 - 3) \tag{5.137}$$

for isothermal deformation. This will be identical with the isothermal reversible work of deformation w per unit volume and hence

$$w = \tfrac{1}{2}NkT(\lambda_1^2 + \lambda_2^2 + \lambda_3^2 - 3) \tag{5.138}$$

(iii) *Limitations and use of the theory*

The statistical theory is remarkable in that it enables the macroscopic deformation behaviour of a rubber to be predicted from considerations of how the molecular structure responds to an applied strain. However, it is important to realize that it is only an approximation to the actual behaviour and has significant limitations. Perhaps the most obvious problem is with the assumption that end-to-end distances of the chains can be described by the Gaussian distribution. This problem has been highlighted earlier in connection with solution properties (Section 3.1) where it was shown that the distribution cannot be applied when the chains become extended. It can be overcome to a certain extent with the use of more sophisticated distribution functions, but the use of such functions is beyond the scope of this present discussion. Another problem concerns the value of N. This will be governed by the number of junction points in the polymer network which can be either chemical (cross-links) or physical (entanglements) in nature. The structure of the chain network in an elastomer has been discussed earlier (Section 4.4.5). There will be chain ends and loops which do not contribute to the strength of the network, but if their presence is ignored it follows that if all network chains are anchored at two cross-links then the density, ρ, of the polymer can be expressed as

$$\rho = NM_c/N_A \tag{5.139}$$

where M_c is the number average molar mass of the chain lengths between cross-links and N_A is the Avogadro number. It follows that N is given by

$$N = \rho N_A/M_c = \rho R/M_c k \tag{5.140}$$

and Equation (5.138) then becomes

$$w = \frac{\rho RT}{2M_c}(\lambda_1^2 + \lambda_2^2 + \lambda_3^2 - 3)$$

This is normally written as

$$w = \tfrac{1}{2}G(\lambda_1^2 + \lambda_2^2 + \lambda_3^2 - 3) \tag{5.142}$$

where G is given by

$$G = \rho RT/M_c \tag{5.143}$$

The parameter G relates w, the work of deformation per unit volume which has dimensions of stress to the extension ratios and so G also has dimensions of stress. It is therefore often referred to as the modulus of the rubber. Inspection of Equation (5.143) shows that G has some interesting properties. As might be expected, G increases as the length of chain between cross-links is reduced. This means that the rubber becomes stiffer as the cross-link density increases and the network becomes tighter. A rather more surprising prediction of the equation, which has been substantiated experimentally, is that, unlike almost every other material, the modulus of a rubber increases as the temperature is increased. This is yet another consequence of the deformation of elastomers being dominated by changes in entropy rather than of internal energy.

5.3.3 *Stress–strain behaviour of elastomers*

The statistical theory allows the stress–strain behaviour of a rubber to be predicted. The calculation is greatly simplified when the observation that rubbers tend to deform at constant volume is taken into account. This means that the product of the extension ratios must be unity

$$\text{i.e.} \quad \lambda_1 \lambda_2 \lambda_3 = 1 \tag{5.144}$$

The form of the stress–strain behaviour depends upon the loading geometry employed. The analysis will be restricted to simple uniaxial tension or compression when specimens are deformed to an extension ratio λ in the direction of the applied stress. It is possible to replace λ_1 by λ but Equation (5.144) requires that $\lambda_2\lambda_3 = 1/\lambda_1$ and so $\lambda_2 = \lambda_3 = 1/\lambda^{1/2}$. Equation (5.142) can therefore be written as

$$w = \tfrac{1}{2}G(\lambda^2 + 2/\lambda - 3) \tag{5.145}$$

for uniaxial tension or compression. If the specimen has an initial cross-sectional area of A_0 and length l_0 and it is extended δl by a uniaxial tensile force, f, then the amount of work δW done *on* the specimen is given by

$$\delta W = f\delta l$$

Equation (5.145) gives an expression for w which is the work per unit volume. The volume of the rubber is $A_0 l_0$ and so

$$\delta w = \delta W/A_0 l_0 = (f/A_0)\,(\delta l/l_0) \tag{5.146}$$

But f/A_0 is the force per unit initial cross-sectional area, i.e. the nominal stress σ_n and $\delta l/l_0 = \delta e \approx \delta \lambda$. A relationship between σ_n and λ can then be derived from Equation (5.145) since rearranging Equation (5.146) and expressing it in a differential form gives

$$\sigma_n = dw/d\lambda = G(\lambda - 1/\lambda^2) \tag{5.147}$$

It is possible to compare the relationship between σ_n and λ predicted by this equation with experimental results by fitting the theoretical curve to experimental data by choosing a suitable value of G. Fig. 5.21(a) shows the results of such an exercise for a sample of vulcanized natural rubber deformed in tension. The value of G chosen was ~ 0.4 MNm^{-2} and it can be seen that the agreement between experiment and theory is good at low elongations ($\lambda < 1.5$). However, the theoretical curve falls below the experimental one as the strain is increased further, but eventually rises above it at elongations of $\lambda > 6$. It can be seen that the theory predicts the basic form of the relationship between σ_n and λ especially at low strains when the Gaussian approximation may be expected to hold well. The lack of agreement at high strains is thought to be due to at least two factors. The Gaussian formula will no longer apply and crystallization can also occur in the rubber (Section 4.4.5). The theory predicts the deformation behaviour rather better in compression. Fig. 5.21(b) shows the variation of σ_n with λ for the same rubber as used in tensile experiment, but in this case deformed into the compressive region ($\lambda < 1$). Good agreement between experiment and theory (Equation 5.147) is found between $\lambda = 0.4$ and 1.3. In this regime the Gaussian formula would be expected to be a better approximation of the behaviour of the chains and crystallization does not occur.

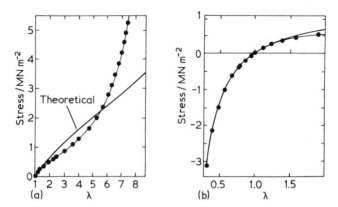

Fig. 5.21 *Relationship between stress and extension ratio λ for a rubber. The theoretical curves for a value of $G = 0.4$ MN m^{-2} are given as heavy lines (a) Extension, (b) Compressive deformation (After Treloar).*

Another aspect of the stress–strain behaviour of elastomers which is not predicted by the statistical theory is the phenomenon of *mechanical hysteresis* which is encountered when a deformed rubber is allowed to relax. This type of behaviour is shown in Fig. 5.22 for a strain-crystallizing rubber. In this case although all the strain is recovered upon removing the load the unloading curve does not follow the same path as the loading curve. The area between the two curves corresponds to the energy dissipated within the cycle. Hysteresis is found to be more prevalent in strain crystallizing and filled rubbers and in certain applications it can lead to undesirable consequences. For example, if a rubber exhibiting a high hysteresis is cycled rapidly between low and high strains the dissipation of energy can lead to a large heat build up which may cause a deterioration in the properties of the rubber.

5.4 Yield in polymers

Determination of the stress–strain behaviour of a material is particularly useful as it gives information concerning important mechanical properties such as Young's modulus, yield strength and brittleness. These are parameters which are vital in design considerations when the material is used in a practical situation. Stress–strain curves can be readily obtained for polymers by subjecting a specimen to a tensile force applied at a constant rate of testing. The stress–strain behaviour of polymers is not fundamentally different from that of conventional materials, the main difference being that polymers show a marked time or rate dependence (cf. viscoelasticity). The situation can be greatly simplified by making measurements at a fixed testing rate, but it is important to bear in mind the underlying time dependence that exists.

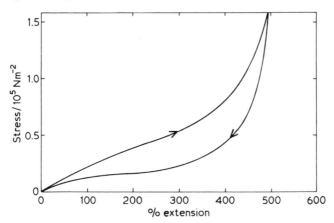

Fig. 5.22 *Illustration of mechanical hysteresis for a strain-crystallizing rubber (After Andrews)*.

5.4.1 *Phenomenology of yield*

Fig. 5.23 shows an idealized stress–strain curve for a ductile polymer sample. In this case the nominal stress σ_n is plotted against the strain e. The change in the cross-section of a parallel-sided specimen is also sketched schematically at different stages of the deformation. Initially the stress is proportional to the strain and Hooke's law is obeyed. The tensile modulus can be obtained from the slope. As the strain is increased the curve decreases in slope until it reaches a maximum. This is conventionally known as the '*yield point*' and the yield stress and yield strain, σ_y and e_y, are indicated on the curve. The 'yield point' for a polymer is rather difficult to define. It should correspond to the point at which permanent plastic deformation takes place, but for polymers a 'permanent set' can be found in specimens loaded to a stress, below the maximum, where the curve becomes non-linear. The situation is further complicated by the observation that even for specimens loaded well beyond the yield strain the plastic deformation can sometimes be completely recovered by annealing the specimen at elevated temperature. In practice, the exact position of the yield point is not of any great importance and the maximum point on the curve suffices as a definition of yield. The value of the yield strain for polymers is typically of the order of 5–10 percent which is very much higher than that of metals and ceramics. Yield in metals normally occurs at strains below 0.1 percent.

During elastic deformation the cross-sectional area of the specimen decreases uniformly as length increases, but an important change occurs at the yield point. The cross-sectional area starts to decrease more rapidly

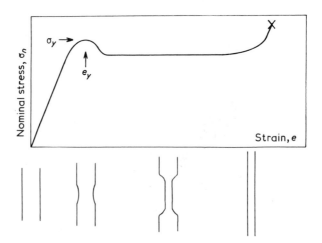

Fig. 5.23 *Schematic representation of the stress–strain behaviour of a ductile polymer and consequent change in specimen dimensions.*

at one particular point along the gauge length as a 'neck' starts to form. The nominal stress falls after yield and settles at a constant value as the neck extends along the specimen. Eventually, when the whole specimen is necked, strain hardening occurs and the stress rises until fracture eventually intervenes. The process whereby the neck extends is known as *cold drawing*. It was originally thought that this was due to local heating of the specimen by the energy expended during deformation. Detailed measurements have shown that a neck can form even at very low strain rates when the heat is easily dissipated.

Polymers differ in their ability to form a stable neck and some do not neck at all. The form of the neck varies from polymer-to-polymer and with the testing conditions for a given polymer. The degree of necking is characterized by the 'draw ratio' which is defined as the length of a fully necked specimen divided by the original length. Since cold drawing takes place at approximately constant volume it is also equal to the ratio of original specimen cross-sectional area to that of the drawn specimen.

The exact form of the stress–strain curve for a particular polymer varies with both the temperature and rate of testing. A series of stress–strain curves for poly(methyl methacrylate) tested at different temperatures is given in Fig. 5.24. The modulus, as given by the initial slope of the curves, increases as the temperature is reduced. On the other hand, the ductility of the polymer decreases and below about 320 K the polymer does not draw and behaves in a brittle manner. The exact temperature at which this *ductile-to-brittle transition* takes place depends upon the rate of testing. Its temperature decreases as the rate of testing increases. There is a similar interdependence of the properties upon temperature and rate as is found for viscoelasticity (Section 5.2). In general, it is found that the effect of increasing the rate of testing upon properties such as the Young's modulus or yield stress is the same as reducing the testing temperature.

5.4.2 *Necking and the Considère construction*

It is clear that some polymers have the ability under certain conditions to form a stable neck and undergo cold-drawing. This ability is shared by certain metals such as mild steel whereas other metals form an unstable neck which continues to thin-down until fracture ensues. The phenomenon of necking and cold drawing can best be considered using the *Considère construction* which involves plotting the true stress σ against normal strain e. This must be contrasted with the curves in Fig. 5.23 and 5.24 where the nominal stress σ_n is plotted against the nominal strain e.

The *true stress* σ is defined as the load P divided by the instantaneous cross-sectional area A and so

$$\sigma = P/A \qquad (5.148)$$

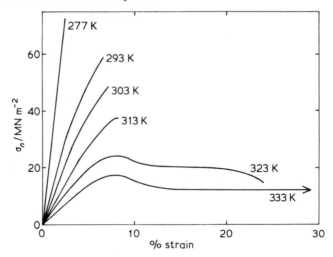

Fig. 5.24 *Variation of the stress–strain behaviour of poly(methyl methacrylate) with temperature (After Andrews).*

It may be assumed that deformation takes place at approximately constant volume and so A can be related to the original cross-sectional area A_0 through the equation

$$Al = A_0 l_0 \tag{5.149}$$

where l and l_0 are the instantaneous and original specimen lengths respectively. The nominal strain is defined as

$$e = (l - l_0)/l_0 \qquad l/l_0 = (1 + e) \tag{5.150}$$

and the nominal stress as

$$\sigma_n = P/A_0 \tag{5.151}$$

Combining Equations (5.148) to (5.151) gives the relationship between the nominal and true stress as

$$\sigma_n = \sigma/(1 + e) \tag{5.152}$$

This means that for finite tensile strains σ will always be greater than σ_n.

When a neck starts to form, the load on the specimen ceases to rise as the strain is increased. This corresponds to the condition that $dP/de = 0$ or $d\sigma_n/de = 0$ (Fig. 5.23). Applying this condition to Equation (5.152) gives

$$\frac{d\sigma_n}{de} = 0 = \frac{1}{(1 + e)}\frac{d\sigma}{de} - \frac{\sigma}{(1 + e)^2}$$

or
$$\frac{d\sigma}{de} = \frac{\sigma}{(1 + e)} \tag{5.153}$$

as the condition for a neck to form. The use of this equation is illustrated in Fig. 5.25. The point at which necking occurs can be determined by drawing a tangent to the curve of true stress against nominal strain from the point $e = -1$. Equation (5.153) will then be satisfied at the point at which the tangent meets the curve. This construction can also be used to determine whether or not the neck will be stable. In the curve drawn in Fig. 5.25(a) only one tangent can be drawn and after the neck forms the sample continues to thin down in the necked region until fracture occurs. A stable neck will only form if a second tangent can be drawn to the curve as shown in Fig. 5.25(b). The point where the second tangent meets the $\sigma-e$ curve corresponds to a minimum in the nominal stress-nominal strain curve (Fig. 5.23) which is necessary for the neck to be stable. The condition for both necking *and* cold-drawing to occur is therefore that $d\sigma/de$ must be equal to $\sigma/(1 + e)$ at *two* points on the $\sigma-e$ curve.

So far the mechanisms whereby a stable neck forms in a polymer have not been considered. In metals it is known to be due to the multiplication of dislocations which lead to a phenomenon known as strain-hardening. The metal becomes 'harder' as the strain is increased and if the strain-hardening takes place to a sufficient extent the material in the neck, which supports a smaller area, does not strain further. The material outside the neck is less deformed and hence 'softer' and continues to deform even though the cross-sectional area is larger and the true stress locally lower. Strain-hardening in polymers is thought to be due principally to the effects of orientation. It has been shown by X-ray diffraction that the molecules are aligned parallel to the stretching direction in the cold-drawn regions of both amorphous and crystalline polymers. Since the anisotropic nature of the chemical bonding in the molecules causes an oriented polymer to be very much stronger and stiffer than an isotropic one, the material in the necked region is capable of supporting a much higher true stress than that outside the neck. Hence, polymers tend to form stable necks and undergo cold-drawing if they do not first undergo brittle fracture.

5.4.3 *Yield criteria*

We have so far in this section on the yield behaviour of polymers only considered tensile deformation. In order to obtain a complete idea of the yield process it is necessary to know under what conditions yield occurs for any general combination of stresses. For example, glassy polymers are usually brittle in tension when the temperature of testing is sufficiently below T_g whereas when they are deformed in compression at similar temperatures they can undergo considerable plastic deformation. Also a knowledge of yield behaviour under general stress systems is important in engineering structures where components are subjected to a variety of

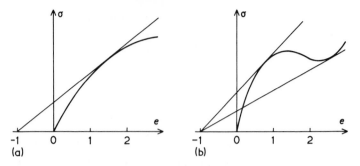

Fig. 5.25 *Schematic curves of true stress, σ against nominal strain, e for polymers showing the Considère construction. (a) Unstable neck, (b) Stable neck.*

tensile, compressive and shear stresses. A description of the conditions under which yield can occur under a general stress system is called a *yield criterion*. Several criteria have been suggested for materials in general, but before they can be described it is necessary to consider how the state of stress on a body can be represented. If the body is isotropic then the stresses can in general be given in terms of a tensor (Section 5.1.1) as

$$\sigma_{ij} = \begin{pmatrix} \sigma_{11} & \sigma_{21} & \sigma_{31} \\ \sigma_{21} & \sigma_{22} & \sigma_{23} \\ \sigma_{31} & \sigma_{23} & \sigma_{33} \end{pmatrix}$$

It is possible to choose a set of three mutually orthogonal axes such that the shear stress are all zero and the stress system can be described in terms of three normal stress such that the tensor becomes

$$\begin{pmatrix} \sigma_1 & 0 & 0 \\ 0 & \sigma_2 & 0 \\ 0 & 0 & \sigma_3 \end{pmatrix}$$

These three principal stresses σ_1, σ_2 and σ_3 are used in the formulation of the different yield criteria which are outlined below.

(i) *Tresca yield criterion*
This was one of the earliest criteria developed to describe the yield behaviour of metals and Tresca proposed that yield occurs when a critical value of the maximum shear stress σ_s is reached. If $\sigma_1 > \sigma_2 > \sigma_3$ then the criterion can be given as

$$\tfrac{1}{2}(\sigma_1 - \sigma_3) = \sigma_s \tag{5.154}$$

In a simple tensile test σ_1 is equal to the applied stress and $\sigma_2 = \sigma_3 = 0$ and so at yield

$$\sigma_s = \sigma_1/2 = \sigma_y/2 \qquad (5.155)$$

where σ_y is the yield stress in tension.

(ii) *Von Mises yield criterion*
Although the relative simplicity of the Tresca criterion is rather attractive it is found that the criterion suggested by von Mises gives a somewhat better prediction of the yield behaviour of most materials. The criterion corresponds to the condition that yield occurs when the shear strain energy in the material reaches a critical value and it can be expressed as a symmetrical relationship between the principal stresses of the form

$$(\sigma_1 - \sigma_2)^2 + (\sigma_2 - \sigma_3)^2 + (\sigma_3 - \sigma_1)^2 = \text{constant} \qquad (5.156)$$

If the case of simple tension is considered then $\sigma_2 = \sigma_3 = 0$ and if the tensile yield stress is again σ_y then the constant is equal to $2\sigma_y^2$. Equation (5.156) can be therefore expressed as

$$(\sigma_1 - \sigma_2)^2 + (\sigma_2 - \sigma_3)^2 + (\sigma_3 - \sigma_1)^2 = 2\sigma_y^2 \qquad (5.157)$$

If the experiment is done in pure shear then $\sigma_1 = -\sigma_2$ and $\sigma_3 = 0$ and Equation (5.157) becomes

$$\sigma_1 = \sigma_y/\sqrt{3} \qquad (5.158)$$

Hence the shear yield stress is predicted to be $1/\sqrt{3}$ times the tensile yield stress. This should be compared with the Tresca criterion which predicts that the shear yield stress is $\sigma_y/2$. On the other hand, it can be readily seen that both criteria predict that the yield stresses measured in uniaxial tension and compression will be equal.

It is useful to represent the yield criteria graphically. This is done for the case of plane stress deformation ($\sigma_3 = 0$) in Fig. 5.26 where the stress axes are normalized with respect to the tensile yield stress. In plane stress the von Mises criterion (Equation 5.157) reduces to

$$\left(\frac{\sigma_1}{\sigma_y}\right)^2 + \left(\frac{\sigma_2}{\sigma_y}\right)^2 - \left(\frac{\sigma_1}{\sigma_y}\right)\left(\frac{\sigma_2}{\sigma_y}\right) = 1 \qquad (5.159)$$

which describes the ellipse shown in Fig. 5.26. It also shows that under certain conditions the applied tensile stress can be larger than σ_y. The Tresca criterion can also be represented on the same plot since when $\sigma_3 = 0$ Equations (5.154) and (5.155) can be combined to give

$$\tfrac{1}{2}(\sigma_1 - \sigma_2) = \sigma_s = \tfrac{1}{2}\sigma_y$$

$$\text{or} \quad \frac{\sigma_1}{\sigma_y} - \frac{\sigma_2}{\sigma_y} = 1 \qquad (5.160)$$

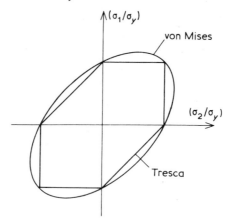

Fig. 5.26 *Schematic representation of the Tresca and von Mises yield criteria for plane stress deformation.*

This equation is applicable in the quadrants where σ_1 and σ_2 have different signs. Where they have the same signs the maximum shear stress criterion requires that

$$\sigma_1/\sigma_y = 1 \quad \text{and} \quad \sigma_2/\sigma_y = 1$$

It can be seen that the Tresca criterion gives a surface consisting of connected straight lines which inscribe the von Mises ellipse.

In practice materials are subjected to triaxial stress systems and the form of the criterion in three-dimensional principal stress space is of interest. They both plot out as infinite cylinders parallel to the [111] direction (cf. crystallography). The Tresca cylinder has a regular hexagonal cross-section and the von Mises one is circular in cross-section. It is found that metals tend to obey the von Mises criterion rather than Tresca although Tresca is often assumed in calculations because of its simpler form.

5.4.4 *Pressure-dependent yield behaviour*

In order to determine which is the most appropriate yield criterion for a particular material it is necessary to follow the yield behaviour by using a variety of different combinations of multiaxial stress. However, with polymers rather unusual features are revealed when the yield stress of the same polymer is measured just in simple uniaxial tension and compression. Both the Tresca and von Mises criteria predict that the yield stress should be the same in both cases and this is what is found for metals. But for polymers the compressive yield stress is usually higher than the tensile one. This difference between the compressive and tensile yield stress can be taken as an indication that the hydrostatic component of the applied stress is exerting an influence upon the yield process. This can be demonstrated more directly by measuring the tensile yield stress under the action of an overall hydrostatic pressure. The results of such measurements on different polymers are shown in Fig. 5.27 and it can be seen that there is a clear increase in σ_y with hydrostatic pressure. In general, it is found that the yield stresses of amorphous polymers show larger pressure dependences than those of crystalline polymers. There is no significant change in the yield stresses of metals at similar pressures, but it is quite reasonable to expect that the physical properties of polymers will change at these relatively low pressures. They tend to have much more lower bulk moduli than metals and so undergo significant volume changes on pressurization.

It is possible to modify the yield criteria described in the last section to take into account the pressure dependence. The hydrostatic pressure p is given by

$$p = \tfrac{1}{3}(\sigma_1 + \sigma_2 + \sigma_3) \tag{5.2}$$

and the effect of hydrostatic stress can be incorporated into Equation (5.156) by writing it as

$$A[\sigma_1 + \sigma_2 + \sigma_3] + B[(\sigma_1 - \sigma_2)^2 + (\sigma_2 - \sigma_3)^2 + (\sigma_3 - \sigma_1)^2] = 1 \tag{5.161}$$

where A and B are constants. It is possible to define A and B in terms of the simple tensile and compressive yield stresses, σ_{yt} and σ_{yc}, since when $\sigma_2 = \sigma_3 = 0$ then

$$A\sigma_{yt} + 2B\sigma_{yt}^2 = 1 \qquad \text{(tension)}$$
$$-A\sigma_{yc} + 2B\sigma_{yc}^2 = 1 \qquad \text{(compression)}$$

and so

$$\left.\begin{array}{l} A = (\sigma_{yc} - \sigma_{yt})/\sigma_{yc}\sigma_{yt} \\ \text{and} \quad B = 1/(2\sigma_{yc}\sigma_{yt}) \end{array}\right\} \tag{5.162}$$

Equation (5.161) can therefore be written as

$$2(\sigma_{yc} - \sigma_{yt})[\sigma_1 + \sigma_2 + \sigma_3]$$
$$+ [(\sigma_1 - \sigma_2)^2 + (\sigma_2 - \sigma_3)^2 + (\sigma_3 - \sigma_1)^2] = 2\sigma_{yc}\sigma_{yt} \tag{5.163}$$

If the magnitude of the tensile yield stress is the same as that for compression the normal von Mises criterion is recovered.

It is possible to modify the Tresca maximum shear-stress criterion in several ways. The simplest way is to make the critical shear stress a function of the hydrostatic pressure p and so σ_s can be expressed by an equation of form

$$\sigma_s = \sigma_s^0 - \mu p \tag{5.164}$$

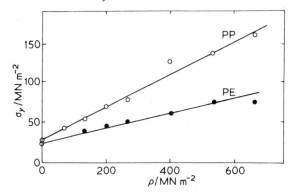

Fig. 5.27 *Variation of yield stress of polyethylene and polypropylene with hydrostatic pressure, p. (Data taken from Mears, Pae and Sauer, J. Appl. Phys.* **40** *(1969) 4229).*

where σ_s^0 is the shear yield stress in the absence of any overall hydrostatic pressure and μ is a material constant. The hydrostatic pressure p is taken to be positive for uniaxial tensile loading and negative in compression. The yield stress in uniaxial tension is therefore given by combining Equations (5.2), (5.155) and (5.164) as

$$\sigma_{yt} = 2\sigma_s^0 - 2\mu\sigma_{yt}/3$$

or $\sigma_{yt} = 2\sigma_s^0/(1 + 2\mu/3)$ (5.165)

and in compression ($\sigma_2 = -\sigma_{yc}$) as

$$\sigma_{yc} = 2\sigma_s^0 + 2\mu\sigma_{yc}/3$$

or $\sigma_{yc} = 2\sigma_s^0/(1 - 2\mu/3)$ (5.166)

Combining these two equations leads to

$$\mu = \frac{3}{2}\left(\frac{\sigma_{yc} - \sigma_{yt}}{\sigma_{yc} + \sigma_{yt}}\right)$$ (5.167)

which shows that $\mu > 0$ when the magnitude of the compressive yield stress is greater than the one measured in tension.

In order to determine which is the most appropriate yield criterion for a particular polymer it is necessary to follow the yield behaviour under a variety of states of stress. This is most conveniently done by working in plane stress ($\sigma_3 = 0$) and making measurements in pure shear ($\sigma_1 = -\sigma_2$) and biaxial tension ($\sigma_1, \sigma_2 > 0$) as well as in the simple uniaxial cases. The results of such experiments on glassy polystyrene are shown in Fig. 5.28. The modified von Mises and Tresca envelopes are also plotted. In both cases they have been fitted to the measured uniaxial tensile and compressive yield stresses, σ_{yt} and σ_{yc}. It can be seen that the von Mises

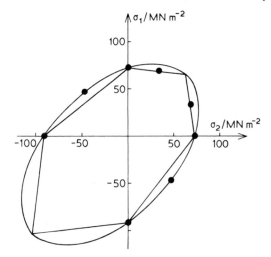

Fig. 5.28 *Modified von Mises and Tresca criteria fitted to experimental data upon the yield behaviour of polystyrene. (Data taken from Whitney and Andrews, J. Polym. Sci.,* **C-16** *(1967) 2981.)*

criterion plots out as distorted ellipse which is inscribed by the modified Tresca hexagon. In general it is found that under most conditions polymers tend to follow a pressure-dependent von Mises criterion rather than the modified Tresca.

In three-dimensional principal stress space the modified Tresca cylinder becomes a hexagonal pyramid and the pressure-dependent von Mises cylinder becomes a cone. The significance of the apices of the pyramid and cone is that they define the conditions for which there can be yielding under the influence of hydrostatic stress alone. This is something which cannot happen for materials which obey the unmodified criteria (Section 5.4.3).

5.4.5 *Rate and temperature dependence*

So far we have considered only deformation which takes place at constant rate and temperature, but plastic deformation, like other aspects of the mechanical behaviour of polymers, has a strong dependence upon the testing rate and temperature. Typical behaviour is illustrated in Fig. 5.29 for a glassy thermoplastic deformed in tension. At given strain-rate the yield stress drops as the temperature is increased and σ_y falls approximately linearly to zero at the glass-transition temperature when the polymer glass becomes a rubber. If the strain-rate is increased and the temperature held constant the yield stress increases [cf. time–temperature superposition (Section 5.2.7)].

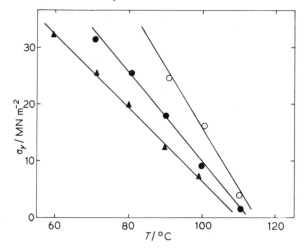

Fig. 5.29 *Variation of the yield stress of poly(methyl methacrylate) with temperature. Strain rates* ▲, 0.002 min^{-1}, ●, 0.02 min^{-1}, ○, 0.2 min^{-1}. *(Data of Langford, Whitney and Andrews reported by Ward.)*

The behaviour of semi-crystalline polymers is rather similar with the main difference being that the yield stress drops to zero at the melting temperature of the crystals rather than at the T_g. Between T_g and T_m the non-crystalline areas are rubbery and the material gains its strength from the crystalline regions which reinforce the rubbery matrix. If the temperature of a semi-crystalline polymer is reduced below T_g it behaves more like a glassy polymer and the crystals do not have such a significant strengthening effect.

5.4.6 *Craze yielding*

Consideration of the plastic deformation of polymers has, so far, been concerned only with 'shear yielding'. This occurs at constant volume and can take place uniformly throughout the sample. Certain polymers, particularly thermoplastics in the glassy state, are capable of undergoing a localized form of plastic deformation known as *crazing*. This is found to take place only when there is an overall hydrostatic tensile stress (i.e. $p >$ 0) and the formation of crazes causes the material to undergo a significant increase in volume. Fig. 5.30 shows a deformed specimen of polycarbonate which has undergone crazing. The crazes appear as small crack-like entities which are usually initiated on the specimen surface and are oriented perpendicular to the tensile axis. Closer examination shows that they are regions of cavitated polymer and so not true cracks, although the cracks which lead to eventual failure of the specimen usually nucleate within pre-existing crazes.

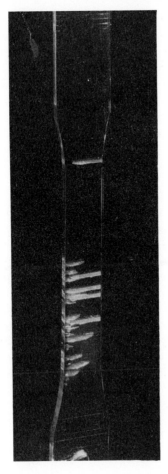

Fig. 5.30 Sample of polycarbonate containing several large crazes. (Kambour, reproduced with permission.)

Crazing can be demonstrated quite simply by rubbing ones fingers over the surface of a sample of glassy polystyrene and bending it to near its point of failure. The crazes appear as a fine surface mist on the tension surface. The crazes are very small, ~1000 Å thick and several microns in lateral dimensions, but they can be seen by the naked eye because they are less dense than the undeformed matrix and so reflect and scatter light. The demonstration also shows how crazing can be made easier by the presence of certain liquids ('crazing agents') such as finger grease. In practice the use of many glassy polymers is often limited by their tendency to undergo crazing at relatively low stresses in the presence of crazing agents. The crazes can mar the appearance of the specimen and lead to eventual catastrophic failure.

As with other mechanical properties, the formation of crazes in polymers depends upon testing temperature and the rate or time-period of loading. This is clearly demonstrated in Fig. 5.31 where the stress required to cause crazing in simple uniaxial tensile specimens of polystyrene is plotted against testing temperature. The craze initiation stress drops as the testing temperature is increased and as the strain-rate is decreased. The data in Fig. 5.31 are for dry crazing. The presence of crazing agents (e.g. finger grease or methanol with polystyrene) causes the craze initiation stress to drop further.

5.4.7 *Craze criteria*

As crazing is a yield phenomenon there have been attempts to establish a *craze criterion* in the same way that Tresca and von Mises criteria have been used for shear yielding. Crazing occurs typically about one-half of the yield stress of the polymer. This is in the 'elastic' region (Fig. 5.23), but the occurrence of crazing does not usually cause any detectable change in slope of the stress–strain curve as the volume fraction of crazed material is initially very low. Early suggestions of craze criteria were that a critical uniaxial tensile stress or strain were required for craze formation, but these were inadequate to describe the behaviour in multiaxial stress systems. Observations upon the way in which polymers craze in uniaxial tension under the action of a superimposed hydrostatic pressure have been made in a similar way to the measurement of shear yielding under hydrostatic pressure described earlier (Fig. 5.27). It is found that superimposed

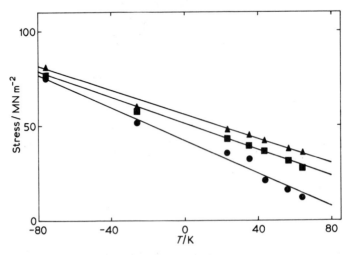

Fig. 5.31 *Dependence of crazing stress upon temperature at different strain-rates for polystyrene. Strain rates,* ●, 0.00067s^{-1}, ■, 0.0267s^{-1}, ▲, 0.267s^{-1},

hydrostatic compression makes crazing more difficult and increases the tensile stress or strain needed to initiate crazes. Hydrostatic tension on the other hand produces a dilatation which helps to open up voids in the structure and hence aids the formation of crazes.

Since crazing involves the opening up of cavitated regions it would be reasonable to expect that it might take place at a critical strain for crazing, e_c. The effect of pressure outlined above could then be incorporated by writing

$$e_c = C + D/p \qquad (5.168)$$

where p is the hydrostatic stress and C and D are time- and temperature-dependent parameters. The maximum tensile strain in an isotropic body subjected to a general state of stress defined by the principal stresses σ_1, σ_2 and σ_3 (where $\sigma_1 > \sigma_2 > \sigma_3$) is given by

$$e_1 = \frac{1}{E}(\sigma_1 - \nu\sigma_2 - \nu\sigma_3) \qquad (5.26)$$

where ν is Poissons ratio and this strain always acts in the direction of the maximum principal stress. The critical strain criterion requires that crazing occurs when e_1 reaches a critical value and so Equation (5.168) can be rewritten to define the criterion in terms of stresses only as

$$\sigma_1 - \nu\sigma_2 - \nu\sigma_3 = X + Y/(\sigma_1 + \sigma_2 + \sigma_3) \qquad (5.169)$$

where X and Y are new time- and temperature-dependent constants. This criterion is useful in that it gives both the conditions under which crazes form and the direction in which the crazes grow (i.e. perpendicular to σ_1). As with yield criteria the easiest way to test craze criteria is by making measurements in plane stress ($\sigma_3 = 0$) where Equation (5.169) becomes

$$\sigma_1 - \nu\sigma_2 = X + Y/(\sigma_1 + \sigma_2) \qquad (5.170)$$

This equation is plotted in Fig. 5.32 and X and Y have been chosen by fitting the curve to experimental data on crazing in polystyrene under biaxial stress. Crazing cannot take place unless there is the overall hydrostatic tension [i.e. ($\sigma_1 + \sigma_2 + \sigma_3) > 0$] which is necessary to allow cavitation to occur. The curves from Equation (5.170) will then be assymptotic to the line where $\sigma_1 = -\sigma_2$. It is instructive to consider the criteria for shear yielding at the same time as craze criteria are examined and so the pressure-dependent von Mises curve from Fig. 5.28 for polystyrene tested under similar conditions is also given in Fig. 5.32. Having both the yield and craze envelopes on the same plot allows a prediction of the type of yielding that may occur under any general state of stress to be made.

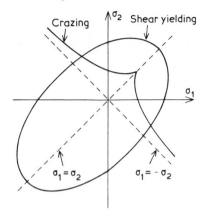

Fig. 5.32 *Envelopes defining crazing and yield for an amorphous polymer undergoing plane stress deformation. (After Sternstein, S. S. and Onachin, L. (1969), ACS Polymer Reprints, **19**, 1117.)*

5.5 Deformation mechanisms

The response of a material to mechanical deformation reflects the microscopic deformation processes which are occurring on a molecular or atomic level. There is now a clear understanding of how this happens in metals. Elastic deformation reflects the displacement of the metal atoms out of their low energy positions in the 'potential well' and plastic deformation occurs by the motion of dislocations through the crystal lattice. In addition methods have been developed of strengthening metals by precipitation hardening. The presence of precipitate particles makes dislocation motion more difficult. Our understanding of the deformation processes which occur in polymers is by no means so well-developed and this aspect of polymer deformation has been somewhat ignored previously in books at this level. There have been important developments over recent years which have increased our knowledge of the mechanisms by which polymers deform and so it is felt that it may be useful at this stage to review the current ideas of the mechanisms of polymer deformation.

The area has been divided up into elastic deformation and plastic deformation (shear yielding and crazing). They have been considered as essentially time-independent processes. It is clear from Section 5.2 that there will always be an underlying time-dependence, but since the discussion is limited to polymers which are generally well away from the viscoelastic region the approximation of time-independent behaviour does not cause too many problems.

5.5.1 *Elastic deformation of single- phase polymers*

It is useful to consider the deformation of single-and multi-phase polymers separately. Unfilled rubbers, polymer glasses and polymer single crystals

TABLE 5.1 *Typical values of Young's modulus for different types of polymers.*

Types of polymer	Young's modulus/Nm^{-2}
Rubber	$\sim 10^6$
Glassy polymer	$\sim 10^9$
Polymer crystal ($\perp$ to c)	$\sim 10^9$
($\parallel$ to c)	$\sim 10^{11}$

can be considered as being essentially single-phase materials and this makes determination of their moduli somewhat more straightforward than for multi-phase systems such as semi-crystalline polymers which can be thought of as consisting of mixtures of the separate phases.

One of the most useful aspects of polymers is the wide range of moduli that can be obtained. Typical values of Young's modulus are given in Table 5.1. *Rubbers* tend to have very low Young's moduli of the order of 10^6 Nm^{-2}. This is because their deformation consists essentially of uncoiling the molecules in a cross-linked network. It is possible to calculate a value of modulus of a rubber using the statistical theory and this gives very good agreement with the experimentally determined values (Section 5.3.2). The moduli of *polymer glasses* are typically of the order of 10^9 Nm^{-2}. The deformation of a polymer glass is somewhat more difficult to model theoretically. The structure is thought to be completely random and so elastic deformation will involve the bending and stretching of the strong covalent bonds on the polymer backbone as well as the displacement of adjacent molecules which is opposed by relatively weak secondary (e.g. van der Waals) bonding.

The most striking feature of the elastic properties of *polymer crystals* is that they are very anisotropic. The moduli parallel to the chain direction are $\sim 10^{11}$ Nm^{-2} which is similar to the values of moduli found for metals (e.g. steel $E \approx 2.1 \times 10^{11}$ Nm^{-2}). However, the moduli of polymer crystals deformed in directions perpendicular to the chain axes are much lower at about 10^9 Nm^{-2}. This large difference in modulus reflects the anisotropy in the bonding in polymer crystals. Deformation parallel to the chain direction involves stretching of the strong covalent bonds and changing the bond angles or if the molecule is in a helical conformation distorting the molecular helix. Deformation in the transverse direction is opposed by only relatively weak secondary van der Waals or hydrogen bonding. Because of the high anisotropy of the bonding in polymer crystals it is possible with knowledge of the crystal structure (Section 4.1.3) and the force constants of the chemical bonds to calculate the theoretical moduli of polymer crystals. An exact calculation is rather complex as it requires detailed knowledge of inter- and intra-molecular interactions, but Treloar has shown how an estimate can be made of the chain-direction modulus of a polymer which crystallizes with its backbone in the form of a planar zig-zag. This involves a relatively simple calculation which assumes that deformation involves only the bending and stretching of the bonds along the molecular backbone.

The model of the polymer chain used in the calculation is shown schematically in Fig. 5.33. It is treated as consisting of n rods of length, l, which are capable of being stretched along their lengths but not of bending, and are joined together by torsional springs. The bond angles are taken initially to be Θ and the angle between the applied force, f, and the individual bonds as initially α. The original length of the chain, L, is then $nl \cos \alpha$ and so the change in length δL on deforming the chain is given by

$$\delta L = n\delta(l \cos \alpha) = n(\cos \alpha \delta l - l \sin \alpha \delta \alpha) \qquad (5.171)$$

It is possible to obtain expressions for δl and $\delta \Theta$ in terms of the applied force f. Consideration of the bond stretching allows δl to be determined. The component of the force acting along the bond direction is $f \cos \alpha$. This can be related to the extension through the force constant for bond stretching k_l which can be determined from infrared or Raman spectroscopy. k_l is the constant of proportionality relating the force along the bond to its extension. It follows therefore that

$$\delta l = f \cos \alpha / k_l \qquad (5.172)$$

The force required to cause valence angle deformation can be determined by using the angular deformation force constant k_θ which relates the change in bond angle $\delta \theta$ to the torque acting around each of the bond angles. This torque is equal to the moment of the applied force about the angular vertices, $\frac{1}{2}fl \sin \alpha$ and so the change in the angle between the bonds is given by

$$\delta \Theta = (fl \sin \alpha)/2k_\theta \qquad (5.173)$$

But since it can be shown by a simple geometrical construction that $\alpha = 90° - \Theta/2$ then it follows that

$$\delta \alpha = -\delta \Theta/2$$

and Equation (5.173) becomes

$$\delta \alpha = -(fl \sin \alpha)/4k_\theta \qquad (5.174)$$

Putting Equations (5.172) and (5.174) into (5.171) gives

$$\frac{\delta L}{f} = n\left[\frac{\cos^2\alpha}{k_l} + \frac{l^2\sin^2\alpha}{4k_\theta}\right] \qquad (5.175)$$

Fig. 5.33 *Model of a polymer chain undergoing deformation (After Treloar, Polymer, 1 (1960) 95).*

The Young's modulus of the polymer is given by

$$E = (f/A)/(\delta L/L) \tag{5.176}$$

where A is the cross-sectional area supported by each chain and so

$$E = \frac{l \cos \alpha}{A} \left[\frac{\cos^2\alpha}{k_l} + \frac{l^2\sin^2\alpha}{4k_\theta} \right]^{-1}$$

or in terms of the bond angle Θ

$$E = \frac{l \sin(\Theta/2)}{A} \left[\frac{\sin^2(\Theta/2)}{k_l} + \frac{l^2\cos^2(\Theta/2)}{4k_\theta} \right]^{-1} \tag{5.177}$$

This equation allows the modulus to be calculated with knowledge of the crystal structure, which gives A, l and Θ, and the spectroscopically measured values of the force constants k_l and k_θ. It predicts a value of E for polyethylene of about 180 GN m^{-2} which is generally rather lower than values obtained from more sophisticated calculations. This is because the above analysis does not allow for intra-molecular interactions which tend to increase the modulus further. The chain-direction modulus does not completely describe the elastic behaviour of a polymer crystal. A complete set of elastic constants c_{ij} (Equation 5.12) are required in order to do this. The number of constants depends upon the crystal symmetry and for an orthorhombic crystal such as polyethylene nine constants are necessary. The measurement of all the elastic constants is very difficult even for good single crystals like the diacetylenes (Section 4.1.6). A set of elastic constants has been calculated for orthorhombic polyethylene and they are given in Table 5.2. They were obtained by Odajima and Maeda who calculated the forces required to stretch the bonds along the polymer backbone and open up the angles between the bonds as in the Treloar model. In addition, inter- and intra-molecular interactions were taken into account.

Table 5.3 gives some calculated and experimentally measured values of chain-direction crystal moduli of several different polymer crystals. Single crystals of polymers are not normally available with dimensions that are suitable to allow them to be tested mechanically. Highly oriented semi-crystalline polymers tend to have moduli of no more than 50 percent of the theoretical values. This is because even though they may possess

TABLE 5.2 *Calculated values of the nine elastic constants for a polyethylene crystal at 298 K. (After Odajima and Maeda, J. Polym. Sci. C* **15** *(1966)55.)*

Stiffness component	$c_{ij}/10^9 Nm^{-2}$
c_{11}	4.83
c_{22}	8.71
c_{33}	257.1
c_{12}	1.16
c_{13}	2.55
c_{23}	5.84
c_{44}	2.83
c_{55}	0.78
c_{66}	2.06

good crystal orientation the presence of an amorphous phase tends to cause a significant reduction in modulus. Estimates of chain-direction moduli can be made experimentally using an X-ray diffraction method. This involves measuring the relative change under an applied stress in the position of any (*hkl*) Bragg reflection which has a component of the chain-direction repeat (usually indexed *c*). The production of 100 percent crystalline macroscopic polymer single crystals by solid-state polymerization (Section 2.5.6) has recently made it possible to measure the moduli of these crystals directly using simple mechanical methods. A stress–strain curve for a polydiacetylene single crystal fibre is given in Fig. 5.34. The

TABLE 5.3 *Calculated and measured values of chain-direction moduli for several different polymer crystals. (After Young, Chapter 7 in 'Developments in Polymer Fracture' edited by E.H. Andrews.)*

Polymer		Calculated E/GN m^{-2}	method	Measured E/GN m^{-2}	method
Polyethylene		182	VFF	240	X-ray
		340	UBFF	358	Raman
		257	UBFF	329	Neutron (CD$_2$)
Polypropylene		49	UBFF	42	X-ray
Polyoxymethylene		150	UBFF	54	X-ray
Polytetrafluoroethylene		160	UBFF	156	X-ray
				222	Neutron
Substituted	(i)	49	VFF	45	Mechanical
Polydiacetylenes	(ii)	65	VFF	61	Mechanical

VFF—Valence force field (e.g. Treloar's method)
UBFF—Urey–Bradley force field (e.g. used to determine Table 5.2)
(i) Phenyl urethane derivative
(ii) Ethyl urethane derivative

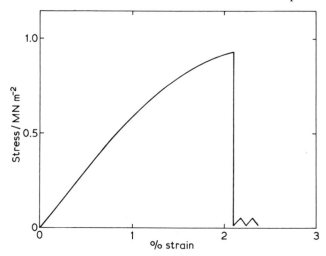

Fig. 5.34 *A stress–strain curve obtained for a polydiacetlyene single crystal fibre.*

modulus calculated from the initial slope of the curve is 61 GN m^{-2} which is very close to the value calculated using the simple bond stretching and bending method of Treloar ($\sim$65 GN m^{-2}). The measured and calculated values of moduli for some polydiacetylene single crystals are also given in Table 5.3

A general picture emerges concerning the values of chain direction moduli of polymer crystals. They tend to be high if the molecule is in the form of a planar zig-zag rather than a helix. For example, polyethylene is stiffer than polyoxymethylene or polytetrafluorethylene which both have molecules in helical conformations (Table 4.1). The helices can be extended more easily than the polyethylene planar zig-zag. Also the presence of large side groups tend to reduce the modulus because they increase the separation of molecules in the crystal. This causes an increase in the area supported by each chain.

Although polymers are generally considered to be relatively low modulus flexible materials it is clear from Table 5.3 that their potential unidirectional moduli are relatively high. This feature has been utilized to a limited extent for many years in polymer fibres. However, conventional fibres do not normally exploit the full potential of the polymer backbone and it is envisaged that future developments in this field will be in producing polymer morphologies which enable this high potential stiffness to be fully utilized. For many purposes it is the *specific modulus* (Young's modulus divided by specific gravity) that is important, especially when light high-stiffness materials are required. Polymers tend to have low specific gravities, typically 0.8–1.5 compared with 7.9 for steel, and so their specific stiffnesses compare even more favourably with metals.

5.5.2 *Elastic deformation of semi-crystalline polymers*

The presence of crystals in a polymer have a profound effect upon its mechanical behaviour. This effect has been known for many years from experience with natural rubber. Fig. 5.35(a) shows the increase in Young's modulus that occurs due to crystallization which can occur during prolonged storage of the rubber at low temperature. There is a hundred-fold increase in modulus when the degree of crystallinity increases from 0 to 25 percent. A similar effect is found in other semi-crystalline polymers. Fig. 5.35(b) shows how the modulus of a linear polyethylene increases as the crystallinity is increased through use of different heat treatments. The degree of crystallinity in linear polyethylene is usually quite high (>60 percent) and the data in Fig. 5.35(b) can be extrapolated to give a modulus of the order of 5 GN m^{-2} for 100 percent crystalline material.

It is apparent from considerations of the structure in Section 4.2 that semi-crystalline polymers are two-phase materials and that the increase in modulus is due to the presence of the crystals. Traditional ideas of the stiffening effect due to the presence of crystals were based upon the statistical theory of rubber elasticity (Section 5.3.2). It was thought that the crystals in the amorphous 'rubber' behaved like cross-links and produced the stiffening through an increase in cross-link density rather than through their own inherent stiffness. Although this mechanism may be relevant at very low degrees of crystallinity it is clear that most semi-crystalline

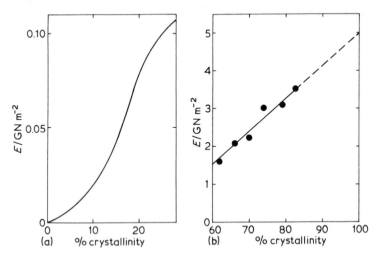

Fig. 5.35 *Variation of Young's modulus, E, with the degree of crystallinity for different polymers. (a) Increase in modulus of rubber with crystallization (Data of Leitner reported by Treloar). (b) Dependence of the modulus of polyethylene upon crystallinity (Data taken from Wang, J. Appl. Phys.,* **44** *(1973) 4052).*

polymers behave as composites and the observed modulus is a reflection of the combined modulus of the amorphous and crystalline regions. The exact way in which this combination of moduli can be represented mathematically is not a trivial problem. For example, spherulitic polymers have elastically isotropic elastic properties, but the individual crystals within the spherulites are themselves highly anisotropic. It is therefore necessary to calculate an effective modulus for the particular type of morphology under consideration. The effective crystal modulus depends not only upon the proportion of crystalline material present, given by the degree of crystallinity, but also upon the size, shape and distribution of the crystals within the polymer sample. It is clear, therefore, that reliable estimates of the properties of the semi-crystalline polymer composite can only be made for specific well-defined morphologies.

A model system which can be used is the diacetylene polymer for which the stress–strain curve was given in Fig. 5.34. By controlling the polymerization conditions it is possible to prepare single crystal fibres which contain both monomer and polymer molecules. The monomer has a modulus of only 9 GN m^{-2} along the fibre axis compared with 61 GN m^{-2} for the polymer and the partly polymerized fibres which contain both monomer and polymer molecules are found to have values of modulus between these two extremes as shown in Fig. 5.36. The variation of the modulus with the proportion of polymer (approximately equal to the conversion) can be predicted by two simple models. The first one due to Reuss assumes that the elements in the structure (i.e. the monomer and polymer molecules) are lined up in series and experience the same stress.

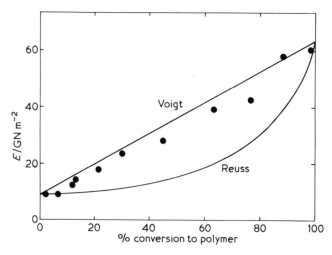

Fig. 5.36 *Dependence of the modulus of polydiacetylene single crystal fibres upon conversion in to polymer.*

The modulus in this case, E_R will be given by an equation of the form

$$1/E_R = V_P/E_p + (1 - V_p)/E_m \qquad (5.178)$$

where V_p is the volume fraction of polymer and E_p and E_m are the moduli of the polymer and matrix (monomer) respectively. The second model due to Voigt assumes that all the elements are lined up in parallel and so experience the same strain. The modulus for the Voigt model, E_V, is given by the rule of mixtures as

$$E_V = E_p V_p + E_m(1 - V_p) \qquad (5.179)$$

Lines corresponding to the predictions of Equations (5.178) and (5.179) are also given in Fig. 5.36. It can be seen that the experimental points lie between the two lines and closest to the Voigt line indicating that the elements are experiencing approximately the same strain in this particular structure. Equations similar to (5.178) and (5.179) can be used to predict the moduli of fibre-reinforced composites and it is necessary that in all similar composite structures that the measured moduli fall either between or on the Voigt and Reuss lines. The agreement that is found between the measured moduli of the diacetylene polymer fibres and the Voigt prediction is due to the structure of the fibres which consist of long polymer molecules lying parallel to the fibre axis and embedded in a monomer matrix. This morphology ensures that the strain will be uniform in the deformed structure.

The situation is considerably more complicated with more random structures such as spherulitic polymers. The measured modulus of such a structure is an average of that of the crystalline and amorphous regions. Even when it is assumed that the amorphous material is isotropic there are problems in determining how to estimate the effective average modulus of the crystals and how to combine this with modulus of the amorphous material to give an overall modulus for a polymer sample. This latter problem can be overcome by extrapolating measurements of modulus as a function of crystallinity to 100 percent crystallinity as shown in Fig. 5.35(b). The structure of such a material can be thought of as a randomly oriented poly-crystalline structure not unlike a poly-crystalline metal. The Reuss and Voigt models can again be applied with their respective assumptions of uniform stress and uniform strain. Using the nine crystal moduli from Table 5.2 the Reuss model predicts a modulus of 4.9 GN m^{-2} compared with 15.6 GN m^{-2} for the Voigt model. Since the extrapolated value is ~5 GN m^{-2} it can be seen that in the case of spherulitic polyethylene the Reuss model gives a better prediction of the observed behaviour.

5.5.3 *Shear yielding in glassy polymers*

Glassy polymers are generally thought of as being rather brittle materials, but they are capable of displaying a considerable amount of ductility below

T_g especially when deformed under the influence of an overall hydrostatic compressive stress. This behaviour is illustrated in Fig. 5.37 where true stress–strain curves are given for an epoxy resin tested in uniaxial tension and compression at room temperature. The T_g of the resin is 100°C and such cross-linked polymers are found to be brittle when tested in tension at room temperature. In contrast they can show considerable ductility in compression and undergo shear yielding. Another important aspect of the deformation is that glassy polymers tend to show 'strain softening'. The true stress drops after yield, not because of necking which cannot occur in compression, but because there is an inherent softening of the material.

The process of shear yielding in epoxy resins is found to be homogeneous. The specimens undergo uniform deformation with no evidence of any localization. The situation is rather similar in other glassy polymers but in certain cases strongly localized deformation is obtained. Fig. 5.38 shows a polarized light-optical micrograph of a section of polystyrene deformed in plane strain compression. Fine deformation bands are obtained in which the shear is highly localized. It is thought that the formation of shear bands may be associated with the strain softening in the material. Once a small region starts to undergo shear yielding it will continue to do so because it has a lower flow stress than the surrounding relatively undeformed regions.

There have been many attempts over the years to explain the process of shear yielding on a molecular level and some of the approaches are outlined below.

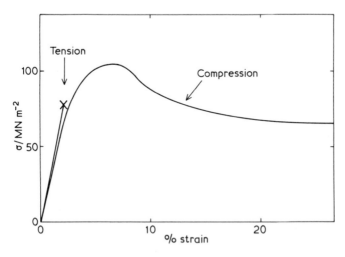

Fig. 5.37 *Stress–strain curves for an epoxy resin deformed in tension or compression at room temperature.*

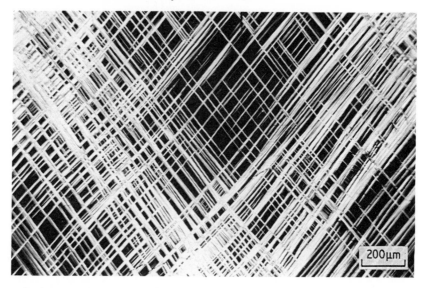

Fig. 5.38 *Shear bands in polystyrene deformed in plane strain compression (Courtesy Dr. R.J. Oxborough)*.

(i) *Stress-induced increase in free volume*

It has been suggested that the one effect of an applied strain upon a polymer is to increase the free volume (Section 4.4.2) and hence increase the mobility of the molecular segments. It follows that this would also have the effect of reducing T_g. Developments of this idea predict that during tensile deformation the overall hydrostatic tension should cause an increase in free volume and so aid plastic deformation. However, there are problems in the interpretation of compression tests where plastic deformation can still take place even though the hydrostatic stress is compressive. Moreover, careful measurements have shown that there is generally a reduction in overall specimen volume during plastic deformation in both tension and compression. It would seem, therefore, that the concept of free volume is not very useful in the explanation of plastic deformation of glassy polymers.

(ii) *Application of the Eyring theory to yield in polymers*

It is possible to think of yield and plastic deformation in polymers as a type of viscous flow, especially since glassy polymers are basically frozen liquids that have failed to crystallize. Eyring developed a theory to describe viscous flow in liquids and it can be readily adapted to describe the behaviour of glassy polymers. The segments of the polymer chain can be thought of as being in a pseudo-lattice and for flow to occur a segment must move to an adjacent site. There will be a potential barrier to overcome,

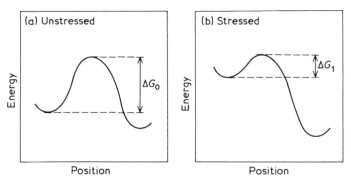

Fig. 5.39 *Schematic illustration of the potential barrier for flow in a glassy polymer.* (a) *Unstressed state with barrier of* ΔG_0. (b) *Stressed state with barrier reduced to* $\Delta G_1 = (\Delta G_0 - \frac{1}{2}\sigma A x)$.

because of the presence of adjacent molecular segments, and the situation is shown schematically in Fig. 5.39(a). The height of the barrier is given by ΔG_0 and the frequency ν_0 at which the segments jump the barrier and move to the new position can be represented by an Arrhenius type equation

$$\nu_0 = B \exp(-\Delta G_0/kT) \tag{5.180}$$

where B is a temperature independent constant and k is Boltzmann's constant. The rate at which segments jump forward over the barrier will be increased by the application of a shear stress σ. This has the effect of reducing the energy barrier by an amount $\frac{1}{2}\sigma A x$ which is the work done in moving the segment a distance, x, through the lattice of unit cross-sectional area A. The new situation is illustrated schematically in Fig. 5.39(b) and the frequency of forward motion increases to become

$$\nu_f = B \exp[-(\Delta G_0 - \frac{1}{2}\sigma A x)/kT]$$

$$\text{or} \quad \nu_f = \nu_0\exp(\sigma A x/2kT] \tag{5.181}$$

There is also the possibility of backward motion occurring. The frequency of backward motion ν_b is given by

$$\nu_b = \nu_0 \exp(-\sigma A x/2kT) \tag{5.182}$$

and the application of stress makes backward motion more difficult. The strain rate $\dot{e}$ of the polymer will be proportional to the difference between the frequencies of forward and backward motion

i.e. $\dot{e} \propto (\nu_f - \nu_b) = \nu_0\exp(\sigma A x/2kT) - \nu_0\exp(-\sigma A x/2kT)$

and this can be written as

$$\dot{e} = K \sinh(\sigma V/2kT) \tag{5.183}$$

where K is a constant and V ($= Ax$) is called the 'activation volume' or 'Eyring volume' and is the volume of the polymer segment which must move to cause plastic deformation. If the stress is high and the frequency of backward motion is small then Equation (5.183) can be approximated to

$$\dot{e} = (K/2)\exp(\sigma V/2kT) \tag{5.184}$$

This approach has met with considerable success in the prediction of the strain-rate dependence of the yield stress of polymers. It is tempting to relate V to structural elements in the polymer. For most glassy polymers it comes out to be the volume of several polymer segments (typically between 2 and 10) although there is no clear correlation between the size of the activation volume and any particular aspect of the polymer structure.

(iii) *Molecular theories of yield*
Recent developments in the interpretation of the yield behaviour of glassy polymers have involved attempts to relate plastic deformation to specific kinds of molecular motion. One particularly successful approach was suggested by Robertson for the case of a planar zig-zag polymer chain. He assumed that at any time the molecular chain segments would be distributed between low-energy *trans* and high-energy *cis* conformations and that the effect of an applied stress would be to cause certain segments to change over from *trans* to *cis* conformations as illustrated in Fig. 5.40. This and other similar molecular theories have lead to accurate predictions of the variation of yield stress with strain-rate and temperature for glassy polymers. The reader is referred to more advanced texts for further details.

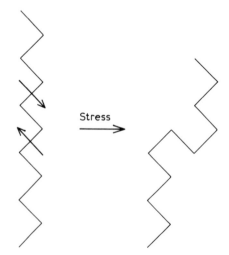

Fig. 5.40 *Schematic illustration of the conformational changes involved in the deformation of a planar zig-zag polymer molecule (After Robertson and Kambour).*

5.5.4 *Crazing in glassy polymers*

The phenomenology of crazing was discussed in section 5.4, but as yet the mechanisms by which crazing occurs have not been considered. Crazes normally nucleate on the surface of a specimen, probably at the site of flaws such as scratches or other imperfections. They tend to be lamellar in shape and grow into the bulk of the specimen from the surface as shown in Fig. 5.41. The flaws on the surface tend to raise the magnitude of the applied stress locally and allow craze initiation to occur.

The basic difference between crazes and cracks is that crazes contain polymer, typically about 50 percent by volume, within their bulk whereas cracks do not. The presence of polymer within crazes was originally deduced from measurements of the refractive index of crazed material, but it can be demonstrated more directly by using electron microscopy. Fig. 5.42 shows a section through a craze in polystyrene viewed at high magnification. It can be clearly seen that there is a sharp boundary between the craze and the polymer matrix and that the craze itself contains many cavities and is bridged by a dense network of fine fibrils of the order of 200 Å in diameter. If a craze is followed along its length to the tip it is found to taper gradually until it eventually consists of a row of voids and microfibrils on the 50 Å scale.

Microscopic examination has given the most important clues as to how crazes nucleate and grow in a glassy polymer. The basic mechanism is thought to be a local yielding and cold-drawing process which takes place in a constrained zone of material. It is known that the local stresses at the tip of a surface flaw or a growing craze are higher than the overall tensile

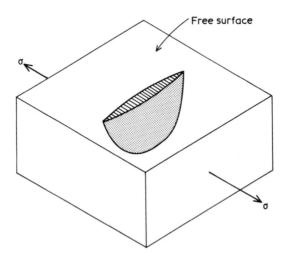

Fig. 5.41 *Schematic diagram of a craze nucleating at the surface of a polymer.*

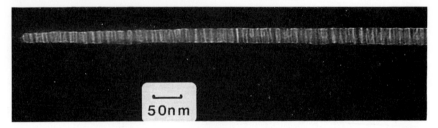

Fig. 5.42 *Transmission electron micrograph of a craze in a thin film of polystyrene (Courtesy of Professor E. Kramer).*

stress applied to the specimen. It is envisaged that yielding takes place locally at these stress concentrations at a stress below that which is required to cause general yielding in the rest of the specimen. If the bulk tensile stress is σ_0 then there will be an overall bulk tensile hydrostatic stress of $p = \sigma_0/3$ but this hydrostatic component will be very much higher at the tip of the flaw or growing craze. These conditions of stress then favour the initiation of a craze at a flaw or the propagation of a preexisting craze. It is envisaged that the fibrils within the craze form by a local yielding and cold-drawing process and the fibrils remain intact because they contain oriented molecules and are consequently stronger than the surrounding uncrazed polymer. The thickening of the craze is thought to take place by uncrazed material at the craze/matrix boundary being deformed into fibrils and so transformed into craze matter. The crazes form as very thin lamellae in a plane perpendicular to the tensile axis because the tensile stress is highest across this plane and the elastic constraint of the surrounding uncrazed material allows the craze to grow only laterally and thicken.

5.5.5 *Plastic deformation of polymer crystals*

There is now clear evidence to suggest that polymer crystals are capable of undergoing plastic deformation in a similar way to other crystalline solids through processes such as slip, twinning and martensitic transformations. It is thought that similar deformation processes occur in both isolated single crystals and in the crystalline regions of semi-crystalline polymers. The main difference between the deformation of polymer crystals and crystals of atomic solids is that the deformation occurs in the polymer such that molecules are not broken and, as far as possible, remain relatively undistorted. This problem of molecular integrity does not arise in atomic solids. Deformation, therefore, generally involves only the sliding of molecules in directions parallel or perpendicular to the chain axis which breaks only the secondary van der Waals or hydrogen bonds. The specific deformation mechanisms will be discussed separately.

(i) *Slip*

The deformation of metal crystals takes place predominantly through slip processes which involve the sliding of one layer of atoms over the other through the motion of dislocations. In fact the ductility of metals stems from there normally being a large number of slip systems (i.e. slip planes and directions) available in metal crystals. Slip is also important in the deformation of polymer crystals, but the need to avoid molecular fracture limits the number of slip systems available. The slip planes are limited to those which contain the chain direction (i.e. $(hk0)$ planes when [001] is the chain direction). The slip directions can in general be any direction which lies in these planes, but normally only slip which takes place in directions parallel or perpendicular to the chain direction is considered in detail. Slip in any general direction in a $(hk0)$ plane will have components of these two specific types. *Chain direction slip* is illustrated schematically in Fig. 5.43. The molecules slide over each other in the direction parallel to the molecular axis. It is known to occur during the deformation of both bulk semi-crystalline polymers and of solution-grown polymer single crystals. It can be readily detected from the change of orientation of the molecules within the crystals as shown in Fig. 5.44 for polytetrafluoroethylene. Initially the molecules are approximately normal to the lamellar surface (Fig. 4.21), but during deformation they slide past each other and become tilted within the crystal. Slip in directions perpendicular to the molecular axes is sometimes called *transverse slip*. It has been shown to occur during the deformation of both melt-crystallized polymers and solution-grown single crystals, but is rather more difficult to detect because it involves no change in molecular orientation within the crystals.

(ii) *Dislocation motion*

It is well established that slip processes in atomic solids are associated with the motion of dislocations. The theoretical shear strength of a metal lattice has been shown (Section 5.1.4) to be of the order of $G/10$ to $G/30$ where G

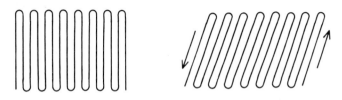

Fig. 5.43 *Schematic representation of chain-direction slip in a chain-folded polymer crystal.*

Fig. 5.44 *A replica of a fracture surface of polytetrafluoroethylene after compressive deformation in the vertical direction.*

is the shear modulus of the crystal. The measured shear yield strengths of many crystalline solids are typically between 10^{-4} G and 10^{-5} G. The difference between the theoretical and experimentally determined values is due to deformation taking place by the glide of dislocations which occurs at stresses well below the theoretical strength. The presence of dislocations in polymer crystals has been discussed previously (Section 4.2.5) and there is no fundamental reason why slip in polymer crystals cannot occur through dislocation motion.

The energy per unit length, $E(l)$, of a dislocation in an isotropic medium can be shown* to be given approximately by

$$E(l) = \frac{Gb^2}{4\pi} \ln\left(\frac{r}{r_0}\right) \tag{5.185}$$

where b is the Burgers vector of the dislocation, r the radius of the crystal and r_0 is the core radius of the dislocation (normally taken $\approx b$). The shear modulus G in Equation (5.185) is that for an isotropic medium. Since the bonding in polymer crystals is highly anisotropic G must be replaced by an effective crystal modulus when the equation is applied to polymer crystals. Inspection of Equation (5.185) shows that for a crystal of a given size, dislocations will have low energy when the shear modulus and b are both

*J. Friedel, *Dislocations*, Pergamon Press, (1964).

small. The total energy of a dislocation will also be reduced if the length of the dislocation line is as short as possible. Since dislocations cannot terminate within a crystal then the dislocations with the shortest length can be obtained in lamellar polymer crystals by having the dislocation line perpendicular to lamellar surface and so parallel to the chain direction. The need to avoid excess deformation of the polymer chains means that *b* should also be parallel to the chain direction. The result of these considerations is that a particularly favourable type of dislocation in lamellar crystals would be expected to be a screw dislocation with its Burgers vector parallel to the chain direction as illustrated schematically in Fig. 4.24.

Such dislocations have been shown to be present in solution-grown polyethylene single crystals and their number is found to increase following deformation (Fig. 5.45). The Burgers vector of such dislocations is the same as the chain direction repeat 2.54 Å (Table 4.1) and the shear modulus for slip in the chain direction will be either $c_{44} = 2.83$ GN m^{-2} or $c_{55} = 0.78$ GN m^{-2} (Table 5.2) depending upon whether the slip plane is (010) or (100). It is found that at room temperature [001] chain direction slip in bulk polyethylene occurs at a stress of the order of 0.015 GN m^{-2} (i.e. ~1/50 to 1/200 of the shear modulus) and calculations have shown that chain direction slip could occur through the thermal activation of the type of dislocations shown in Fig. 4.24 and 5.45. This mechanism is, however, only favourable when the crystals are relatively thin where the length of the dislocation line is small. It is expected that such a dislocation would be favourable in thin crystals of other polymers since they generally have low shear moduli. Most other polymer crystals have larger chain-direction repeats than polyethylene (Table 4.1) especially when the molecule is in a helical conformation. This would give rise to larger values of $E(l)$ through higher Burgers vectors, but it would be expected that in this case the dislocations might split into partials in order to reduce their energy.

(iii) *Twinning*

Twinned crystals of many materials are produced during both growth and deformation. Many mineral crystals are found in a twinned form in nature and deformation twins are often obtained when metals are deformed, especially at low temperatures. The most commonly obtained types of twinned crystals are those in which one part of the crystal is a mirror image of another part. The boundary between the two regions is called the *twinning plane*. The particular types of twins that form in a crystal depend upon the structure of the crystal lattice and deformation twins can be explained in terms of a simple shear of the crystal lattice.

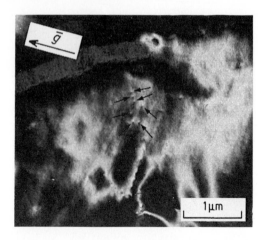

Fig. 5.45 *Dark-field electron micrograph of a polyethylene crystal containing screw disloca-tions with Burgers vectors parallel to the chain direction (normal to the crystal surface). The dislocations can be seen arrowed as dots with black and white contrast. (Courtesy of Dr J. Petermann.)*

As in the case of slip, there is no fundamental reason why polymer crystals cannot undergo twinning and it is found that twinning frequently occurs when crystalline polymers are deformed. The types of twins that are obtained are formed in such a way that the lattice is sheared without either breaking or seriously distorting the polymer molecules. A good example of the formation of twins in polymer crystals is obtained by deforming solid-state polymerized polydiacetylene single crystals (Section 2.5.6). Fig. 5.46 shows a micrograph of such a crystal viewed at 90° to the chain direction. The striations on the crystal define the molecular axis and it can be seen that the molecules kink over sharply at the twin boundary. Accurate measurements have shown that there is a 'mirror image' orientation of the crystal on either side of the boundary. This clearly indicates that the crystal contains a twin. A schematic diagram of the molecular arrangements on either side of the twin boundary is also given in Fig. 5.46. The angle that the molecules can bend over and still retain the molecular structure can be calculated from a knowledge of the crystal structure and exact agreement is found between the measured and calculated angles. The result of the twinning process is that the molecules on every successive plane in the twin are displaced by one unit of *c* parallel to the chain direction. This is clearly identical to the result of chain-direction slip taking place and it has now become recognized that this type of twinning could be important in the deformation of many types of polymer crystals.

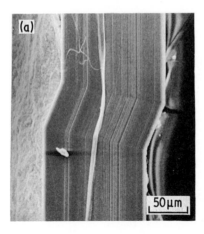

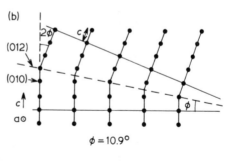

Fig. 5.46 *A large twin in a polydiacetylene single crystal.* (a) *Scanning electron micrograph of twin;* (b) *Schematic diagram of twin on a molecular level.* (*Young, Bloor, Batchelder and Hubble, J. Mater, Sci.,* **13** *(1978) 62, reproduced with permission.*)

Another type of twinning that has been found to occur in polymer crystals involves a simple shear in directions perpendicular to the chain axis and in this case does not bend the molecules. An example of this type of twinning is shown schematically in Fig. 5.47 for the polyethylene unit cell projected parallel to the chain direction. Two particular twinning planes (110) and (310) have been found and the twinning has been reported after deformation of both solution-grown single crystals and melt-crystallized polymer. The occurrence of twinning can be detected from measurements of the rotation of the *a* and *b* crystal axes about the molecular *c* axis since the twinning causes rotation of the crystal lattice.

(iv) *Martensitic transformations*

This is a general term which covers transformations which bear some similarity to twinning, but involve a change of crystal structure. The name comes from the well-known, important transformation which occurs when austenitic steel is rapidly cooled and forms the hard tetragonal metastable phase called 'martensite'. These transformations can also occur on the deformation or cooling of many crystalline materials and all of these martensitic transformations share the similar feature that they take place with no long-range diffusion.

Martensitic transformations have not been widely reported in polymers, but an important one occurs on the deformation of polyethylene. The normal crystal structure of both solution-grown single crystals and melt-crystallized polymer is orthorhombic, but after deformation extra Bragg reflections are found in diffraction patterns and these have been

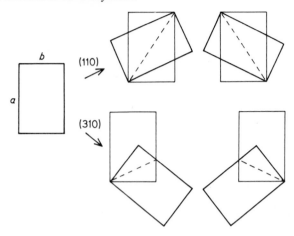

Fig. 5.47 *Schematic representation of (110) and (310) twinning in orthorhombic polyethylene. The unit cells are viewed projected parallel to the chain direction. (After Frank, Keller and O'Conner, Phil. Mag. 3 (1958) 64.)*

explained in terms of the formation of a monoclinic form (Table 4.1). The transformation takes place by means of a simple two-dimensional shear of the orthorhombic crystal structure in a direction perpendicular to c. The chain axis remains virtually unchanged by the transformation at about 2.54 Å, but in the monoclinic cell (Table 4.1) it is conventionally indexed as b.

5.5.6 *Plastic deformation of semi-crystalline polymers*

Melt-crystallized polymers have a complex structure consisting of amorphous and crystalline regions in a spherulitic microstructure (Fig. 4.14). They display their most useful properties in the temperature range $T_g < T < T_m$ and they are normally designed such that the temperature of operation comes within this range. Below the T_g they are brittle and above T_m they behave as viscous liquids or rubbers. The deformation of semi-crystalline polymers can be considered on several levels of structure. Fig. 5.48 shows how the spherulitic microstructure changes with deformation. The spherulites become elongated parallel to the stretching direction as the strain is increased and at high strains the spherulitic microstructure breaks up and a fibre-like morphology is obtained. Clearly the change in appearance of the spherulitic structure must be reflecting changes which are occurring on a finer structural level within the crystalline and amorphous areas. It is thought that when the spherulites are deforming homogeneously, during the early stages of deformation, the crystalline regions deform by a combination of slip, twinning and martensitic transformations as outlined in the last section. In addition it is thought that

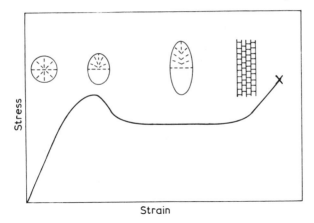

Fig. 5.48 *Schematic illustration of the change in spherulite morphology during the cold-drawing of a semi-crystalline polymer (After Magill).*

because the amorphous phase is in a rubbery state it allows deformation to take place by the shear of crystals relative to each other. During these early and intermediate stages of deformation the crystals become distorted but remain essentially intact. Towards the later stages when the spherulitic morphology is lost there appears to be a complete break-up of the original crystalline microstructure and a new fibre-like structure is formed. The mechanisms whereby this happens are not fully understood but one possible method by which this could take place is illustrated in Fig. 5.49. It is assumed for simplicity that the undeformed polymer has crystals which are stacked with the molecules regularly folded. Deformation then takes place by slip and twinning etc. until eventually the crystals start to crack and chains are pulled out. At sufficiently high strains the molecules and crystal blocks become aligned parallel to the stretching direction and a fibrillar structure is formed. Earlier theories of cold-drawing suggested that the structural changes occurred because the heat generated during deformation caused melting to occur and the material subsequently recrystallized on cooling. Although it is clear that there is a small rise in temperature during deformation this is not sufficient to cause melting, and at normal strain rates the heat is dissipated relatively quickly.

Polymers oriented by drawing are of considerable practical importance because they can display up to 50 percent of the theoretical chain direction modulus of the crystals. This short-fall in modulus is caused by imperfections within the crystals and the presence of voids and noncrystalline material. The value of modulus obtained depends principally upon the crystal structure of the polymer and the extent to which a sample is drawn (i.e. the draw-ratio). The maximum draw ratio that can be obtained in turn

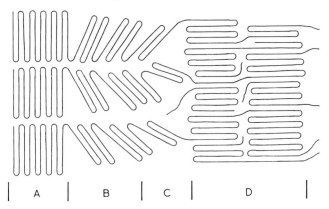

Fig. 5.49 *Schematic representation of the deformation which takes place on a molecular level during the formation of a neck in a semi-crystalline polymer. (A) Idealized representation of undeformed structure, (B) Crystals deforming by slip, twinning and martensitic transformations, (C) Crystals break up and molecules are pulled out, (D) Fibrillar structure forms.*

depends upon the temperature of drawing and the average molar mass and molar mass distribution of the polymer. There is now considerable interest is optimizing these variables to obtain ultra-high modulus, oriented polymers.

5.6 Fracture

Fracture can be defined simply as the creation of new surfaces within a body through the application of external forces. The failure of engineering components through fracture can have catastrophic consequences and much effort has been put in by materials scientists and engineers in developing materials and designing structures which are resistant to premature failure through fracture. It is possible to classify materials as being either 'brittle' like glass which shatters readily or 'ductile' like pure metals such as copper or aluminium which can be deformed to high strains before they fail. Polymers are found to display both of these two types of behaviour depending upon their structure and the conditions of testing. In fact, the same polymer can sometimes be made either brittle or ductile by changing the temperature of rate of testing. Over the years some polymers have been advertized as being 'unbreakable'. Of course, this is very misleading as every material can only stand limited stresses. On the other hand, domestic containers made from polymers such as polypropylene or low-density polyethylene are remarkably tough and are virtually indestructible under normal conditions of use.

Polymers are capable of undergoing fracture in many different ways and the stress–strain curves of several different polymers tested to failure are

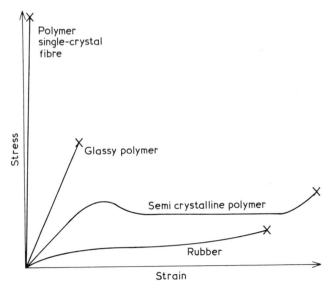

Fig. 5.50 *Schematic stress–strain curves of different types of polymers, drawn approximately to scale.*

given schematically in Fig. 5.50. Glassy polymers tend to be rather brittle, failing at relatively low strains with very little plastic deformation. On the other hand, semi-crystalline polymers are normally more ductile, especially between T_g and T_m, and undergo cold-drawing before ultimate failure. Cross-linked rubbers are capable of being stretched elastically to high extensions, but a tear will propagate at lower strains if a pre-existing cut is present in the sample. Polymer single-crystal fibres break at very high stresses, but because their moduli are high the stresses correspond to strains of only a few percent. Because of this wide variation in fracture behaviour it is convenient to consider the different fracture modes separately.

5.6.1 Ductile–Brittle transitions

Many materials are ductile under certain testing conditions, but when these conditions are changed such as, for example, by reducing the temperature they become brittle. This type of behaviour is also encountered in steels and is very common with polymers. Fig. 5.51(a) shows how the strength of poly(methyl methacrylate) varies with testing temperature. At low temperatures the material fails in a brittle manner, but when the testing temperature is raised to just above room temperature the polymer undergoes general yielding and brittle fracture is suppressed. This type of brittle–ductile transition has been explained in terms of brittle fracture

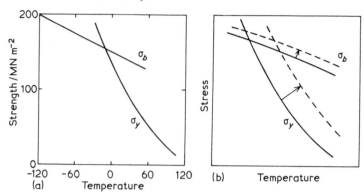

Fig. 5.51 *Ductile-to-brittle transitions.* (a) *Variation with temperature of brittle strength measured in flexure,* σ_b *and yield strength measured in tension,* σ_y *for poly(methyl methacrylate).* (b) *Effect of strain-rate upon the transition. Curves for a higher strain-rate are drawn as dashed lines (After Ward).*

and plastic deformation being independent processes which have a different dependence on temperature as shown schematically in Fig. 5.51(b). Since the yield stress increases more rapidly as the temperature is reduced, at a critical temperature it becomes higher than the stress required to cause brittle fracture. The temperature at which this occurs corresponds to that of the *brittle–ductile transition*, T_b, and it is envisaged that the process which occurs during deformation is that which can take place at the lowest stress, i.e. brittle fracture at $T < T_b$ and general yielding at $T > T_b$.

With polymers there is the added complication of there being a strong dependence of the mechanical properties upon strain rate as well as upon temperature. For example, nylon can be cold-drawn at room temperature when relatively low strain rates are used, but it becomes brittle as the strain-rate is increased. It is thought that this is because the curves of yield stress and brittle stress against temperature are moved to the right by increasing the strain rate and so the temperature of the brittle–ductile transition is increased (Fig. 5.51(b)).

It is known that the temperature at which the brittle–ductile transition occurs for a particular polymer is sensitive to both the structure of the polymer and the presence of surface flaws and notches. The temperature of the transition is generally raised by crosslinking and increasing the crystallinity. Both of these factors tend to increase the yield stress without affecting the brittle stress significantly. The addition of plasticizers to a polymer has the effect of reducing yield stress and so lowers T_b. Plasticization is widely used in order to toughen polymers such as poly(vinyl chloride) which would otherwise have a tendency to be brittle at room temperature.

The presence of surface scratches, cracks or notches can have the effect of both reducing the tensile strength of the material and also increasing the temperature of the brittle–ductile transition. In this way the presence of a notch can cause a material to fail in a brittle manner at a temperature at which is would otherwise be ductile. As one might expect this phenomenon can play havoc with any design calculations which assume the material to be ductile. This type of 'notch sensitivity' is normally explained by examining the state of stress that exists at the root of the notch. When an overall uniaxial tensile stress is applied to the body there will be a complex state of triaxial stress at the tip of the crack or notch. This can cause a large increase (up to three times) in the yield stress of the material at the crack tip as shown in Fig. 5.52. The brittle fracture stress will be unaffected and so the overall effect is to raise T_b. If the material is used between T_b for the bulk and T_b for the notched specimen it will be prone to suffer from 'notch brittleness'.

It is tempting to relate the temperature at which the ductile–brittle transition takes place to either the glass transition or secondary transitions (Section 5.2.6) occurring within the polymer. In some polymers such as natural rubber or polystyrene T_b and T_g occur at approximately the same temperature. Many other polymers are ductile below the glass transition temperature (i.e. $T_b < T_g$). In this case it is sometimes possible to relate T_b to the occurrence of secondary low-temperature relaxations. However, more extensive investigations have shown that there is no general correlation between the brittle–ductile transition and molecular relaxations. This may not be too unexpected since these relaxations are detected at low strains whereas T_b is measured at high strains and depends upon factors such as the presence of notches which do not affect molecular relaxations.

5.6.2 *Brittle fracture and flaws*

One of the most useful approaches that can be used to explain the fracture of brittle polymers is the theory of brittle fracture developed by Griffith to

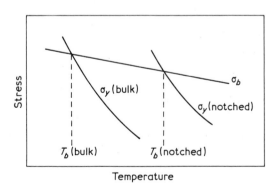

Temperature

Fig. 5.52 *Schematic representation of the effect of notching upon the ductile-to-brittle transition (After Andrews).*

account for the fracture behaviour of glass. The theory is concerned with the thermodynamics of initiation of cracks from pre-existing flaws in a perfectly elastic brittle solid. These flaws could be either scratches or cracks which both have the effect of causing a stress concentration. This means that the local stress in the vicinity of the crack tip is higher than that applied to the body as a whole. The theory envisages that fracture starts at this point and the crack then continues to propagate catastrophically.

The effect of the presence of a crack in a body can be determined by considering an elliptical crack of major and minor axes a and b in a uniformly loaded infinite plate as illustrated in Fig. 5.53. If the stress at a great distance from the crack is σ_0 then the stress, σ, at the tip of the crack is given by*

$$\sigma = \sigma_0(1 + 2a/b) \tag{5.186}$$

The ratio σ/σ_0 defines the stress-concentrating effect of the crack and is three for a circular hole ($a = b$). It will be even larger when $a > b$. The equation can be put in a more convenient form using the radius of curvature ρ of the sharp end of the ellipse which is given by

$$\rho = b^2/a$$

Equation (5.186) then becomes

$$\sigma = \sigma_0[1 + 2(a/\rho)^{1/2}] \tag{5.187}$$

If the crack is long and sharp ($a \gg \rho$), σ will be given approximately by

$$\sigma = \sigma_0 2(a/\rho)^{1/2} \tag{5.188}$$

Inspection of these equations shows that the effect of a sharp crack will be to cause a large concentration of stress which is a maximum at the tip of the crack, the point from which the crack propagates.

Griffith's contribution to the argument concerned a calculation of the energy released by putting the sharp crack into the plate and relaxing the surrounding material, and relating this to the energy required to create new surface. The full calculation requires an integration over the stress field around the crack but an estimate of the energy released in the creation of new surface can be made by considering that the crack completely relieves the stresses in a circular zone of material of radius a around the crack as shown in Fig. 5.53. If the plate has unit thickness then the volume of this region is $\sim \pi a^2$. Since the material is assumed to be linearly elastic then the strain energy per unit volume in an area deformed to a stress of σ_0 is $\frac{1}{2}\sigma_0^2/E$, where E is the modulus. If it is assumed that the circular area is completely relieved of stress when the crack is present then the energy

---

*See J.G. Williams, *Stress Analysis of Polymers*.

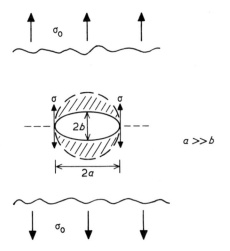

Fig. 5.53 *Model of an elliptical crack of length 2a in a uniformly loaded infinite plate.*

released is $\frac{1}{2}\pi\, a^2\sigma_0^2/E$ per unit thickness. The surface energy of the crack is $4a\gamma$ per unit thickness, where γ is the surface energy of the material. The extra factor of 2 appears because the crack of length $2a$ has two surfaces. It is postulated that the crack will propagate if the energy gained by relieving stress is greater than that needed to create new surface and this can be expressed mathematically as

$$\frac{\partial}{\partial a}\left(-\frac{\pi\sigma_0^2 a^2}{2E} + 4a\gamma\right) > 0 \tag{5.189}$$

or at the point of fracture

$$\sigma_0 = \sigma_f = \left(\frac{4E\gamma}{\pi a}\right)^{1/2} \tag{5.190}$$

where σ_f is the fracture stress. This equation is similar to that obtained by Griffith who showed by a better calculation of the stresses around the crack that

$$\sigma_f = \left(\frac{2E\gamma}{\pi a}\right)^{1/2} \tag{5.191}$$

This equation is the usual form in which the Griffith criterion is expressed, but it is only strictly applicable when plane stress conditions prevail. In plane strain it becomes

$$\sigma_f = \left[\frac{2E\gamma}{\pi(1-\nu^2)a}\right]^{1/2} \tag{5.192}$$

where ν is Poisson's ratio.

The Griffith theory was tested initially for glass and it has since been applied to the fracture of other brittle solids. The main predictions are borne out in practice. The theory predicts that there should be a proportional relationship between the fracture strength and $a^{-1/2}$ and so in a uniform stress field once a crack starts to propagate it should continue to do so since the stress required to drive it drops as the crack becomes longer.

The theory provides a good qualitative prediction of the fracture behaviour of brittle polymers such as polystyrene and poly(methyl methacrylate). Fig. 5.54 shows how the strength of uniaxial tensile specimens of these two polymers varies with the length of artificially induced cracks. There is a clear dependence of σ_f upon $a^{-1/2}$ as predicted by the Griffith theory but detailed investigations have shown that there are significant discrepancies. For example, since the Young's moduli of the polymers are known, the values of γ can be calculated from the curves in Fig. 5.54 for the two polymers using Equation (5.191). The value of γ for poly(methyl methacrylate) is found to be ~210 Jm^{-2} and that for polystyrene ~1700 Jm^{-2}. These values are very much larger than the surface energies of these polymers which are thought to be of the order of 1 Jm^{-2}. The values of γ determined for other materials are also larger than the theoretical surface energies (see for example Table 5.4). The discrepancy arises because the Griffith approach assumes that the material behaves elastically and does not undergo plastic deformation. It is known that even if a material appears to behave in a brittle manner and does not undergo general yielding there is invariably a small amount of plastic deformation, perhaps only on a very local level, at the tip of the crack. The energy absorbed during plastic deformation is very much higher than the surface energy and this is reflected in the measured values of γ for most materials being much larger than the theoretical ones.

The Griffith approach assumes that for any particular material the fracture stress is controlled by the size of the flaws present in the structure. The strength of any particular sample can be increased by reducing the size of these flaws. This can be easily demonstrated for glass in which flaws are introduced during normal handling and so it normally has a low fracture stress. The strength can be greatly increased by etching the glass in hydrofluoric acid which removes most of the flaws. The reduction of the size of artificially induced flaws clearly also causes an increase in the strength of brittle polymers (Fig. 5.54). However, this rise in strength does not go on indefinitely and when the flaw size is reduced below a critical level, about 1 mm for polystyrene and 0.07 mm for poly(methyl methacrylate) at room temperature, σ_f becomes independent of flaw size. The materials, therefore, behave as if they contain natural flaws of these critical sizes and so the introduction of smaller artificial flaws does not

TABLE 5.4 *Approximate values of fracture surface energy γ determined using the Griffith theory.*

Material	γ/Jm^{-2}
Inorganic glass	7–10
Poly(methyl methacrylate)	200–400
Polystyrene	1000–2000
Epoxy resins	50–200
MgO	8–10
High-strength aluminium	~8000
High-strength steel	~25 000

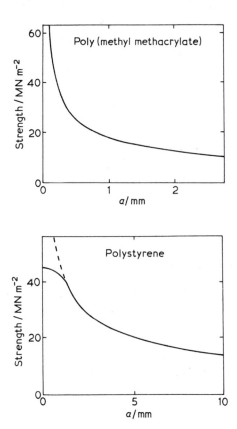

Fig. 5.54 *Dependence of fracture strength of poly(methyl methacrylate) and polystyrene upon the crack length, a. (After Berry, Chapter 2 in Fracture VII, edited by H. Liebowitz, Academic Press, 1972.)*

affect the strength. The exact size of the natural or inherent flaws can be determined from the fracture strength of an unnotched specimen, the modulus and the value of γ measured from the curves in Fig. 5.54. The relatively low inherent flaw size of poly(methyl methacrylate) accounts for the observation that in the unnotched state it has a higher strength than polystyrene even though polystyrene has a higher value of γ (Table 5.4).

Since polystyrene behaves as if it contains natural flaws of the order of 1 mm it might be expected that they could be detected in undeformed material but this is not the case. The inherent flaws are in fact crazes which form during deformation. When an unnotched specimen of polystyrene is deformed in tension large crazes of the order of 1 mm in length are seen to form whereas in the case of poly(methyl methacrylate) the crazes are much smaller. Fracture then takes place through the breakdown of one of these crazes which becomes transformed into a crack. On the other hand, when specimens containing cracks of greater than the inherent flaw size are deformed general crazing does not occur and failure takes place at the artificial flaw.

5.6.3 *Linear elastic fracture mechanics*

The derived value of γ is usually greater than the true surface energy because other energy-absorbing processes occur and so it is normal to replace 2γ in Equation (5.191) with $\mathcal{G}_c$ which then represents the total work of fracture. The equation then becomes

$$\sigma_f = \left(\frac{E\mathcal{G}_c}{\pi a}\right)^{1/2} \tag{5.193}$$

and this equation can be thought of as a prediction of the critical stress and crack length at which a crack will start to propagate in an unstable manner. Failure would be expected to occur when the term $\sigma\sqrt{\pi a}$ reaches a value of $\sqrt{E\mathcal{G}_c}$. The term $\sigma\sqrt{\pi a}$ can be considered as the driving force of crack propagation and so the stress intensity factor, K, is defined as

$$K = \sigma\sqrt{\pi a} \tag{5.194}$$

and the condition for crack propagation to occur is that K reaches a critical value, K_c given by

$$K_c = \sqrt{E\mathcal{G}_c} \tag{5.195}$$

The critical stress intensity factor, K_c, is sometimes called the fracture toughness since it characterizes the resistance of a material to crack propagation.

K_c and $\mathcal{G}_c$ are the two main parameters used in *linear elastic fracture mechanics*. Equations (5.194) and (5.195) can only be strictly applied if the material is linearly elastic and any yielding is confined to a small region in the vicinity of the crack tip. These restrictions can then only normally be applied to glassy polymers undergoing brittle fracture and the application of linear elastic fracture mechanics to the failure of brittle polymers has allowed significant progress to be made in the understanding of polymer fracture processes. In the case of more ductile polymers and rubbers generalized versions of the Griffith equation must be applied.

The understanding of brittle fracture is also greatly improved by using more sophisticated test-pieces than the simple uniaxial tensile specimens. It is possible to design test pieces such that cracks can be grown in a stable manner at controlled velocities. In the case of the uniaxial tensile specimen once a crack starts to propagate the crack runs away very rapidly. This can be overcome by having a specimen in which the effective stress driving the

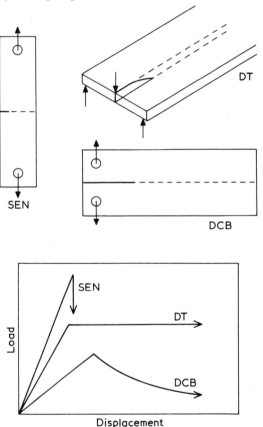

Fig. 5.55 *Schematic representation of different fracture mechanics specimens used with brittle polymers and load–displacement curves for each specimen. SEN, single-edge notched; DT, double torsion; DCB, double-cantilever beam.*

crack drops as the crack grows longer. Fig. 5.55 shows typical fracture mechanics test pieces. In the double-torsion specimen the crack is driven by twisting the two halves of one end of the specimen in opposite directions and the stress distribution is such that for a given load applied at the end of the specimen the stress-intensity factor, K, is independent of the crack length. This must be contrasted with the unaxial tensile specimen where, for a given applied stress, K increases rapidly with increasing crack length. It is possible to have specimens where K actually drops as the crack length increases such as in the case of the double cantilever beam which is also shown in Fig. 5.55. The effect of this variation of K with crack length upon load–displacement curves obtained from poly(methyl methacrylate) specimens when the specimen ends are displaced at a constant rate are also given in Fig. 5.55. In all three cases the load is initially approximately proportional to the displacement. For the simple tensile specimen the load drops immediately the crack starts to grow and the crack runs rapidly in an uncontrolled manner to the end of the specimen. When the crack starts to grow in the double-torsion specimen it propagates at a constant load (and also constant velocity) until it reaches the end of the specimen. Finally, in the double cantilever beam the crack propagates at a decreasing load and velocity when the specimen ends are displaced at a constant rate.

The value of K_c at which crack propagation occurs in a specimen of finite size depends upon the applied load, crack length and specimen dimensions. It is only given by Equation (5.194) for a crack of length $2a$ in an infinite plate. At other times it must be determined from an elastic analysis of the particular specimen used and this so-called 'K-calibration' has only been done for certain specific specimens. The formulae for K which have been determined for the three specimens in Fig. 5.55 are given in Table 5.5. It should be noted that for the double-torsion specimen the crack length, a does not appear in the formula.

It is sometimes more convenient to use $\mathscr{G}_c$ rather than K_c to characterize the fracture behaviour of a material especially since $\mathscr{G}_c$ is an energy parameter and so can be related more easily to particular energy-absorbing fracture processes. If the value of K_c is known then $\mathscr{G}_c$ can be obtained through Equation (5.195) but care is required with polymers since E depends upon the rate of testing and it may not be obvious as to what is the appropriate strain-rate for the fracture experiment. It is possible to calculate $\mathscr{G}_c$ without any prior knowledge of K_c or E for any general type of fracture specimen using a compliance calibration technique. This involves measurement of $(\partial C/\partial a)$, the variation of specimen compliance C with crack length, the compliance in this case being defined as the ratio of the displacement of the ends of the specimens to the applied load, P. When crack propagation occurs $\mathscr{G}_c$

TABLE 5.5 *Expression for K_c for different fracture mechanics specimens used in the study of the fracture of brittle polymers. (After Young, Chapter 6 in 'Developments in Polymer Fracture' ed. E.H. Andrews).*

Geometry of specimen	K_c
Single-edge notched (SEN)	$\dfrac{Pa^{1/2}}{bt}[1.99 - 0.41(a/b) + 18.7(a/b)^2 \\ - 38.48(a/b)^3 + 53.85(a/b)^4]$
Double torsion (DT)	$PW_m\left[\dfrac{3(1 + \nu)}{Wt^3 t_n}\right]^{1/2} \quad (W/2 \gg t)$
Double cantilever beam (DCB)	$\dfrac{2P}{t}\left[\dfrac{3a^3}{h^3} + \dfrac{1}{h}\right]^{1/2}$

P = applied load
b = specimen breadth
t = sheet thickness
a = crack length
h = distance of specimen edge from fracture plane
t_n = thickness of sheet in plane of crack
W = specimen width
W_m = length of moment arm
ν = Poisson's ratio

is given by*

$$\mathcal{G}_c = \frac{P^2}{2t}\left(\frac{\partial C}{\partial a}\right) \tag{5.196}$$

where t is the specimen thickness.

5.6.4 *Crack propagation in poly(methyl methacrylate)*

The slow growth of cracks in poly(methyl methacrylate) is an ideal application of linear elastic fracture mechanics to the failure of brittle polymers. Cracks grow in a very well-controlled manner when stable test-pieces such as the double-torsion specimen are used. In this case the crack will grow steadily at a constant speed if the ends of the specimen are displaced at a constant rate. The values of K_c or $\mathcal{G}_c$ at which a crack propagates depends upon both the crack velocity and the temperature of testing, another result of the rate and temperature of the mechanical properties of polymers. This behaviour is demonstrated clearly in

*See for example, J.F. Knott, *Fundamentals of Fracture Mechanics*, Butterworths, London (1973).

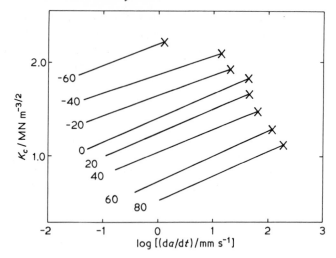

Fig. 5.56 *Dependence of K_c upon crack velocity for crack growth in poly(methyl methacrylate) at different temperatures (in °C). The values of K_c^* in each case are indicated by a cross. (Data taken from Marshall, Coutts and Williams, J. Mater, Sci.,* **9** *(1974) 1409).*

Fig. 5.56. The value of K_c is a unique function of crack velocity at a given temperature with K_c increasing as the crack velocity increases and the temperature of testing is reduced. The data in Fig. 5.56 are presented in the form of a log–log plot and they can be fitted to an empirical equation of the form

$$K_c = A(da/dt)^n \qquad (5.197)$$

where A is a temperature dependent constant and n is an exponent equal to the slope of the lines (~0.07). At a given temperature the value of K_c does not continue to rise indefinitely but reaches a maximum value K_c^*. This corresponds to a point of instability above which the crack grows rapidly at a lower value of K_c. This instability has been explained in terms of an adiabatic-isothermal transition. It is thought that at velocities below the instability, cracks grow isothermally because the heat can be dissipated from the tip of the growing crack. Whereas at velocities above the transition the crack grows too rapidly for heat to be dissipated and adiabatic conditions prevail at the crack tip. The consequent rise in temperature softens the polymer and so reduces the value of K_c for propagation.

A consequence of the dependence of K_c upon crack velocity for poly(methyl methacrylate) is that the polymer is prone to time-dependent failure during periods of prolonged loading. Since polymers are often used under stress for long periods of time this can be a serious problem. A typical use might be as windows in the pressurized cabins of air-craft. The time-to-failure, t_f, for a specimen under stress can be calculated by integrating Equation (5.197). If the specimen contains flaws of size a_0 and is subjected to a constant stress σ_0 then a crack will start to grow slowly at a velocity governed by Equation (5.197). If it is postulated that failure occurs when the crack reaches a critical length a^* then the equation can be rearranged to give

$$\int_0^{t_f} dt = \int_{a_0}^{a^*} \frac{A^{1/n}}{K_c^{1/n}} da \tag{5.198}$$

If the flaw is small and so can be considered to be in an infinite plate then since K is given by $\sigma\sqrt{\pi a}$ Equation (5.198) can be written as

$$t_f = \int_{a_0}^{a^*} A^{1/n}\sigma_0^{-1/n}\pi^{-1/2n}a^{-1/2n} da \tag{5.199}$$

and integrating gives

$$t_f = \frac{A^{1/n}(\sigma_0^2\pi)^{-1/2n}}{(1-1/2n)}\left[a^{*(1-1/2n)} - a_0^{(1-1/2n)}\right]$$

Since $a^* \gg a_0$ and $n \ll 1$ (Fig. 5.56) then the term in a^* can be neglected and so the time-to-failure will be given approximately by

$$t_f = \frac{A^{1/n}a_0^{(1-1/2n)}}{(\sigma_0^2\pi)^{1/2n}(1/2n - 1)} \tag{5.200}$$

The time-to-failure can therefore be determined from a knowledge of σ_0, a_0 and the variation of K_c with crack velocity. This particular equation is for a small crack in an infinite plate. Similar equations can be derived for other specimen geometries.

So far the mechanisms of crack propagation have not been considered but they are important since the observed fracture behaviour is a reflection of deformation processes taking place at the crack tip. There is accumulated evidence to show that in poly(methyl methacrylate), and probably many other glassy polymers, crack propagation takes place through the breakdown of a craze at the tip of the growing crack. One way that this can be demonstrated indirectly is from the presence of craze debris on the polymer fracture surfaces which gives rise to interference colours when the surfaces are viewed in reflected light. Fig. 5.57 shows the geometry of the crack/craze entity. It is envisaged that propagation occurs through the growth of a crack with a single craze at its tip and a steady-state situation is reached whereby the rate of craze growth equals that of craze breakdown and the crack/craze entity propagates gradually through the specimen. The formation of craze and the subsequent breakdown absorbs

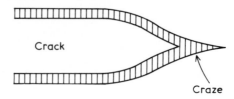

Fig. 5.57 *Schematic representation of a crack growing through a single craze in poly(methyl methacrylate).*

a considerable amount of energy and it is these processes which account for the fracture energies being much higher than the energy required just to create new surface (Table 5.4).

The mechanisms of crack propagation in poly(methyl methacrylate) are particularly amenable to analysis. The situation is not so simple for other polymers. For example, in polystyrene there is usually multiple crazing in the vicinity of the crack tip. In tougher polymers such as polycarbonate general yielding as well as crazing often takes place at the crack tip. In these cases crack propagation does not occur in such a well-controlled manner as in poly(methyl methacrylate) and it is more difficult to analyse.

5.6.5 *Tearing of rubbers*

It is impossible to apply linear elastic fracture mechanics to polymers which fail at high strains and when the deformation is not linear elastic. There are, however, certain cases where the Griffith approach (Section 5.6.2) can be generalized and failure criteria established. Such a case is in the tearing of rubber where the growth of a crack can be considered as being a balance between strain energy in the material and the tearing energy $\mathcal{T}$ of the rubber. This case is particularly easy to analyse because the material behaves in a reasonably (non-linear) elastic manner and the energy losses are confined to regions in the vicinity of the crack tip. The situation can best be analysed by taking a specific example such as the pure shear test-piece of Rivlin and Thomas which is shown in Fig. 5.58. It is possible to calculate for this type of specimen the 'energy release rate', $-d\mathscr{E}/dA$, which is the amount of energy released per unit increase in crack area, A.

When the specimen is deformed the strain energy will not be the same in the different zones of the specimen and so they must be considered separately. Region A is unstressed and so the strain energy $\mathscr{E}_A$ will be zero. In region B around the crack tip the stress field is complex and the strain energy is difficult to determine. Region C will be in pure shear and so the energy in this region will be given by $\mathscr{E}_c = WV_C$ where W is the energy density (i.e. energy per unit volume) of the rubber in pure shear at the relevant strain and V_C is the volume of region C. Region D will not be in

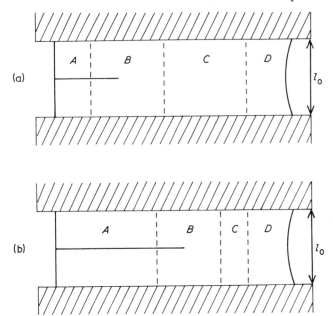

Fig. 5.58 *Pure shear test-piece used in the study of the tearing of rubbers. (a) Specimen before crack propagation; (b) After crack propagation. (After Andrews, Chapter 9, in Polymer Science ed. Jenkins, North-Holland, 1972.)*

pure shear because of its proximity to the ends of the specimen and so $\mathscr{E}_D$ will be difficult to calculate. The strain energy in the different regions can be summarized as follows

$$\mathscr{E}_A = 0, \qquad \mathscr{E}_B = ?$$
$$\mathscr{E}_C = WV_C, \qquad \mathscr{E}_D = ?$$

The lack of knowledge of $\mathscr{E}_B$ and $\mathscr{E}_D$ can be overcome by calculating only the *change* in energy when crack propagation occurs. If the specimen grips are held at constant displacement and the crack propagates a distance Δa then the net result is to increase region A at the expense of region C and so since region A is unstrained the change in specimen strain energy $\Delta\mathscr{E}$ is

$$\Delta\mathscr{E} = -W\Delta V_C \qquad (5.201)$$

where ΔV_C is the reduction in the volume of C due to crack growth. The size of regions B and D are unchanged and so the fact that their strain energies are not known does not matter. Holding the grips at constant displacement ensures that no work is done by or on the specimen and so the overall change in stored energy is given by Equation (5.201). Since rubbers shear at approximately constant volume ΔV_C can be determined from either the unstrained or strained specimen dimensions and is given by

$$\Delta V_C = l_0 t_0 \Delta a \tag{5.202}$$

where l_0 and t_0 are the initial specimen length and thickness. The energy change is then given by

$$\Delta \mathscr{E} = -W l_0 t_0 \Delta a \tag{5.203}$$

and so

$$\frac{-\partial \mathscr{E}}{\partial a} = W l_0 t_0 \tag{5.204}$$

Since the crack has two surfaces then the change in crack area ΔA is given by

$$\Delta A = 2 t_0 \Delta a$$

and so

$$\frac{-\partial \mathscr{E}}{\partial A} = \tfrac{1}{2} W l_0 \tag{5.205}$$

The Griffith criterion for brittle solids assumes that crack propagation occurs when the release of elastic strain-energy during an increment of crack growth is greater than the corresponding increase in surface energy due to the creation of new surface. This can be expressed in a similar way for the case of rubber as

$$\frac{-\partial \mathscr{E}}{\partial A} \geq \mathscr{T} \tag{5.206}$$

where $\mathscr{T}$ is the tearing energy or the characteristic energy for unit area of crack propagation in the rubber. (Analogous to γ in the Griffith theory or $\mathscr{G}_c/2$ in linear elastic fracture mechanics). In terms of a critical stored energy density W_c the criterion for crack growth in the pure shear test piece is

$$W_c = 2\mathscr{T}/l_0 \tag{5.207}$$

It is possible to determine W_c from the state of strain in the sample and the stress/strain characteristics of the material.

The quantity $-\partial \mathscr{E}/\partial A$ can be determined for testing geometries other than the pure shear specimen. The values of $-\partial \mathscr{E}/\partial A$ for a variety of other specimens commonly used to study the tearing of rubbers are given with schematic diagrams of the specimens in Fig. 5.59. It is found that cracks will propagate at a given crack velocity and temperature at the same value of $-\partial \mathscr{E}/\partial A$ in all of these types of specimen. This is strong evidence that $\mathscr{T}$ can be thought of as a true material property. It is also found that $\mathscr{T}$ varies

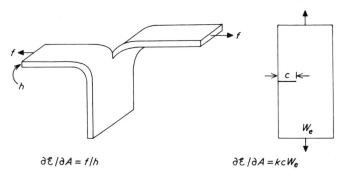

$$\partial\mathcal{E}/\partial A = f/h \qquad\qquad \partial\mathcal{E}/\partial A = kcW_e$$

Fig. 5.59 *Trouser and edge-crack test pieces used in the study of the tearing of rubbers. Expressions for* $-\partial\mathcal{E}/\partial A$ *are given in each case. (After Andrews, Chapter 9 in Polymer Science ed. Jenkins, North-Holland, 1972.)*

with crack velocity and temperature for rubbers in a similar way that K_c depends upon these variables for glassy polymers such as poly(methyl methacrylate) (Fig. 5.56). This is neatly summarized in the three-dimensional plot in Fig. 5.60 for the tearing of a styrene–butadiene rubber. The value of $\mathcal{T}$ increases with increasing crack velocity and decreasing temperature. This variation with rate and temperature can be represented in the form of a single WLF plot (Section 5.2.7) which enables the value of $\mathcal{T}$ to be determined for any velocity or temperature. At any given temperature the variation of $\mathcal{T}$ with crack velocity can be represented by an equation of the form

$$\mathcal{T} = B(da/dt)^n \qquad\qquad\qquad (5.208)$$

where n has a value of about 0.25 for the SBR rubber used and B is a temperature dependent constant. This equation can be used to predict the time-dependent failure of the SBR rubber in the same way that Equation (5.197) is used with poly(methyl methacrylate).

The values of $\mathcal{T}$ for rubbers are typically between 10^3 and $10^5\,\mathrm{Jm^{-2}}$ which is very much higher than the energies required to break bonds. The extra energy is not used up in crazing or plastic deformation but is dissipated when material in the vicinity of the crack tip is deformed and relaxed viscoelastically as the crack propagates. If the cracks are propagated very slowly at high temperatures these losses can be reduced and the values of $\mathcal{T}$ become closer to those expected for bond breaking. The behaviour shown in Fig. 5.60 is only encountered if the rubber is unfilled and does not crystallize. In these more complex rubbers crack propagation tends to be unstable with the crack moving in a series of jumps and the plots of $\mathcal{T}$ against crack velocity and temperature tend to be more complicated than that in Fig. 5.60. This difference in behaviour is due to the presence of filler

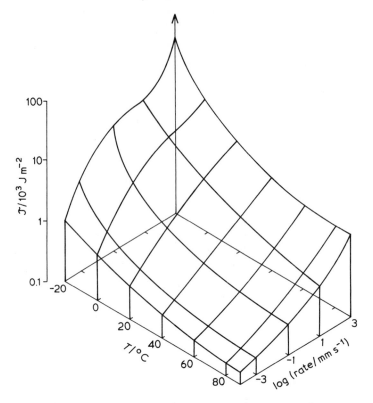

Fig. 5.60 *Dependence of tearing energy, $\mathcal{T}$ upon rate and temperature for a gum styrene–butadiene rubber. (After Greensmith and Thomas, J. Polym. Sci.,* **18**, *(1955) 189).*

particles and crystallization modifying the viscoelastic properties of the rubber.

5.6.6 *Fatigue*

Many materials suffer failure under cyclic loading at stresses well below those they can sustain during static loading. This phenomenon is known as *fatigue* and the development of fatigue cracks in the metallic components of aircraft and other structures has lead to disastrous consequences. It is perhaps less well-known that polymers can also undergo fatigue fracture. Fig. 5.61 gives the dependence of the failure stress of a variety of polymers upon the number, N, of cycles during fatigue loading. The polymers had been cycled at a frequency of 0.5 Hz between zero stress and the stress indicated. The curves have a characteristic sigmoidal shape and tend to flatten out when N becomes sufficiently large, suggesting a fatigue limit. This is stress to which the material can be cycled without causing failure. Similar so-called $S-N$ curves have been found for metals, but one

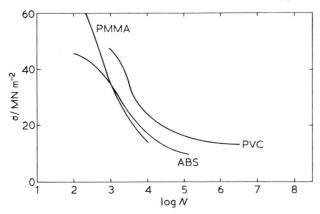

Fig. 5.61 *Dependence of failure stress of a variety of engineering polymers upon the number of cycles accumulated during a dynamic fatigue test at 20°C. The polymers used were poly(methyl methacrylate) (PMMA), rigid poly(vinyl chloride) (PVC) and an acrylonitrile-butadiene-styrene copolymer (ABS). (After Bucknall, Gotham and Vincent, Chapter 10 in Polymer Science ed. Jenkins, North-Holland, 1972.)*

important difference with polymers is the strong dependence of the fatigue behaviour upon the testing frequency. There is often a large rise in specimen temperature, especially at high stresses and high testing frequencies, which causes failure to take place through thermal softening. The temperature rise is due to the energy dissipated by the viscoelastic energy loss processes not being conducted away sufficiently rapidly because of the low thermal conductivity of the polymer.

Fatigue failure occurs through a process of initiation and propagation of cracks. In the propagation phase the fatigue crack grows by a small amount during each cycle. This behaviour can be monitored by measuring the amount of crack growth per cycle, da/dN as a function of the applied stress. It is found that for many brittle polymers the propagation rate is related to the range of stress-intensity factor ΔK by an equation of the form

$$da/dN = C(\Delta K)^m \qquad (5.209)$$

where m is a constant of the order of 4 and C is a temperature-dependent constant for the particular polymer. This type of relationship is also found for many metals. The exact form of behaviour for a particular polymer is found to depend upon the testing frequency and the mean value of K used. Analogous behaviour is also found for non-crystallizing rubbers where the crack growth rate can be related to $\mathcal{T}$ through an equation of the form

$$da/dN = D\mathcal{T}^m \qquad (5.210)$$

where D and m are constants which depend on the type of rubber and testing conditions used.

5.6.7 *Environmental fracture*

The Achilles heel of many polymers is their tendency to fail at relatively low stress levels through the action of certain hostile environments. This type of failure falls into several different categories.

(a) One of the most widely studied forms of environmental failure is that of rubbers in the presence of *ozone*. The ozone reacts at the surface of unsaturated hydrocarbon elastomers and even when the materials are subjected to relatively low stresses cracks can nucleate and grow catastrophically through the material. It is found that ozone cracking takes place above a critical value of tearing energy $\mathcal{T}_0$. This can be of the order of 0.1 Jm^{-2} which is close to the true surface energy of the polymer and well below the tearing energies of rubbers in the absence of a hostile environment.

(b) Another type of environmental failure which occurs through surface interactions is the failure of polyolefins such as high density polyethylene in the presence of *surfactants* such as certain detergents. It is found that when this material is held under constant stress in these environments its behaviour changes from ductile failure at high stresses over short time periods to brittle fracture at low stresses after longer periods of time.

(c) Glassy polymers in particular are prone to failure through the action of certain organic liquids which act as *crazing and cracking* agents. Failure can occur at stresses and strains well below those required in the absence of the agents. The detailed behaviour is rather complex and it is difficult to generalize from one system to another. It is thought that in most cases the failure occurs through the initial formation of crazes although the crazes may not be particularly stable and may transform directly into cracks. Two particular mechanisms have been proposed to account for the action of crazing and cracking agents in glassy polymers and in many cases it is thought that both mechanisms may apply. One suggestion is that the presence of the liquid lowers the surface energy of the polymer and makes the formation of new surface during crazing easier. Another is that the organic liquid swells the polymer and lowers its T_g allowing deformation and crazing to take place at lower stresses and strains.

The effect of certain liquids with different solubility parameters, δ, upon the critical strain for crazing and cracking poly(2,6-dimethyl-1,4-phenylene oxide) is shown in Fig. 5.62. In certain cases the crazes are not stable and they transform directly into cracks. The minimum in the curve corresponds to liquids which have the highest solubilities in the polymer, when the values of δ are matched for the polymer and solvent (Section 3.1.8). It is clear, therefore, that the crazing and cracking in this particular system is due principally to a reduction in the T_g by swelling. However, this is not the

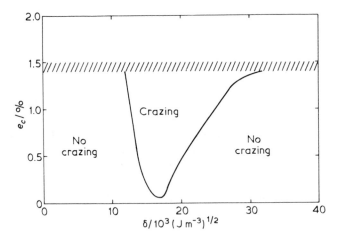

Fig. 5.62 *Dependence of critical strain for crazing, e_c, upon solubility parameter δ for poly(2,6-dimethyl-1,4-phenylene oxide) in various liquids. The shaded area gives e_c for the polymer in air. (After Kambour and Robertson, Chapter 11 in Polymer Science, ed. Jenkins, North-Holland, 1972.)*

whole story since in the liquids with high or low values of δ which produce no swelling at all, the polymer fails by crazing whereas when tested at constant strain-rate in air it undergoes cold-drawing with very little crazing. It is thought that these liquids tend to reduce the surface energy and make the formation of holes and voids during crazing easier.

5.6.8 *Molecular failure processes*

In this final section we will examine the processes which take place on a molecular level when a polymer is fractured. It is clear that, in general, failure will occur through a combination of molecular rupture and the slippage of molecules past each other, but the extent to which these two processes occur is not easy to determine except for particular specific structures.

The rupture of primary bonds is likely to be the principal mechanism of failure during the fracture of polymer single crystals deformed in tension parallel to the chain direction. Fig. 5.63 shows a side view of a polydiacetylene single crystal fibre that has been fractured in tension. The stress–strain curve of a similar crystal is given in Fig. 5.34. Examination of the fractured end of the crystal shows clearly that molecular fracture has taken place. Since the covalent bonds along the polymer chain are very strong the material has a very high fracture strength. Fig. 5.64 shows a plot of the fracture strength of the polymer as a function of fibre diameter. The fracture strength increases as the diameter is reduced. This is because the

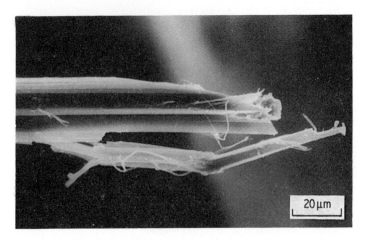

Fig. 5.63 *Scanning electron micrograph of a fractured polydiacetylene single-crystal fibre.*

thinner fibres contain less strength-reducing flaws (e.g. Griffith theory). The narrowest fibres have a strength of the order of 1.5 GN m^{-2}. The modulus of the polymer is about 60 GN m^{-2} and so the maximum fracture strength is of the order of $E/40$. Theoretical considerations (Section 5.1.5) show that the highest fracture strength might be expected to be about $E/10$ and most materials are, because of flaws, well below this value. The values of fracture strengths of the polydiacetylene single crystals are therefore close to the theoretical estimates and compare favourably with those of other high-strength materials.

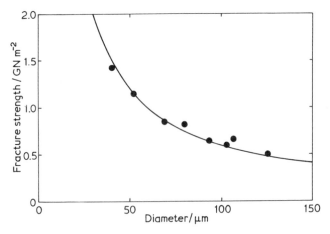

Fig. 5.64 *Dependence of fracture strength of polydiacetylene single-crystal fibres upon fibre diameter.*

Another group of materials in which molecular fracture must occur during crack growth is cross-linked polymers such as rubbers. The cross-linking points do not allow molecular pull-out to occur and bond breakage must then take place. It can be demonstrated that this is indeed the failure mechanism by determining the tearing energy $\mathcal{T}_0$ of a rubber in the absence of any viscoelastic energy losses. This is found to be close to that expected if failure takes place through bond breakage.

There is now a good deal of interest in using techniques which are capable of monitoring directly the occurrence of molecular fracture during the deformation of polymer. The most important method in this respect is electron spin resonance spectroscopy (e.s.r.) (Section 3.6.3) which detects the production of free radicals produced through the fracture of covalent bonds. Specimens can be fractured within the cavity of the spectrometer and molecular fracture is manifest be the generation of an e.s.r. signal. The technique suffers from certain limitations such a difficulty in being able to get a sufficiently large volume of material into the cavity to produce a strong enough signal and the drop in signal strength due to the decay of radicals through combination and other reactions. The radicals are found to be more stable at low temperature where the mobility of the chains is reduced. It is found that for many polymers only weak signals are obtained because very little molecular fracture takes place and failure occurs primarily through molecular slippage. The strongest signals are obtained from cross-linked polymers and crystalline polymer fibres where the cross-links and crystals act as anchorage points and reduce the amount of slippage that can occur.

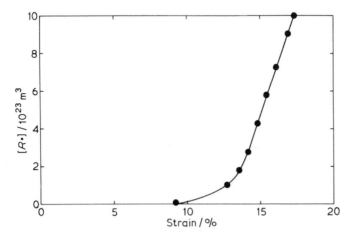

Fig. 5.65 *Dependence of radical concentration,* $[R\cdot]$ *upon the strain for nylon 6 deformed at room temperature. (Data taken from Becht and Fischer, Kolloid-Z.u.Z. Polymere,* **229**, *(1969) 167.)*

Fig. 5.65 shows how the radical concentration increases with strain as nylon 6 fibres are deformed in tension. An appreciable signal starts to build up above a strain of about 10 percent, well before the specimen fractures. The molecules in the fibre are aligned and strained to a certain extent before deformation. The build up in e.s.r. signal is thought to be due to the failure of the most highly stressed molecules within the structure before overall fracture of the specimen takes place.

Further reading

Andrews, E.H. (1968), *Fracture in Polymers*, Oliver and Boyd Ltd, London.

Arridge, R.G.C. (1975), *Mechanics of Polymers*, Clarendon Press, Oxford.

Haward, R.N. (Ed.) (1973), *The Physics of Glassy Polymers*, Applied Science Publishers, London.

Jenkins, A.D. (Ed.) (1972), *Polymer Science*, North-Holland Publishing Co., London.

Kambour, R.P. (1973), *Crazing and Fracture in Thermoplastics*, Journal of Polymer Science, Macromolecular Reviews, Vol. 7.

Kelly, A. (1966), *Strong Solids*, Clarendon Press, Oxford.

Kelly, A. and Groves, G.W. (1970), *Crystallography and Crystal Defects*, Longman, London.

Kinloch, A.J. and Young, R.J. (1983), *Fracture Behaviour of Polymers*, Applied Science, London.

Treloar, L.R.G. (1958), *The Physics of Rubber Elasticity*, Clarendon Press, Oxford.

Ward, I.M. (1971), *Mechanical Properties of Solid Polymers*, Wiley-Interscience, London.

Williams, J.G. (1973), *Stress Analysis of Polymers*, Longman, London.

Problems

5.1 Mechanical models are often used to describe the viscoelastic behaviour of polymers. The Maxwell model which consists of a spring and dashpot in series predicts that the variation of strain, e, with time, t, for a viscoelastic material can be described by an equation of the form

$$\frac{de}{dt} = \frac{1}{E}\frac{d\sigma}{dt} + \frac{\sigma}{\eta} \qquad \begin{array}{l} E = \text{modulus of spring} \\ \eta = \text{viscosity of dashpot} \end{array}$$

where σ is the applied stress. The Voigt model which consists of a spring and dashpot in parallel leads to an equation of the form

$$\frac{de}{dt} = \frac{\sigma}{\eta} - \frac{Ee}{\eta}$$

(i) Derive equations describing the variation of strain with time for both models subjected to a constant stressing-rate (i.e. $d\sigma/dt$ = const.).

(ii) Derive equations describing the variation of stress with time when both models are subjected to a constant strain-rate (de/dt = const.).

5.2 A viscoelastic polymer which can be assumed to obey the Boltzmann superposition principle is subjected to the following loading history. At time, $t = 0$, a tensile stress of 10 MN m^{-2} is applied and maintained for 100 s. The stress is then removed instantaneously. If the creep compliance of the material is given by

$$J(t) = J_0(1 - \exp(-t/\tau_0))$$

where $J_0 = 2$ m^{2}GN^{-1} and $\tau_0 = 200$ s, what is the net creep strain after (a) 100 s and (b) 200 s?

5.3 A ring-shaped seal, made from a viscoelastic material, is used to seal a joint between two rigid pipes. When incorporated in the joint the seal is held at a fixed compressive strain of 0.2. Assuming that the seal can be treated as a Maxwell model, determine the time before the seal begins to leak under an internal fluid pressure of 0.3 MN m^{-2}. It can be assumed that the relaxation time τ_0 of the material is 300 days and the short-term (instantaneous) modulus of the material is 3 MN m^{-2}.

5.4 A viscoelastic polymer is subjected to an oscillating sinusoidal load applied at an angular frequency, ω. Assuming that the variation of the strain, e, and stress, σ, with time, t, can be represented by equations of the form

$$e = e_0\sin \omega t$$

$$\sigma = \sigma_0\sin(\omega t + \delta)$$

where δ is the phase lag between the stress and strain, show that the energy dissipated per cycle of the deformation ΔU can be given by

$$\Delta U = \sigma_0 e_0 \pi \sin \delta$$

5.5 The isothermal reversible work of deformation w per unit volume of a rubber is given by the statistical theory of rubber elasticity as

$$w = \tfrac{1}{2}G(\lambda_1^2 + \lambda_2^2 + \lambda_3^2 - 3)$$

where λ_1, λ_2, and λ_3 are the principal extension ratios. Using this equation deduce the stress–strain relations for the general two-dimensional strain of a sheet of rubber and hence show that the stress–strain relation for an equal two-dimensional extension of λ is

$$\sigma_t = G(\lambda^2 - 1/\lambda^4)$$

where σ_t is the true stress acting on planes normal to the plane of stretch.

5.6 If a polymer obeys a pressure-dependent von Mises yield criterion of the form

$$A[\sigma_1 + \sigma_2 + \sigma_3] + B[(\sigma_1 - \sigma_2)^2 + (\sigma_2 - \sigma_3)^2 + (\sigma_3 - \sigma_1)^2] = 1$$

where σ_1, σ_2 and σ_3 are the principal stresses and A and B are constants, show that the material will yield under the action of hydrostatic stress alone and calculate the hydrostatic stress required to cause yield, in terms of the yield stresses of the material in tension and compression.

5.7 Polystyrene is capable of undergoing either crazing or shear yielding during deformation depending upon the state of stress imposed. It is found that it undergoes shear yielding at a stress of 73 MN m^{-2} in uniaxial tension and at 92 MN m^{-2} in uniaxial compression at a particular strain-rate and temperature. Under the same conditions it is found to undergo crazing at a stress of 47 MN m^{-2} in uniaxial tension and under a biaxial tensile stress of 45 MN m^{-2}. Assuming that polystyrene obeys a pressure-dependent von Mises yield criterion and a critical strain crazing criterion, determine the conditions of stress at which crazing and shear yielding will take place simultaneously for plane stress deformation. (Poisson's ratio for polystyrene may be taken as 0.33).

5.8 A sample of an oriented semi-crystalline polymer is deformed in tension in an X-ray diffractometer. The position of the (002) peak is found to change as the stress on the sample is increased as indicated below

stress/MN m^{-2}	(Bragg angle)/degrees
0	37.483
40	37.477
80	37.471
120	37.466
160	37.460
200	37.454

Determine the Young's modulus of the crystals in the polymer in the chain-direction assuming that the stress on the crystals is equal to the stress applied to the specimen as a whole. (The radiation used was CuKα λ = 1.542 Å).

5.9 It is thought that polyethylene is capable of undergoing twinning through a kinking of the molecules in a similar way to which twinning takes place in polydiacetylene single crystals. The most likely planes on which the kinking can occur are thought to be (200), (020) and (110). Draw molecular sketches of each of these twins viewed perpendicular to the plane of shear and hence or otherwise determine

(a) the twinning plane in each case;
(b) the angle through which the molecules bend;
(c) the direction of shear for each twin;
(d) the twin involving the lowest shear.

(Polyethylene is orthorhombic with a = 7.40 Å, b = 4.93 Å, c = 2.537 Å.)

5.10 The fracture strength of an un-notched parallel-sided sheet of a brittle polymer of thickness 5 mm and breadth 25 mm is found to be 85 MN m^{-2}. If the critical stress intensity factor, K_c, for the polymer is determined from a separate experiment on a notched specimen to be 1.25 MN m$^{-3/2}$ calculate the inherent flaw size for the polymer. (The value of K_c for a single-edge notched sheet specimen may be determined from the appropriate expression in Table 5.5).

5.11 It is found that the rate of growth of a crack in a non-crystallizing rubber

(da/dt) can be given by an equation of the form

$$da/dt = q\mathcal{T}^n$$

where $\mathcal{T}$ is the tearing energy, and q and n are constants for a given set of testing conditions. Determine the time-to-failure, t_f, for a sheet of rubber held at constant extension and containing a small edge-crack of initial length a_0, given that for the sheet of rubber

$$\mathcal{T} = KaW$$

where K is a constant and W is the stored energy density in the rubber at the strain imposed.

Answers to problems

2.1 ω-amino carboxylic acid $H_2\,NRC{-}OH$ with $C{=}O$

2 molecules:

$$H_2\,NRC{-}OH + H_2\,NRC{-}OH \longrightarrow H_2\,NRC{-}NHRC{-}OH + H_2O$$

Therefore in general:

$$n\,H_2\,NRC{-}OH \longrightarrow H({-}NHRC{-})_n OH + (n-1)H_2O$$

Calculation

2nd order kinetics $\dfrac{1}{c} = k't + \dfrac{1}{c_o}$ therefore plot $\dfrac{1}{c}$ versus t

3rd order kinetics $\dfrac{1}{c^2} = 2kt + \dfrac{1}{c_o^2}$ therefore plot $\dfrac{1}{c^2}$ versus t

Plot both and see which gives best straight line.

t/h	t/s	$c/\text{mol dm}^{-3}$	$c/\text{mol m}^{-3}$	$\dfrac{1}{c} / \text{m}^3\,\text{mol}^{-1}$	$\left(\dfrac{1}{c}\right)^2 / \text{m}^6\,\text{mol}^{-2}$
0	0	3.10	3.10×10^3	0.32×10^{-3}	0.10×10^{-6}
0.5	1800	1.30	1.30×10^3	0.77×10^{-3}	0.59×10^{-6}
1.0	3600	0.83	0.83×10^3	1.20×10^{-3}	1.45×10^{-6}
1.5	5400	0.61	0.61×10^3	1.64×10^{-3}	2.68×10^{-6}
2.0	7200	0.48	0.48×10^3	2.08×10^{-3}	4.34×10^{-6}
2.5	9000	0.40	0.40×10^3	2.50×10^{-3}	6.25×10^{-6}
3.0	10800	0.34	0.34×10^3	2.94×10^{-3}	8.65×10^{-6}

From the graphs it can be seen that the 2nd order plot gives the best straight line. Therefore the reaction was *2nd order* and so a catalyst *was* used. Rate constant $k' = 2.43 \times 10^{-7}\ \text{m}^3\,\text{mol}^{-1}\,\text{s}^{-1}$

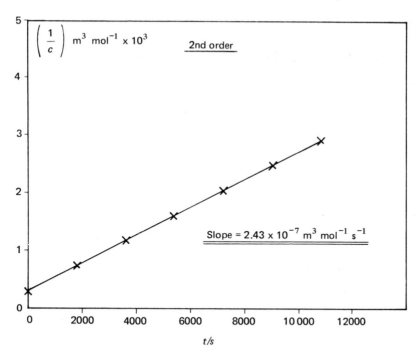

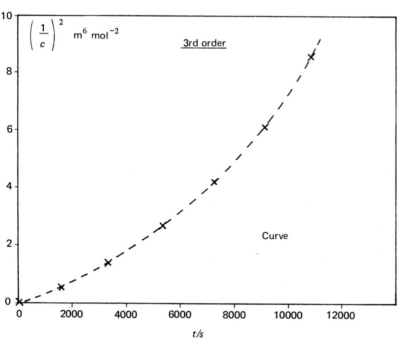

Extents of reaction, p

$$p = (c_o - c)/c_o$$

Therefore

$$\frac{1}{(1-p)} - 1 = c_o k' t$$

Therefore

$$p = 1 - (c_o k' t + 1)^{-1}$$

$$k' = 2.43 \times 10^{-7} \text{ m}^3 \text{ mol}^{-1} \text{ s}^{-1}$$

$$c_o = 3.1 \times 10^3 \text{ mol m}^{-3}$$

$t = 1$ hour $= 3600$ s   therefore $p = 0.731$

$t = 5$ hours $= 18\,000$s therefore $p = 0.931$

2.2   The polymer produced will be

$$[-O(CH_2)_{14}\overset{\overset{\displaystyle O}{\|}}{C}-]_n$$

The molar mass of the repeat unit M_0 is given by

$$(15 \times 12) + (28 \times 1) + (2 \times 16) = 240 \text{ g mol}^{-1}$$

If $\bar{M}_n = 24\,000$ therefore $\bar{x}_n = \bar{M}_n/M_0 = 100$

But $\bar{x}_n = 1/(1-p)$ therefore $p = 1 - 1/\bar{x}_n$

Thus if $\bar{x}_n = 100$ $p = 0.990$

Percentage conversion $= 99.0\%$

2.3   Reaction between 1.2 moles of dicarboxylic acid, 0.4 moles of glycerol (triol), 0.6 moles of ethylene glycol (diol).

Statistical theory

$$p_G = [1 + \gamma(f-2)]^{-1/2}$$

f = functionality of branching unit = 3

$$\gamma = \frac{\text{number of } -\text{OH groups on branch units}}{\text{total number of } -\text{OH groups}}$$

$$= (0.4 \times 3)/[(0.6 \times 2) + (0.4 \times 3)] = 0.5$$

Therefore

$$p_G = [1 + (0.5 \times 1)]^{-1/2} = 0.816$$

Modified Carothers equation

$$p_G = 2/f_{av}$$

Average functionality $= f_{av} = \dfrac{(2 \times 1.2) + (0.4 \times 3) + (2 \times 0.6)}{1.2 + 0.4 + 0.6}$

Thus $f_{av} = 4.8/2.2$ and so $p_G = 0.917$

Observed value of $p_G = 0.866$ and lies between the two theoretical values.

Statistical theory ignores rings and loops which do not aid gelation and so gives a value of p_G which is too low. *Modified Carothers* gives a value of p_G which is too high because long molecules greater than the average cause gelation between $\bar{x}_n \to \infty$.

2.4   Ester interchange reaction starting with 1 kg of $\bar{M}_n = 10\,000$ g mol^{-1} and 1 kg of $\bar{M}_n = 30\,000$ g mol^{-1}. If the polymers have the ideal distribution of molar mass

For $\bar{M}_n = 10\,000$, $\bar{M}_w = 20\,000$

For $\bar{M}_n = 30\,000$, $\bar{M}_w = 60\,000$

$\bar{M}_n$ and $\bar{M}_w$ before interchange

$$(\bar{M}_n)_{mix} = \frac{\Sigma n_i M_i}{\Sigma n_i} = \frac{n_1 (\bar{M}_n)_1 + n_2 (\bar{M}_n)_2}{n_1 + n_2}$$

But $n_1 = \dfrac{1000}{10\,000} = 0.1$ mol

and $n_2 = \dfrac{1000}{30\,000} = 0.0333$ mol

Therefore

$$(\bar{M}_n)_{mix} = \frac{(0.1 \times 10\,000) + (0.0333 \times 30\,000)}{0.13\,333} = 15\,000 \text{ g mol}^{-1}$$

$$(\bar{M}_w)_{mix} = \Sigma w_i M_i \quad \text{and} \quad w_1 = w_2 = 0.5$$

Therefore

$$(\bar{M}_w)_{mix} = 0.5 \times 20\,000 + 0.5 \times 60\,000$$

Therefore

$$(\bar{M}_w)_{mix} = 40\,000 \text{ g mol}^{-1}.$$

$\bar{M}_n$ and M_w after interchange

During the interchange reaction, the total number of molecules remains constant. The molar mass distribution adjusts itself to become the most probable distribution. Therefore total number of moles ($n_1 + n_2$) remains constant. Therefore

$\bar{M}_n$ is the same as that of the mixture

Hence $(\bar{M}_n)_{int} = 15\,000$ g mol^{-1}

   $(\bar{M}_w)_{int} = 30\,000$ g mol^{-1}

2.5   Polymerization of (CH$_2$=CHCN) with AZBN

Initiation

$$(CH_3)_2\underset{\underset{CN}{|}}{C}-N\equiv N-\underset{\underset{CN}{|}}{C}(CH_3)_2 \longrightarrow 2(CH_3)_2\underset{\underset{CN}{|}}{C}\cdot + N_2$$

$$(CH_3)_2\underset{\underset{CN}{|}}{C}\cdot + CH_2=\underset{\underset{CN}{|}}{CH} \longrightarrow (CH_3)_2\underset{\underset{CN}{|}}{C}-CH_2-\underset{\underset{CN}{|}}{CH}\cdot$$

Propagation

$$(CH_3)_2\underset{\underset{CN}{|}}{C}-CH_2-\underset{\underset{CN}{|}}{CH}\cdot + CH_2=\underset{\underset{CN}{|}}{CH} \longrightarrow (CH_3)_2\underset{\underset{CN}{|}}{C}-CH_2-\underset{\underset{CN}{|}}{CH}-CH_2-\underset{\underset{CN}{|}}{CH}\cdot$$

In general

$$(CH_3)_2\underset{\underset{CN}{|}}{C}-(CH_2-\underset{\underset{CN}{|}}{CH})_i\cdot + CH_2=\underset{\underset{CN}{|}}{CH} \longrightarrow (CH_3)_2\underset{\underset{CN}{|}}{C}-(CH_2-\underset{\underset{CN}{|}}{CH})_{i+1}\cdot$$

Termination by combination

$$(CH_3)_2\underset{\underset{CN}{|}}{C}-(CH_2-\underset{\underset{CN}{|}}{CH})_i\cdot + \cdot(\underset{\underset{CN}{|}}{CH}-CH_2)_j-\underset{\underset{CN}{|}}{C}(CH_3)_2$$

$$\longrightarrow (CH_3)_2\underset{\underset{CN}{|}}{C}-(CH_2-\underset{\underset{CN}{|}}{CH})_{i+j}-\underset{\underset{CN}{|}}{C}(CH_3)_2$$

Calculation

The mechanism of termination can be calculated by comparing the activity of 1 mole of polystyrene with that of 1 mole of AZBN.

$\bar{x}_n = 10\,000$ for polystyrene

$M_o = (8 \times 12) + (8 \times 1) = 104$ g mol^{-1}

Therefore molar mass of polystyrene $= 1.04 \times 10^6$ g mol^{-1}

0.001 kg = 1 g of polystyrene has an activity of 6×10^3 cs^{-1}

Therefore 1 mole has an activity of

$6 \times 10^3 \times 1.04 \times 10^6 = 6 \times 10^9$ cs^{-1} mol^{-1}

This is exactly the same as the activity of the AZBN initiator.

Therefore 1 mole of polystyrene contains 1 mole of AZBN

and therefore 1 molecule of polystyrene contains 1 molecule of AZBN.

But each AZBN molecule breaks into 2 radicals and so each growing chain contains ½ molecule of AZBN. Hence the mechanism of termination must be *combination*.

2.6 Breakdown of benzoyl peroxide

$$I \xrightarrow{k_i} 2R \cdot$$

1st order rate equation

$$-\frac{d[I]}{dt} = k_i[I]$$

Integrate rate equation

$$[I] = [I]_0 \exp(-k_i t) \ ([I]_0 \text{ initial concentration})$$

$$[I] = 0.5[I]_0 \text{ when } t = t_{1/2} \text{ (half-life)}$$

Therefore $0.5[I]_0 = [I]_0 \exp(-k_i t_{1/2})$

$([I]_0$ cancel)

Therefore $t_{1/2} = -\ln(0.5)/3.4 \times 10^{-6} = 2.04 \times 10^5 \text{s} = 56.63$ h

After 1 hour

$$[I]/[I]_0 = \exp(-3.4 \times 10^{-6} \times 3600) = 0.988$$

Hence [I] changes very little after 1 hour and thus the approximation that [I] = constant (p. 39) is justified.

2.7 Rate of polymerization is defined as the rate of consumption of monomer

$$= \frac{-d[M]}{dt}$$

Where

$$-\frac{d[M]}{dt} = \frac{k_p k_i^{1/2} [M][I]^{1/2}}{k_t^{1/2}}$$

The initial rate of polymerization is obtained by using the initial reactant concentrations

$$[M] = 0.2 \text{ mol dm}^{-3}$$

$$[I] = 0.05 \text{ mol dm}^{-3}$$

Therefore

$$-\frac{d[M]}{dt} = \frac{2.0 \times 10^3 \times (1.5 \times 10^{-5})^{1/2} \times 0.2 \times 0.05^{1/2}}{(7.8 \times 10^8)^{1/2}}$$

And thus

$$-\frac{d[M]}{dt} = 1.24 \times 10^{-5} \text{ mol dm}^{-3} \text{ s}^{-1}$$

or

$$-\frac{d[M]}{dt} = 1.24 \times 10^{-2} \text{ mol m}^{-3} \text{ s}^{-1}$$

2.8   1. $[M_i{}^\bullet] = (k_i[I]/k_t)^{1/2}$ independent of $[M]$

(a) increasing $[M]_0$ has no effect
(b) increasing $[I]_0$ by 4 increases $[M_i{}^\bullet]$ by $(4)^{1/2} = 2$

   2. $-d[M]/dt = k_p k_i^{1/2}[M][I]^{1/2}/k_t^{1/2}$

(a) increasing $[M]_0$ by 4 increases $-d[M]/dt$ by 4
(b) increasing $[I]_0$ by 4 increases $-d[M]/dt$ by $(4)^{1/2} = 2$

   3. $\bar{x}_n = ak_p[M]/2(k_i k_t)^{1/2}[I]^{1/2}$

(a) increasing $[M]_0$ by 4 increases $\bar{x}_n$ by 4
(b) increasing $[I]_0$ by 4 increases $\bar{x}_n$ by $(4)^{-1/2}$
   i.e. decreases $\bar{x}_n$ by factor of 2.

2.9   For free radical addition polymerization

$$-\frac{d[M]}{dt} = \frac{k_p k_i^{1/2}}{k_t^{1/2}}[M][I]^{1/2} \text{ at steady state}$$

where $k_i = 5.6 \times 10^{14} \exp\left(\dfrac{-126 \times 10^3}{RT}\right)$

$k_p = 2.2 \times 10^7 \exp\left(\dfrac{-34 \times 10^3}{RT}\right)$

$k_t = 2.6 \times 10^9 \exp\left(\dfrac{-10 \times 10^3}{RT}\right)$

Therefore

$$\frac{-d[M]}{dt} = 2.2 \times 10^7 \left(\frac{5.6 \times 10^{14}}{2.6 \times 10^9}\right)^{1/2} \exp\left(\frac{-92 \times 10^3}{RT}\right)[M][I]^{1/2}$$

$$= K \exp\left(\frac{-92 \times 10^3}{RT}\right)$$

where K is a temperature independent constant.

$60°C \equiv 333$ K

$$\left(\frac{-d[M]}{dt}\right)_{60°C} = K \exp\left(\frac{-92 \times 10^3}{333R}\right) = \text{rate}_{60°C}$$

$70°C \equiv 343$ K

$$\left(\frac{-d[M]}{dt}\right)_{70°C} = K \exp\left(\frac{-92 \times 10^3}{343R}\right) = \text{rate}_{70°C}$$

Therefore

$$\frac{\text{rate}_{70°C}}{\text{rate}_{60°C}} = \frac{\exp(-92 \times 10^3/343R)}{\exp(-92 \times 10^3/333R)} = 2.63$$

where $R = 8.314$ J mol^{-1} K^{-1}

Therefore rate at $70°C$ is 2.63 times rate at $60°C$.

$$\bar{x}_n = \frac{ak_p[M]}{2(k_i k_t)^{1/2}[I]^{1/2}}$$

Therefore

$$\bar{x}_n = \underbrace{\frac{a[M]A_p}{2[I]^{1/2}(A_i A_t)^{1/2}}}_{} \exp\left(\frac{-34 + 126/2 + 10/2}{10^{-3} RT}\right)$$

All constant with variations in T

Thus

$$\frac{(\bar{x}_n)_{70°C}}{(\bar{x}_n)_{60°C}} = \frac{\exp(34 \times 10^3/343R)}{\exp(34 \times 10^3/333R)} = 0.699$$

where $R = 8.314$ J mol^{-1} K^{-1}

Therefore $\bar{x}_n$ at $70°C$ is 0.699 of that at $60°C$

Overall rate equation when depolymerization takes place is

$$\frac{-d[M]}{dt} = k_p[M]\, \Sigma[M_i{}^{\bullet}] - k_{dp}\Sigma[M_i{}^{\bullet}]$$

Ceiling temperature, T_c is when $-d[M]/dt = 0$ i.e. when $k_p[M] = k_{dp}$ where

$[M] = 1$ mol dm^{-3}

$$k_p = 2.2 \times 10^7 \exp\left(\frac{-34 \times 10^3}{RT}\right) \text{ mol dm}^{-3} \text{ s}^{-1}$$

$$k_{dp} = 10^{13} \exp\left(\frac{-122 \times 10^3}{RT}\right) \text{ s}^{-1}$$

Ceiling temperature, T_c, is given by

$$2.2 \times 10^7 \exp\left(\frac{-34 \times 10^3}{RT_c}\right) \times 1 = 10^{13} \exp\left(\frac{-122 \times 10^3}{RT_c}\right)$$

Therefore

$$T_c = \frac{-88 \times 10^3}{8.314 \times \ln(2.2 \times 10^{-6})}$$

Therefore

$$T_c = 812.5 \text{ K } (539°C)$$

2.10 This is a living polymer system

$$\bar{x}_n = [M]_0/[I]_0$$

Therefore

$$\bar{x}_n = \frac{10^3}{0.5} = 2 \times 10^3$$

There is a range of molar mass as the active centres are formed gradually throughout the reaction. Therefore, the molecules grow to different lengths.

2.11 Styrene in liquid ammonia initiated by KNH_2

$$\frac{-d[M]}{dt} = \frac{k_p k_i}{k_{tr}} \frac{[NH_2^-][M]^2}{[NH_3]}$$

but

$$\frac{k_p k_i}{k_{tr}} = \frac{A_p A_i}{A_{tr}} \exp\left(\frac{E_{tr} - E_p - E_i}{RT}\right)$$

$$T = -33.5°C \text{ i.e. } \left(\frac{-d[M]}{dt}\right)_{-33.5°C} \propto \exp\left(\frac{42.0 - 25.2 - 54.6}{10^{-3} R \times 240}\right)$$

$$T = -70°C \text{ i.e. } \left(\frac{-d[M]}{dt}\right)_{-70°C} \propto \exp\left(\frac{42.0 - 25.2 - 54.6}{10^{-3} R \times 203}\right)$$

Therefore

$$\frac{rate_{-33.5°C}}{rate_{-70°C}} = \frac{\exp(-18.94)}{\exp(-22.39)} = 33$$

i.e. rate at $-33.5°C$ is 33 times greater than rate at $-70°C$.

$$\bar{x}_n = \frac{k_p}{k_{tr}} \frac{[M]}{[NH_3]}$$

but $\dfrac{k_p}{k_{tr}} \propto \exp\left(\dfrac{E_{tr} - E_p}{RT}\right)$

$T = -33.5°C$ i.e. $(\bar{x}_n)_{-33.5°C} \propto \exp\left(\dfrac{42.0 - 25.2}{10^{-3}\,R \times 240}\right)$

$T = -70°C$ i.e. $(\bar{x}_n)_{-70°C} \propto \exp\left(\dfrac{42.0 - 25.2}{10^{-3}\,R \times 203}\right)$

Therefore

$$\dfrac{(\bar{x}_n)_{-33.5°C}}{(\bar{x}_n)_{-70°C}} = \dfrac{\exp(8.42)}{\exp(9.95)} = 0.215$$

i.e. $\bar{x}_n$ at $-33.5°C$ is 0.215 of that at $-70°C$

Chapter 3

3.1 (a) $\bar{M}_n = \dfrac{\Sigma N_i M_i}{\Sigma N_i}$, $N_1 = N_2 = N_3 = N$

Therefore $\bar{M}_n = \dfrac{(N \times 10\,000 + N \times 30\,000 + N \times 100\,000)}{3N}$

(N's cancel)

Therefore $\bar{M}_n = 46\,667$ g mol^{-1}

$$\bar{M}_w = \dfrac{\Sigma N_i M_i^2}{\Sigma N_i M_i}$$

$$\bar{M}_w = \dfrac{N \times 10\,000^2 + N \times 30\,000^2 + N \times 100\,000^2}{N \times 10\,000 + N \times 30\,000 + N \times 100\,000}$$

(N's cancel)

Therefore $\bar{M}_w = 78\,571$ g mol^{-1}

(b) $\bar{M}_n = 1/\Sigma(w_i/M_i)$, $w_1 = w_2 = w_3 = \tfrac{1}{3}$

Therefore $\bar{M}_n = 1/(\tfrac{1}{3}/10\,000 + \tfrac{1}{3}/30\,000 + \tfrac{1}{3}/100\,000)$

Therefore $\bar{M}_n = 20\,930$ g mol^{-1}

$\bar{M}_w = \Sigma w_i M_i$

$\bar{M}_w = \tfrac{1}{3} \times 10\,000 + \tfrac{1}{3} \times 30\,000 + \tfrac{1}{3} \times 100\,000$

Therefore $\bar{M}_w = 46\,667$ g mol^{-1}

(c) $w_1 = 0.145$, $w_2 = 0.855$

Therefore $\bar{M}_n = 1/\Sigma(w_iM_i) = 1/(0.145/10^4 + 0.855/10^5)$

Therefore $\bar{M}_n = 43\,384$ g mol^{-1}

$$\bar{M}_w = \Sigma w_iM_i = 0.145 \times 10\,000 + 0.855 \times 100\,000$$

Therefore $\bar{M}_w = 86\,950$ g mol^{-1}

$\bar{M}_w/\bar{M}_n \sim 2$ but the mixture does *not* have the 'Ideal Distribution'.

3.2 10^{-3} kg of polyester is neutralized with 0.012×10^{-3} mol of alkali (monofunctional)

i.e. 10^{-3} kg of polyester contains 1.2×10^{-5} mol of $-COOH$ groups. But since each polymer molecule contains one unreacted $-COOH$ group then 10^{-3} kg of polyester contains 1.2×10^{-5} mol of polymer molecules.

Therefore

$$\bar{M}_n = 10^{-3}/1.2 \times 10^{-5} \text{ kg mol}^{-1}$$

i.e. $\bar{M}_n = 83.33$ kg mol^{-1} = $83\,333$ g mol^{-1}

3.3   Since the polymerization takes place with termination by combination then

1 mole of polymer contains 1 mole of AZBN

Activity of 1 mole of AZBN   = 2.5×10^8 cs^{-1}
Activity of 0.001 kg of polymer   = 3.2×10^3 cs^{-1}
Activity of 1 mole of polymer   = 2.5×10^8 cs^{-1}

Therefore

$$0.001 \text{ kg polymer contains } \frac{3.2 \times 10^3}{2.5 \times 10^8} \text{ mol} = 1.28 \times 10^{-5} \text{ mol}$$

Therefore

$$\text{molar mass of polymer} = \frac{0.001}{1.28 \times 10^{-5}} = 78.125 \text{ kg mol}^{-1}$$

Therefore

$$\bar{M}_n = 78\,125 \text{ g mol}^{-1}$$

3.4 $$\Omega_c = \frac{(N_1 + N_2)!}{N_1! N_2!}$$

$$S_c = k \ln \Omega_c = \Delta S_m$$

Therefore

$$\Delta S_m = k \ln \left[\frac{(N_1 + N_2)!}{N_1! N_2!} \right]$$

$$= k \ln (N_1 + N_2)! - \ln N_1! - \ln N_2!$$

Now $\ln N! = N \ln N - N$ (for large N)

$$\Delta S_m = k[(N_1 + N_2) \ln (N_1 + N_2) - (N_1 + N_2) - N_1 \ln N_1$$
$$+ N_1 - N_2 \ln N_2 + N_2]$$

(N's cancel)

Therefore

$$\Delta S_m = -k\{N_1 \ln [N_1/(N_1 + N_2)] + N_2 \ln [N_2/(N_1 + N_2)] \}$$

Now $X_1 = N_1/(N_1 + N_2)$, etc. (mole fractions)

and $N_1 = N_A n_1$, etc. (where N_A is the Avagadro number)

$$N_A k = R$$

Thus $\Delta S_m = -R[n_1 \ln X_1 + n_2 \ln X_2]$

For an ideal solution, $\Delta H_m = 0$

Thus $\Delta G_m = -T \Delta S_m$

Therefore

$$\Delta G_m = RT[n_1 \ln X_1 + n_2 \ln X_2]$$

3.5   Given $\mu_1 - \mu_1^0 = -\Pi \bar{V}_1$

but $\mu_1 - \mu_1^0 = \Delta \bar{G}_1 = \left(\frac{\partial \Delta G_m}{\partial n_1} \right)_{n_2}$

and $\Delta G_m = RT[n_1 \ln X_1 + n_2 \ln X_2]$

Now $\left(\frac{\partial \Delta G_m}{\partial n_1} \right) = RT \left\{ \frac{\partial}{\partial n_1} \left[n_1 \ln \left(\frac{n_1}{n_1 + n_2} \right) \right] + \frac{\partial}{\partial n_1} \left[n_2 \ln \left(\frac{n_2}{n_1 + n_2} \right) \right] \right\}$

Therefore

$$\Delta \bar{G}_1 = RT \left\{ \frac{\partial}{\partial n_1} [n_1 \ln n_1 - n_1 \ln (n_1 + n_2)] + \frac{\partial}{\partial n_1} [n_2 \ln n_2 - \ln (n_1 + n_2)] \right\}$$

Therefore

$$\Delta \bar{G}_1 = RT\left[\frac{n_1}{n_1} + \ln n_1 - \frac{n_1}{(n_1 + n_2)} - \ln (n_1 + n_2) + 0 - \frac{n_2}{(n_1 + n_2)}\right]$$

i.e. $$\Delta \bar{G}_1 = RT \ln \left[\frac{n_1}{(n_1 + n_2)}\right] = RT \ln X_1$$

But $$\Pi = -\frac{\Delta \bar{G}_1}{\bar{V}_1}$$

Therefore

$$\Pi = -\frac{RT \ln X_1}{\bar{V}_1}, \text{ osmotic pressure of an ideal solution.}$$

3.6 $$\mu_1 - \mu_1^0 = \Delta \bar{G}_1 = \left(\frac{\partial \Delta G_m}{\partial n_1}\right)_{n_1}$$

Given $$\Delta G_m = RT[n_1 \ln \phi_1 + n_2 \ln \phi_2 + \chi n_1 \phi_2]$$

Remember

$$\phi_1 = n_1/(n_1 + xn_2), \quad \phi_2 = xn_2/(n_1 + xn_2), x = v_2/v_1$$

Therefore

$$\Delta \bar{G}_1 = RT\left\{\frac{\partial}{\partial n_1} [n_1 \ln n_1 - n_1 \ln (n_1 + xn_2)]\right.$$

$$\left. + \frac{\partial}{\partial n_1} [n_2 \ln xn_2 - n_2 \ln (n_1 + xn_2)] + \frac{\partial}{\partial n_1}\left[\chi n_1 \frac{xn_2}{(n_1 + xn_2)}\right]\right\}$$

Therefore

$$\frac{\Delta \bar{G}_1}{RT} = \frac{n_1}{n_1} + \ln n_1 - \frac{n_1}{n_1 + xn_2} - \ln (n_1 + xn_2) - \frac{n_2}{(n_1 + xn_2)}$$

$$+ n_2 \chi x \left[\frac{-n_1}{(n_1 + xn_2)^2} + \frac{1}{(n_1 + xn_2)}\right]$$

i.e. $$\frac{\Delta \bar{G}_1}{RT} = \ln \phi_1 + \frac{(n_1 + xn_2 - n_1 - n_2)}{(n_1 + xn_2)} + \chi \phi_2^2$$

i.e. $$\frac{\Delta \bar{G}_1}{RT} = \ln (1 - \phi_2) + \left(\frac{xn_2}{(n_1 + xn_2)}\right)\left(1 - \frac{1}{x}\right) + \chi \phi_2^2$$

Therefore

$$\Pi = -\frac{\Delta \bar{G}_1}{\bar{V}_1} = -\frac{RT}{\bar{V}_1}\left[\ln (1 - \phi_2) + \left(1 - \frac{1}{x}\right)\phi_2 + \chi \phi_2^2\right]$$

3.7

Contour length $= nl \sin \theta/2$

For a C—C chain $l = 1.54$ Å, $\theta = 109.5°$

(a)

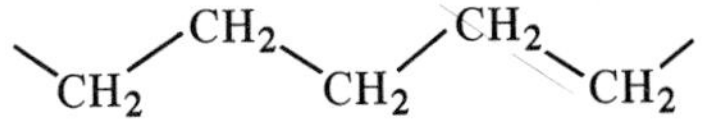

$M = 14\,000$ g mol^{-1}

There is one —CH_2— per link

Therefore

$$n = \frac{14\,000}{14} = 1\,000$$

Contour length $= 1\,000 \times 1.54 \times \sin 54.75°$

$$= 1257.6 \text{ Å}$$

(b)

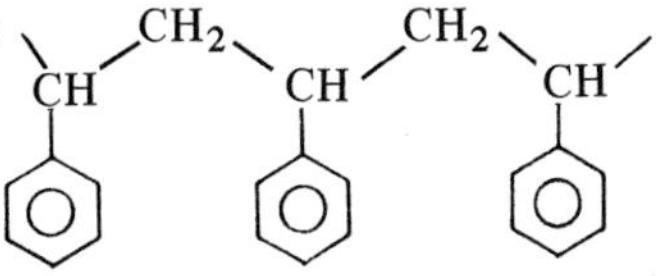

$M = 140\,000$ g mol^{-1}

Each —CH—CH_2 group involves 2 links

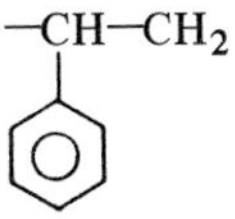

Therefore

$$n = \frac{140\,000}{52}$$

$$\text{Contour length} = \frac{140\,000}{52} \times 1.54 \times \sin 54.75°$$

$$= 3385.9 \text{ Å}$$

$(\overline{r^2})^{1/2} = ln^{1/2}(1 - \cos\theta/1 + \cos\theta)$, $M = 119\,980$ g mol^{-1}

$$n = \frac{119\,980}{14}, l = 1.54 \text{ Å}, \theta = 109.5°$$

Therefore

$$(\overline{r^2})^{1/2} = 1.54 \times \left(\frac{119\,980}{14}\right)^{1/2} \left(\frac{1 - \cos 109.5}{1 + \cos 109.5}\right)^{1/2}$$

Therefore

$$(\overline{r^2})^{1/2} = 201.7 \text{ Å (cf. } nl \sin\theta/2 = 10\,777.9 \text{ Å)}$$

3.8 $W(r) = (\beta/\pi^{1/2})^3 \exp(-\beta^2 r^2)4\pi r^2$

where $\beta = (1/l)(3/2n)^{1/2}$

   (a) Most probable value is when $\dfrac{\partial}{\partial r}(W(r)) = 0$

i.e. $\dfrac{\partial}{\partial r}[(\beta/\pi^{1/2})^3 \exp(-\beta^2 r^2)4\pi r^2] = 0$

Therefore

$$\exp(-\beta^2 r^2)2r - \exp(-\beta^2 r^2)2r^3\beta^2 = 0$$

or $r = 1/\beta$

Therefore most probable value of $r = l(2n/3)^{1/2}$

   (b) $\overline{r^2} = \displaystyle\int_0^\infty W(r)r^2\,dr \bigg/ \int_0^\infty W(r)\,dr$

But $\displaystyle\int_0^\infty W(r)\,dr = 1$ since the molecule must have some length.

Therefore

$$\overline{r^2} = \int_0^\infty (\beta/\pi^{1/2})^3 \exp(-\beta^2 r^2)4\pi r^4\,dr$$

NB $\displaystyle\int_0^\infty r^4 \exp(-\beta^2 r^2)\,dr = (3/8)\pi^{1/2}\beta^{-5}$

Therefore

$$\overline{r^2} = \frac{\beta^3}{\pi^{3/2}}\frac{3}{8}\pi^{1/2}\beta^{-5}4\pi$$

Cancelling throughout gives

$$(\overline{r^2})^{1/2} = \left(\frac{3}{2}\right)^{1/2}\frac{1}{\beta} \quad \text{but} \quad \beta = \frac{1}{l}\left(\frac{3}{2n}\right)^{1/2}$$

Therefore $(\overline{r^2})^{1/2} = ln^{1/2}$

   (c) Mean value of $r = \overline{r} = \displaystyle\int_0^\infty W(r)\,r\,dr\bigg/\int_0^\infty W(r)\,dr$

But $\displaystyle\int_0^\infty W(r)\,dr = 1$

And therefore

$$\bar{r} = \int_0^\infty (\beta/\pi^{1/2})^3 \exp(-\beta^2 r^2) 4\pi r^3 \, dr$$

NB $$\int_0^\infty r^3 \exp(-\beta^2 r^2) dr = \frac{1}{2} (\beta^2)^{-2}$$

Therefore

$$\bar{r} = \frac{\beta^3}{\pi^{3/2}} \frac{4\pi}{2} \frac{1}{\beta^4} = \frac{2}{\pi^{1/2}} \frac{1}{\beta}$$

But $$\beta = \frac{1}{l} \left(\frac{3}{2n}\right)^{1/2}$$

Therefore

$$\bar{r} = \frac{2}{\pi^{1/2}} l \left(\frac{2n}{3}\right)^{1/2} = 0.921 \, ln^{1/2}$$

3.9 $\pi = \rho g h = 861.8 \times 9.81 \times 10^{-3} \, h$ (osmotic pressure in mm solvent)

$= 8.454 \, h \, \text{N m}^{-2}$

$c/\text{kg m}^{-3}$	2.56	3.80	5.38	7.80	8.68
$h/\text{mm toluene}$	3.25	5.45	8.93	15.78	18.56
$\Pi \, \text{N m}^{-2}$	27.48	46.08	75.50	133.41	156.91
$(\Pi/c)/\text{kg}^{-1} \, \text{N m}$	10.73	12.13	14.03	17.10	18.08

$$\frac{\Pi}{c} = \frac{RT}{\bar{M}_n} + Bc + Cc^2 + \dots$$

So plot Π/c versus c (see graph)

$$\text{Intercept} = \frac{RT}{\bar{M}_n} = 7.75 \, \text{N m kg}^{-1}$$

Therefore

$$\bar{M}_n = \frac{RT}{7.75} = \frac{8.314 \times 298}{7.75} \left(\frac{\text{J K kg}}{\text{mol K N m}}\right)$$

Therefore

$$\bar{M}_n = 319.69 \, \text{kg mol}^{-1} \cong 320\,000 \, \text{g mol}^{-1}$$

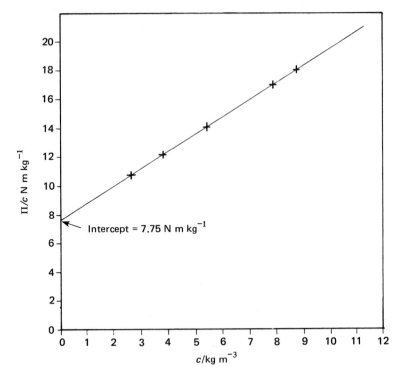

Slope = second virial coefficient = B

$$\beta = \left(\frac{19.7 - 7.7}{10} \right) \frac{N\ m\ m^3}{kg^2}$$

Therefore

$$B = 1.2\ N\ m^4\ kg^{-2}$$

3.10 $\bar{M}_n \longrightarrow$ osmotic pressure, Π

$\Pi = \rho gh = 903 \times 9.81 \times 10^{-3}\ h$ (osmotic pressure in mm solvent)

$= 8.858\ h\ N\ m^{-2}$

$c/\text{kg m}^{-3}$	1.30	2.01	3.01	5.49	6.62
$h/\text{mm solvent}$	7.08	11.15	17.10	33.20	41.10
$\Pi/\text{N m}^{-2}$	62.72	98.77	151.48	294.10	364.08
$(\Pi/c)/\text{kg}^{-1}\ \text{N m}$	48.24	49.14	50.32	53.57	55.00

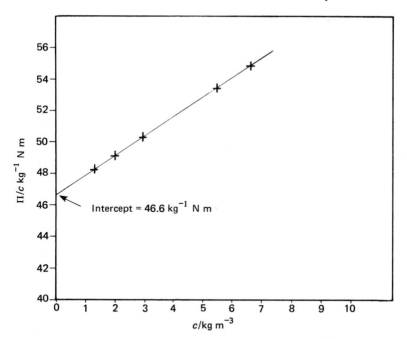

Plot Π/c versus c (see graph)

$$\text{Intercept} = \frac{RT}{\bar{M}_n} = 46.6 \text{ kg}^{-1} \text{ N m}$$

Therefore

$$\bar{M}_n = \frac{8.314 \times 298.2}{46.6} \left(\frac{\text{Nm K kg}}{\text{K mol N m}} \right)$$

Therefore

$$\bar{M}_n = 52.20 \text{ kg mol}^{-1} = 53\ 200 \text{ g mol}^{-1}$$

$\bar{M}_w \longrightarrow$ light scattering

$$\frac{K(1 + \cos^2 \theta)c}{R_\theta} = \frac{1}{\bar{M}_w}(1 + 2Bc)$$

where $\quad K = \dfrac{2\pi^2 \bar{n}_0^2}{N_A \lambda^4} \left(\dfrac{\mathrm{d}\bar{n}}{\mathrm{d}c} \right)$

$$= \frac{2 \times 3.142^2 \times 1.513^2 \times 1.11^2 \times 10^{-8}}{6.023 \times 10^{23} \times 4.358^4 \times 10^{-28}} \left(\frac{\text{m}^6 \text{ mol}}{\text{kg}^2 \text{ m}^4} \right)$$

Therefore

$$K = 2.563 \times 10^{-5} \text{ m}^2 \text{ mol kg}^{-2}$$

But $\theta = \pi/2$

Therefore

$$\cos \theta = 0 \quad \text{and} \quad \frac{K(1 + \cos^2 \theta)c}{R_\theta} = \frac{Kc}{R_{90}}$$

$c/\text{kg m}^{-3}$	1.30	2.01	3.01	5.49	6.62
$10^2 R_\theta/\text{m}^{-1}$	0.383	0.558	0.767	1.180	1.325
$(Kc/R_{90})/\text{mol kg}^{-1}$	0.008 69	0.009 23	0.010 05	0.0119	0.0128

Plot Kc/R_{90} versus c (see graph).

Intercept $= 1/\bar{M}_w = 0.0077 \text{ mol kg}^{-1}$

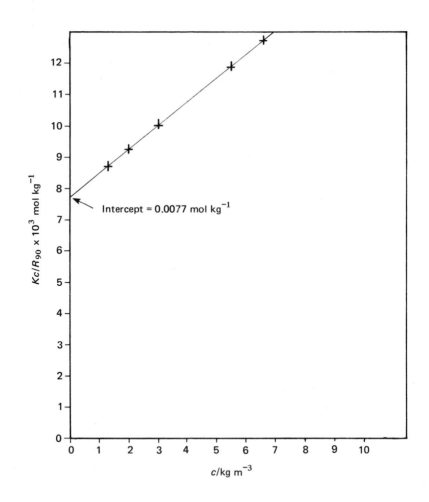

Therefore

$$\bar{M}_w = 129.9 \text{ kg mol}^{-1} = 129\,900 \text{ g mol}^{-1}$$

Ratio $\bar{M}_w/\bar{M}_n = 2.44$

Shape and size of coil could be determined if interference effects had been found.

3.11 $\bar{M}_w$ and $(\overline{r^2})^{1/2}$ can be determined from a Zimm plot which is a plot of:

$$\frac{K(1 + \cos^2 \theta)c}{R_\theta} \text{ versus } \sin^2 \left(\frac{\theta}{2} \right) + 0.1c$$

where c is in kg m^{-3}

$$K = 2\pi^2 \bar{n}_0^2 \left(\frac{d\bar{n}}{dc} \right)^2 \frac{1}{N_A \lambda^4}$$

Thus $\quad K = \dfrac{2\pi^2 \times 1.502^2 \times 1.06^2 \times 10^{-8}}{6.023 \times 10^{23} \times 546.1^4 \times 10^{-36}} = 9.3407 \times 10^{-6} \text{ m}^3 \text{ mol/kg}^{-2}$

Table of values of $K(1 + \cos^2 \theta)c/R_\theta \times 10^3$ mol/kg^{-1}

	θ			
c/kg m^{-3}	30°	60°	90°	120°
2.0	3.00	3.26	3.51	3.78
1.5	2.58	2.77	3.08	3.37
1.0	2.16	2.38	2.68	2.95
0.5	1.81	2.04	2.39	2.63

Table of values of $\sin^2 \theta/2 + 0.1c$ kg mol^{-1}

	θ			
c/kg m^{-3}	30°	60°	90°	120°
2.0	0.267	0.45	0.70	0.95
1.5	0.217	0.40	0.65	0.90
1.0	0.167	0.35	0.60	0.85
0.5	0.117	0.30	0.55	0.80
0	0.067	0.25	0.50	0.75 Values used for extrapolation

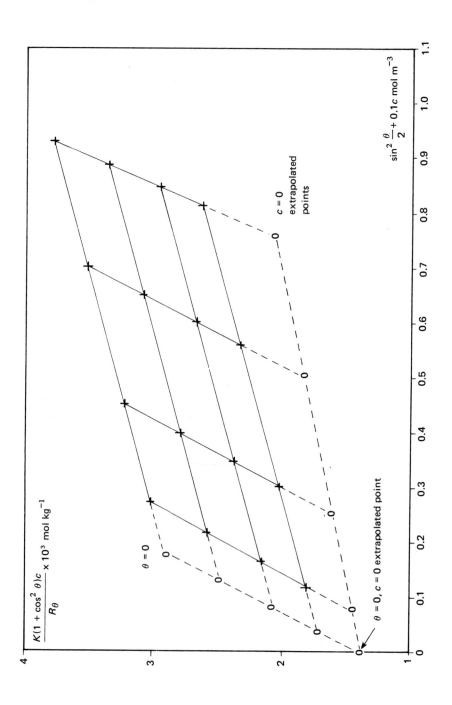

Intercept (see graph) at $(c = 0, \theta = 0) = \dfrac{1}{\bar{M}_w}$

Therefore

$$\frac{1}{\bar{M}_w} = 1.38 \times 10^{-3} \text{ mol kg}^{-1}$$

Therefore

$$\bar{M}_w = 725 \text{ kg mol}^{-1} = 725\,000 \text{ g mol}^{-1}$$

Slope of $c = 0$ line $= 1 \times 10^3$ mol kg^{-1}

$$\text{Slope} = (\bar{r^2}) \times K'/\bar{M}_w = \frac{8\pi^2 \bar{r^2} \bar{n_0^2}}{9\lambda^2 \bar{M}_w} = 9.15 \times 10^{10} \bar{r^2}$$

Therefore

$$\bar{r^2} = 1.09 \times 10^{-14} \text{ m}^2$$

and $(\bar{r^2})^{1/2} = 1.05 \times 10^{-7}$ m $= 1050$ Å

NB values vary depending upon measurement of slopes.

3.12 Mark–Houwink equation is $[\eta] = KM^a$

Therefore

$$\log[\eta] = \log K + a \log M$$

$M/\text{g mol}^{-1}$	$\log M$	$[\eta]/\text{m}^3 \text{ kg}^{-1}$	$\log[\eta]$
76 000	4.881	0.0382	−1.418
135 000	5.130	0.0592	−1.228
163 000	5.212	0.0696	−1.157
336 000	5.526	0.1054	−0.977
444 000	5.647	0.1292	−0.889
556 000	5.745	0.165	−0.783
850 000	5.929	0.221	−0.656

Slope $= a = 0.73$ (see graph)

$\log K$ is the intercept at $\log M = 0$

Therefore

$$K = 1.05 \times 10^{-5} \text{ m}^3 \text{ kg}^{-1}$$

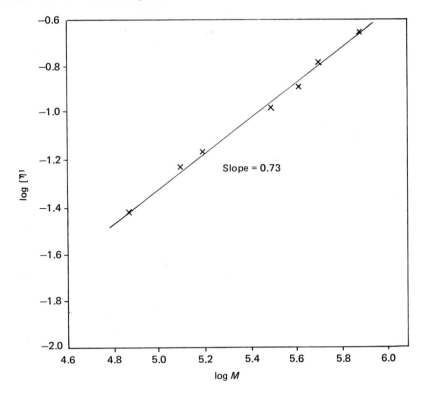

$c/\text{kg m}^{-3}$	t/s	$t_2 - t_1$	$(t_2 - t_1)/t_1 c$
0	100		
1.0	113.5	13.5	0.1350
1.3	118	18.0	0.1385
2.0	128	28.0	0.1400
4.0	160	60.0	0.1500

$$[\eta] = \lim_{c \to 0} (t_2 - t_1)/t_1 c$$

Therefore

$$[\eta] = 0.13 \text{ m}^3 \text{ kg}^{-1},$$

$$\log M = 5.61$$

Therefore

$$M = 407\ 380 \text{ g mol}^{-1}$$

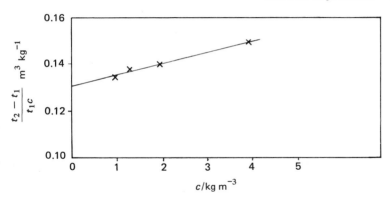

Chapter 4

4.1

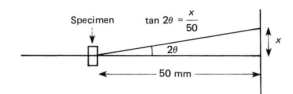

Ring radii 19.7, 22.2, 36.6 mm $\lambda = 1.542$ Å

Braggs Law $\lambda = 2d \sin \theta$ (where d = plane spacings)

x/mm	2θ	θ	$\sin \theta$	d/Å
19.7	21.5	10.75	0.186	4.13
22.2	23.9	11.97	0.207	3.71
36.6	36.2	18.10	0.311	2.48

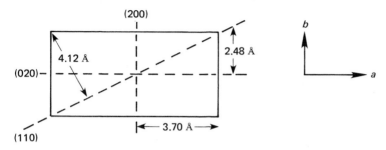

Therefore planes are (110), (200) and (020)

4.2

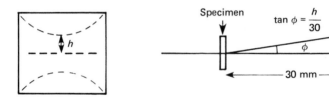

Height of 1st order line = 22.9 mm

$$\tan \phi = \frac{22.9}{30}$$

Therefore $\phi = 37.3°$

(a) The crystal plane spacing, c, can be determined using the same theory as a diffraction grating

i.e. $\sin \phi = \dfrac{\lambda}{c}$ (1st order)

Therefore

$$c = \frac{\lambda}{\sin \phi} = \frac{1.54}{0.606} = 2.54 \text{ Å}$$

(b) For 2nd order line

$$\sin \phi = \frac{2\lambda}{c} = \frac{3.08}{2.54} = 1.21$$

But $\sin \phi \leqslant 1$ hence the 2nd order line cannot be obtained.

X-rays with a wavelength of less than $2.54/2 = 1.27$ Å would have to be used to obtain 2nd order lines.

(c)

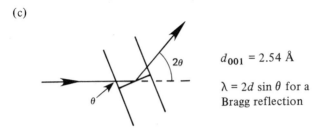

$d_{001} = 2.54$ Å

$\lambda = 2d \sin \theta$ for a Bragg reflection

Specimen tilted to the angle θ from vertical where

$$\theta = \sin^{-1}\left(\frac{\lambda}{2d_{001}}\right) = 17.65°$$

4.3 Mass fraction crystallinity $x_c = \dfrac{\rho_c}{\rho}\left(\dfrac{\rho - \rho_a}{\rho_c - \rho_a}\right)$

$\rho_c = 1000 \text{ kg m}^{-3}$ $\rho_a = 865 \text{ kg m}^{-3}$

Linear: $\rho = 970 \text{ kg m}^{-3}$

$$x_c = \frac{1000}{970}\left(\frac{970 - 865}{1000 - 865}\right) = 0.802$$

i.e. the polymer is 80.2% crystalline.

Branched: $\rho = 917 \text{ kg m}^{-3}$

$$x_c = \frac{1000}{917}\left(\frac{917 - 865}{1000 - 865}\right) = 0.420$$

i.e. the polymer is 42.0% crystalline.

Branched molecules do not crystallize as easily as linear ones.

4.4 T_m^o is determined by plotting T_m versus T_c

T_m^o is given by the extrapolation of the data to $T_m = T_c$ (see graph)

Therefore

$$T_m^o = 349 \text{ K}$$

Bragg equation $n\lambda = 2d \sin \theta$

At low angles the position of the maxima is given by the Bragg angle θ

T_c/K	θ	$d/\text{Å}$	$l = 0.45\,d$	$\Delta T/\text{K}$	$\Delta T^{-1}/\text{K}^{-1}$
270	0.44	100.4	45.2	79	0.0126
280	0.39	113.1	50.9	69	0.0145
290	0.33	133.7	60.2	59	0.0169
300	0.27	164.0	73.5	49	0.0204
310	0.22	200.5	90.2	39	0.0256
320	0.17	259.5	116.8	29	0.0345
330	0.11	401.1	180.5	19	0.0526

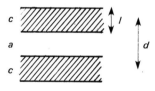

$l \sim d \times$ degree of crystallinity

$$l \sim \frac{2\gamma_e T_m^o}{\Delta H_v \Delta T}$$

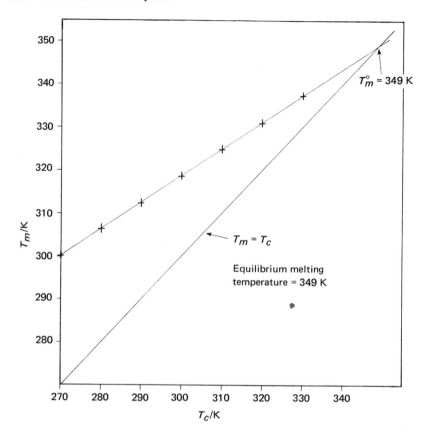

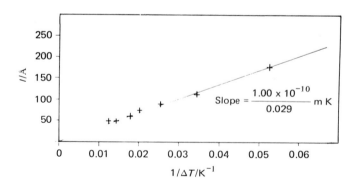

Therefore plot l versus $\dfrac{1}{\Delta T}$

$$\text{Slope} = \frac{2\gamma_e T^\circ_m}{\Delta H_v} = 3.45 \times 10^{-7} \text{ m K}$$

Therefore

$$\gamma_e = \frac{6.9 \times 10^{-7} \times 1.5 \times 10^8}{2 \times 349} \quad \frac{\text{m K J m}^{-3}}{\text{K}}$$

Therefore

$$\gamma_e = 0.074 \text{ J m}^{-2}$$

4.5 Plot $\bar{n}$ versus T and look for a change in slope (see graph).

i.e. $T_g = 89^\circ\text{C}$

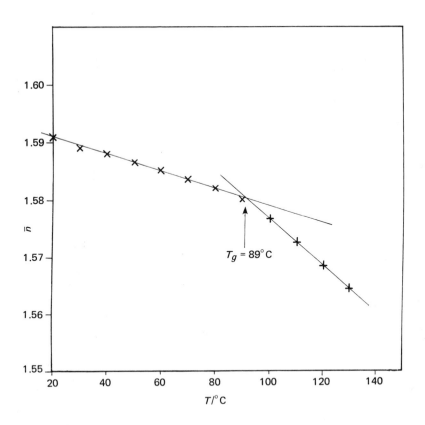

4.6 All the polymers are isomers of each other

(a) $(-CH_2-CH_2-)_n$

$T_g \sim 150$ K
Flexible molecule
therefore rotates easier

$(-CH_2-CH(CH_3)-)_n$

$T_g \sim 250$ K
$-CH_3$ side-group
hinders rotation

(b) $(-CH_2-CH(C(=O)OCH_3)-)_n$

$T_g = 283$ K
Bulky side group
hinders rotation

$(-CH_2-CH(O-C(=O)CH_3)-)_n$

$T_g = 305$ K
Bulky part of side-
group further from chain

(c) $(-CH_2-CH(CH_2CH_3)-)_n$

$T_g = 249$ K
Side group long
and bulky therefore
rotation difficult

$(-CH(CH_3)-CH(CH_3)-)_n$

$T_g = 200$ K
Side groups
shorter therefore
rotation easier

(d) $(-CH_2-CH_2-O-)_n$

$T_g = 232$ K
Flexible main chain
– easy rotation

$(-CH_2-CH(OH)-)_n$

$T_g = 358$ K
$-OH$ side group
rotation difficult

(e) $(-CH_2-CH(C(=O)OC_2H_5)-)_n$

$T_g = 249$ K
Side group long
and flexible

$(-CH_2-C(CH_3)(C(=O)OCH_3)-)_n$

$T_g = 378$ K
Side-groups very bulky
therefore rotation difficult

4.7 For a glassy polymer

$$T_g = T_g^{\infty} - \frac{\rho x N_A \theta}{\alpha_f \bar{M}_n} \quad \text{where } x = \text{number of ends per chain}$$

Therefore

$$T_g = T_g^{\infty} - \frac{Kx}{\bar{M}_n} \quad \text{if } K \text{ is constant.}$$

Linear polymer $\bar{M}_n = 2300$, $T_g = 121°C$, $x = 2$

Therefore

$$121 = T_g^\infty - \frac{2K}{2300} \tag{1}$$

Linear polymer $\bar{M}_n = 9000$, $T_g = 153°C$, $x = 2$

Therefore

$$153 = T_g^\infty - \frac{2K}{9000} \tag{2}$$

Branched polymer $\bar{M}_n = 5200$, $T_g = 115°C$, $x = ?$

$$115 = T_g^\infty - \frac{xK}{5200} \tag{3}$$

$(2) - (1)$ gives

$$32 = 2K\left(\frac{1}{2300} - \frac{1}{9000}\right)$$

Therefore

$$K = 49\,433 \text{ mol g}^{-1}$$

Substituting in (1) gives

$$T_g^\infty = 121 + \left(2 \times \frac{49\,433}{2300}\right)$$

Therefore

$$T_g = 164°C$$

Substituting in (3) gives

$$115 = 164 - \left(x \times \frac{49\,433}{5200}\right)$$

Therefore

$$x = 5.15 \text{ (number of ends)}$$

Therefore the average number of branches = 3.15

Chapter 5
5.1
(i) Constant stressing rate

$$\frac{d\sigma}{dt} = R = \text{constant}$$

Maxwell model: $\dfrac{de}{dt} = \dfrac{1}{E}\dfrac{d\sigma}{dt} + \dfrac{\sigma}{\eta}$

Therefore

$$\frac{de}{dt} = \frac{R}{E} + \frac{Rt}{\eta} \quad \text{since } \sigma = \int R \, dt$$

Voigt model: $\dfrac{de}{dt} = \dfrac{\sigma}{\eta} - \dfrac{Ee}{\eta}$

Differentiating

$$\frac{d^2e}{dt} = \frac{1}{\eta}\frac{d\sigma}{dt} - \frac{E}{\eta}\frac{de}{dt}$$

Therefore

$$\frac{d^2e}{dt^2} + \frac{E}{\eta}\frac{de}{dt} - \frac{R}{\eta} = 0 \quad \text{since } R = \frac{d\sigma}{dt}$$

This is a differential equation which can be solved in one of several ways. One is the 'convolution integral' method outlined by J. G. Williams in *Stress Analysis of Polymers*. The simplest method is to 'guess' the result as follows:

$$e = \frac{R}{E}\left\{ t - \tau_0 \left[1 - \exp\left(-\frac{t}{\tau_0} \right) \right] \right\} \quad \text{where } \tau_0 = \frac{\eta}{E}$$

Therefore

$$\frac{de}{dt} = \frac{R}{E}\left[1 - \exp\left(-\frac{t}{\tau_0} \right) \right]$$

and $\dfrac{d^2e}{dt^2} = \dfrac{R}{E\tau_0} \exp\left(-\dfrac{t}{\tau_0} \right)$

The equation then becomes

$$\frac{d^2e}{dt^2} + \frac{E}{\eta}\frac{de}{dt} - \frac{R}{\eta} = \frac{R}{E\tau_0}\exp\left(-\frac{t}{\tau_0} \right) + \frac{E}{\eta}\frac{R}{E}\left[1 - \exp\left(-\frac{t}{\tau_0} \right) \right] - \frac{R}{\eta}$$

$$= \frac{R}{E}\frac{E}{\eta}\exp\left(-\frac{t}{\tau_0} \right) + \frac{R}{\eta} - \frac{R}{\eta}\exp\left(-\frac{t}{\tau_0} \right) - \frac{R}{\eta} = 0$$

Cancelling throughout the relation between e and t becomes

$$e = \frac{R}{E}\left\{ t - \tau_0 \left[1 - \exp\left(-\frac{t}{\tau_0} \right) \right] \right\}$$

(ii) Constant strain rate

$$\frac{de}{dt} = S = \text{constant}$$

Maxwell model: $\dfrac{de}{dt} = \dfrac{1}{E}\dfrac{d\sigma}{dt} + \dfrac{\sigma}{\eta}$

Therefore

$$S = \frac{1}{E}\frac{d\sigma}{dt} + \frac{\sigma}{\eta} \quad \text{or} \quad \frac{d\sigma}{(S\eta - \sigma)} = \frac{E}{\eta}dt$$

Integrating gives

$$\ln(S\eta - \sigma) = -\frac{Et}{\eta} + A$$

where A is a constant of integration.

Since when $t = 0$, $\sigma = 0$ therefore $A = \ln(S\eta)$

Hence $\ln\left[\frac{(S\eta - \sigma)}{S\eta}\right] = -\frac{Et}{\eta}$

Therefore

$$\sigma = S\eta\left[1 - \exp\left(-\frac{t}{\tau_0}\right)\right] \quad \text{where } \tau_0 = \frac{\eta}{E}$$

Voigt model: $\dfrac{de}{dt} = \dfrac{\sigma}{\eta} - \dfrac{Ee}{\eta} = S$

Rearranging gives

$$\sigma = Ee + S\eta$$

Therefore

$$\sigma = ESt + S\eta \quad \text{since} \quad S = \frac{e}{t}$$

5.2

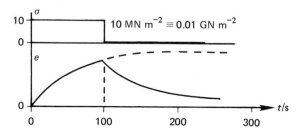

$$J(t) = J_0\left[1 - \exp\left(-\frac{t}{\tau_0}\right)\right]$$

$$J_0 = 2 \text{ m}^2 \text{ GN}^{-1} \quad \text{and} \quad \tau_0 = 200 \text{ s}$$

(a) $t = 100$ s Thus $J(100) = J_0\left[1 - \exp\left(-\frac{100}{200}\right)\right]$

Therefore $J(100) = 2(1 - e^{-0.5}) = 0.787 \text{ m}^2 \text{ GN}^{-1}$

But
$$J = \frac{e}{\sigma} \quad \text{therefore} \quad e = J\sigma$$

Therefore $e(100) = 0.01 \times 0.787 = 0.007\,87$

(b) If stress was not removed the strain after 200 s would have been:
$$e(200) = 0.01 \times 2 \times (1 - e^{-1}) = 0.012\,64$$

The removal of stress is equivalent to the addition of a negative strain of $0.007\,87$ after $(200 - 100)$s.

Therefore residential strain $= 0.012\,64 - 0.007\,87$

Strain after 200 s $= 0.004\,77$

5.3   Maxwell model for stress relaxation
$$\sigma = \sigma_0 \exp\left(-\frac{t}{\tau_0}\right)$$

σ_0 = strain $\times$ short-term modulus

Therefore
$$\sigma_0 = 0.2 \times 3 = 0.6 \text{ MN m}^{-2}$$
$$\text{Leaking pressure} = 0.3 = 0.6 \exp\left(-\frac{t}{300}\right)$$

Therefore
$$t = 208 \text{ days}$$

5.4 $e = e_0 \sin \omega t$

$\sigma = \sigma_0 \sin(\omega t + \delta)$

Therefore
$$\sigma = \sigma_0 \sin \omega t \cos \delta + \sigma_0 \cos \omega t \sin \delta$$

$$\Delta U = \int \sigma \, de$$

In one cycle
$$\Delta U = \int_0^{2\pi/\omega} \sigma \frac{de}{dt} \, dt$$

But
$$\frac{de}{dt} = e_0 \omega \cos \omega t$$

Therefore

$$\Delta U = \int_0^{2\pi/\omega} e_0\omega \cos \omega t(\sigma_0 \sin \omega t \cos \delta + \sigma_0 \cos \omega t \sin \delta)dt$$

$$= e_0\omega\sigma_0 \int_0^{2\pi/\omega} (\cos \delta \sin \omega t \cos \omega t + \sin \delta \cos^2 \omega t)dt$$

Therefore

$$\Delta U = e_0\omega\sigma_0 \frac{\pi \sin \delta}{\omega} = e_0\sigma_0\pi \sin \delta$$

5.5 $w = \frac{1}{2}G(\lambda_1^2 + \lambda_2^2 + \lambda_3^2 - 3)$

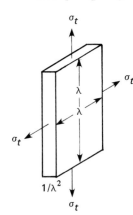

For two dimensional straining volume is constant

Therefore $\lambda_1\lambda_2\lambda_3 = 1$

$$\lambda_1 = \lambda_2 = \lambda, \quad \lambda_3 = \frac{1}{\lambda^2}$$

Therefore $w = \frac{1}{2}G\left(2\lambda^2 + \frac{1}{\lambda^4} - 3\right)$

Therefore

$$dw = 2G\left(\lambda - \frac{1}{\lambda^5}\right)d\lambda$$

The forces per unit area in the *unstrained* state are f_1, f_2 and f_3 in each direction.
Then the work done by these forces will be in general:

$$dw = f_1 d\lambda_1 + f_2 d\lambda_2 + f_3 d\lambda_3$$

But $f_1 = f_2 = f$ and $f_3 = 0$

Therefore

$$dw = 2f d\lambda = 2G\left(\lambda - \frac{1}{\lambda^5}\right)d\lambda$$

Therefore

$$f = G\left(\lambda - \frac{1}{\lambda^5}\right)$$

These forces are related to the true stresses i.e. the forces on the strained area σ_1, σ_2 and σ_3 by

$$f_1 = \sigma_1\lambda_2\lambda_3, \text{ etc. but } \lambda_1\lambda_2\lambda_3 = 1$$

Therefore

$$f_1 = \frac{\sigma_1}{\lambda_1} \quad \text{or} \quad f = \frac{\sigma_t}{\lambda}$$

where σ_t is the two-dimensional true stress.

Therefore

$$\sigma_t = G\left(\lambda - \frac{1}{\lambda^5}\right)\lambda$$

Therefore

$$\sigma_t = G\left(\lambda^2 - \frac{1}{\lambda^4}\right)$$

5.6   Pressure-dependent von Mises yield criterion where

$$A[\sigma_1 + \sigma_2 + \sigma_3] + B[(\sigma_1 - \sigma_2)^2 + (\sigma_2 - \sigma_3)^2 + (\sigma_3 - \sigma_1)^2] = 1$$

Let the tensile yield stress $= \sigma_{yt}$ and compressive yield stress $= \sigma_{yc}$

Therefore the criterion can be written as

$$A\sigma_{yt} + 2B\sigma_{yt}^2 = 1 \quad \text{(tension)}$$
$$-A\sigma_{yc} + 2B\sigma_{yc}^2 = 1 \quad \text{(compression)}$$

Therefore

$$A = (\sigma_{yc} - \sigma_{yt})/\sigma_{yc}\sigma_{yt}$$

and $B = 1/(2\sigma_{yc}\sigma_{yt})$

Hence the criterion becomes

$$2(\sigma_{yc} - \sigma_{yt})[\sigma_1 + \sigma_2 + \sigma_3] + [(\sigma_1 - \sigma_2)^2 + (\sigma_2 - \sigma_3)^2 + (\sigma_3 - \sigma_1)^2]$$
$$= 2\sigma_{yc}\sigma_{yt}$$

Let the critical hydrostatic pressure to cause yielding be p_c.

Then $p_c = \frac{1}{3}(\sigma_1 + \sigma_2 + \sigma_3)$

and $\sigma_1 = \sigma_2 = \sigma_3$

The criterion becomes

$$2(\sigma_{yc} - \sigma_{yt})[3p_c] = 2\sigma_{yc}\sigma_{yt}$$

Therefore

$$p_c = \frac{\sigma_{yc}\sigma_{yt}}{3(\sigma_{yc} - \sigma_{yt})}$$

5.7 Plane stress deformation $\sigma_3 = 0$

Yield criterion: $A[\sigma_1 + \sigma_2] + B[(\sigma_1 - \sigma_2)^2 + \sigma_2^2 + \sigma_1^2] = 1$

Craze criterion: $\sigma_1 - \nu\sigma_2 = X + Y/(\sigma_1 + \sigma_2)$

From **5.6** $A = (\sigma_{yc} - \sigma_{yt})/\sigma_{yc}\sigma_{yt}$

and $B = 1/(2\sigma_{yc}\sigma_{yt})$

But $\sigma_{yc} = 92$ MN m^{-2} and $\sigma_{yt} = 73$ MN m^{-2}

Therefore

$A = 2.829 \times 10^{-3}$ m^2 MN^{-1}

and $B = 7.495 \times 10^{-5}$ m^4 MN^{-2}

Crazing: $\sigma_1 = 47$ MN m^{-2}, $\sigma_2 = 0$

and $\sigma_1 = 45$ MN m^{-2}, $\sigma_2 = 45$ MN m^{-2}

Therefore

$47 = X + Y/47$

$45 - (0.33 \times 45) = X + Y/(45 + 45)$

Therefore

$X = 11.73$ MN m^{-2}

$Y = 1657.6$ MN^2 m^{-4}

Therefore the two equations become (in MN m^{-2})

Shear yielding:

$2.829 \times 10^{-3}(\sigma_1 + \sigma_2) + 2 \times 7.445 \times 10^{-5}(\sigma_1^2 + \sigma_2^2 - \sigma_1\sigma_2) - 1 = 0$

Crazing:

$\sigma_1^2 - 0.33\sigma_2^2 + \sigma_1\sigma_2(1 - 0.33) - 11.73(\sigma_1 + \sigma_2) - 1657.6 = 0$

Crazing and shear yielding will take place simultaneously at values of σ_1 and σ_2 which satisfy both of the above equations. I have not been able to solve the problem analytically. However it can be solved by

(a) Plotting the two functions graphically (Fig. 5.32)
(b) Numerically by computer.

Both methods have been tried and give

$\sigma_1 = 59$ MN m^{-2} and $\sigma_2 = -30$ MN m^{-2}

The shear yielding equation is symmetrical in σ_1 and σ_2. However, the crazing equation is not symmetrical and is different with σ_1 and σ_2 interchanged. Hence shear yielding and crazing will also take place when

$\sigma_1 = -30$ MN m^{-2} and $\sigma_2 = 59$ MN m^{-2}

5.8   The chain direction repeat in polyethylene is 2.54 Å

Therefore

$$d_{002} = 1.270 \text{ Å} \quad \text{(unstrained)}$$

The modulus E = stress/strain = σ/e

Strain, $e = \Delta d/d$ \quad but \quad $\lambda = 2d \sin \theta$

Therefore

$$d = \lambda/2 \sin \theta$$

σ/MN m^{-2}	θ/degrees	d/Å	Δd/Å	$\Delta d/d$	σ/e MN m^{-2}
0	37.483	1.267 00	0	0	
40	37.477	1.267 17	0.000 17	0.000 134	298 507
80	37.471	1.267 34	0.000 34	0.000 268	298 507
120	37.466	1.267 49	0.000 49	0.000 387	310 077
100	37.460	1.267 66	0.000 66	0.000 521	307 101
200	37.454	1.267 83	0.000 83	0.000 655	305 343

Therefore the modulus $\sim$ 300 GN m^{-2}

5.9   This problem is taken from:

Bevis, M. (1978), *Coll. Polym. Sci.*, **256**, 234 and

Young, R. J. *et. al.*, (1979) *J. Polym. Sci: Polym. Phys. Ed.*, **17**, 1325.

General sketch of twins in plane of shear:

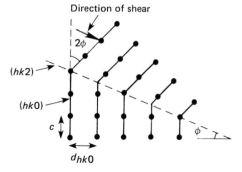

The (*hk*0) planes bend by an angle of 2ϕ across the twin boundary.

(a) If the planes of molecules that bend or kink are of the type $(hk0)$ the twinning plane is $(hk2)$ and hence:

Plane of molecules	Twinning plane
(200)	(202)
(020)	(022)
(110)	(112)

(b) The angle is 2ϕ and from the sketch it can be seen that:

$$\tan \phi = c/2d_{hk0} \qquad (c = 2.537 \text{ Å})$$

$(hk0)$	$d_{hk0}/\text{Å}$	$\tan \phi$	2ϕ	$(hk2)$
200	3.700	0.343	$37.8°$	202
020	2.465	0.515	$54.5°$	022
110	4.103	0.309	$34.4°$	112

(c) The direction of shear lies in the twinning plane, $(hk2)$ and in the plane of shear.

$(hk0)$	$(hk2)$	Plane of shear	Direction of shear
200	202	(010)	$[20\bar{2}]$
020	022	(100)	$[02\bar{2}]$
110	112	Irrational	$\sim [11\bar{2}]$

The plane of shear and direction of shear are irrational for the (112) twin.

(d) The magnitude of the shear, s, is given by

$$s = 2 \tan \phi$$

Therefore the twin with the lowest shear is (112).

5.10

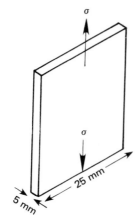

$\sigma_f = 85 \text{ MN m}^{-2}$

$K_c = 1.25 \text{ MN m}^{-3/2}$

The sheet contains 'inherent flaws' equivalent to small cracks of length, a_0. The equation for K_c in Table 5.5 for a single-edge notched specimen can therefore be used.

i.e. $K_c = \dfrac{Pa^{1/2}}{bt} [1.99 - 0.41(a/b) + 18.7(a/b)^2 - 38.48(a/b)^3 + 53.85(a/b)^4]$

As the flaws are very small $a \ll b$

Therefore

$$K_c \sim Pa_0^{1/2} \times 1.99/bt$$
$$b = 25 \text{ mm}, t = 5 \text{ mm}$$

But $\sigma_f = P/bt = 85 \text{ MN m}^{-2}$

Therefore

$$P = 85 \times 25 \times 5 \times 10^{-6} = 0.010\,625 \text{ MN}$$

Therefore

$$a_0^{1/2} = 1.25/1.99 \times 85 \text{ m}^{1/2}$$

Therefore

$$a_0 = 55 \times 10^{-6} \text{ m} = 55 \text{ } \mu m$$

5.11 For the rubber

$$\mathscr{T} = KaW$$

and $\quad \dfrac{da}{dt} = q\mathscr{T}^n = q(KaW)^n$

If a sheet is held at constant extension

$$dt = \frac{1}{qK^n W^n} \times \frac{da}{a^n}$$

The crack grows from a_0 to a where a approaches the sheet width in time t_f

then $\quad \displaystyle\int_0^{t_f} dt = \frac{1}{qK^n W^n} \int_{a_0}^a \frac{da}{a^n}$

Therefore

$$t_f = \frac{1}{qK^n W^n (n-1)} \left[\frac{1}{a_0^{n-1}} - \frac{1}{a^{n-1}} \right]$$

But n is generally large and $a \gg a_0$

Hence

$$t_f = \frac{1}{qK^n W^n (n-1)a_0^{n-1}}$$

Index